INTRODUCTION TO BIOTECHNOLOGY

Edited by

Natassia Correa

Toronto Academic Press

INTRODUCTION TO BIOTECHNOLOGY

Editor: **Natassia Correa**

Toronto Academic Press
4164 Lakeshore Road
Burlington ON L7L 1A4
Canada
www.tap-books.com

Email: orders@arclereducation.com

ISBN: 978-1-77956-292-0

ABOUT THE EDITOR

Natássia Corrêa is a biologist with a background in Molecular Biology, Cellular Biology, and Biotechnology. She is currently a public servant at the Federal University of Uberlândia, Minas Gerais, Brazil. Her academic journey includes earning her Master's and Doctorate degrees in Molecular Biology from the Federal University of Minas Gerais. In addition to working in the biotechnology industry, she completed her postdoctoral studies at FIOCRUZ and served as a faculty member at the Federal University of Paraná. Dr. Natássia Corrêa has always valued the diversity of her learning experiences, driven by a passion for new knowledge, as evidenced by her diverse and in-depth curriculum in Molecular and Cellular Biology.

Table of Contents

List of Figures

List of Tables

PREFACE

The concept of "biotechnology" was first introduced by Karl Ereky, a Hungarian engineer, back in 1919. He used this term to describe a scientific area and methods that allow us to make products from basic materials using living organisms. Biotechnology is a broad field that involves working directly with living cells or using substances derived from these cells. This work is done for the benefit of humans and involves a variety of tools and technologies. Essentially, biotechnology merges biology (the science of living things) with engineering (the field of designing and building things), using living organisms, their cells, or parts of them to create products and services.

The roots of biotechnology stretch back centuries, starting when humans first began to breed plants and animals selectively to improve desirable traits. Early uses of biotechnology include the use of microorganisms to brew beer, make wine, and produce cheese and bread. Over time, this field has significantly evolved. Today, biotechnology involves manipulating living organisms to create useful substances for use in medicine, agriculture, and industry. Traditional biotechnology refers to older practices like making bread and cheese using organisms, while modern biotechnology focuses on using parts of cells, like DNA and proteins, to develop new medical treatments, improve crops, and solve other problems.

This book aims to guide readers through the vast potential applications of biotechnology. It covers how biotechnology is used to create new medicines, diagnostic tools, and treatments such as monoclonal antibodies, stem cells, and gene therapy. It also explores agricultural biotechnology for crop improvement, uses in pollution control through bioremediation, applications in industrial and marine settings, and the development of biomaterials. Additionally, it addresses the ethical and safety considerations that come with certain biotechnological products.

This textbook is primarily intended for undergraduate students pursuing degrees in biotechnology, life sciences, and related disciplines. It aims to bridge the gap between complex biological processes and the practical applications of biotechnology in healthcare, agriculture, environmental science, and industrial processes.

Covering a broad range of topics, from the molecular and cellular foundations of biotechnology to the latest advancements and ethical considerations, the book seeks to equip readers with a solid understanding of the key concepts and techniques that underpin the biotechnology sector. Through detailed explanations, real-world examples, and case studies, students are introduced to the vast potential of biotechnology.

CHAPTER

1 Introduction to Biotechnology

LEARNING OBJECTIVES

After studying this chapter, you will be able to:

- Understand the overview of biotechnology;
- Explain the basic applications of biotechnology.

INTRODUCTORY EXAMPLE

The Definition of Biotechnology:

Biotechnology is the youngest of the sciences and is increasing in knowledge at an unprecedented rate. It is the fastest-growing technical discipline and has probably gained more information per year than any other field of science. Advances in biotechnology even outpace new developments in computer science. Because of the rapid advance, biotechnology is called a revolutionary science that outpaces the ability of people to keep up with an understanding of applications in society. The term biotechnology was first used by Hungarian engineer Karoly Ereky in 1919. His use of the term varies somewhat from its meaning today. Ereky used biotechnology to describe the industrial production of pigs by feeding them sugar beets as an inexpensive large-scale source of nutrients. He then generalized the term to all areas of industry in which commercial products are created from raw materials with the aid of organisms. Ereky predicted a biochemical age that rivaled the societal impacts of the Stone and Iron Ages.

The science of biotechnology is an amalgamation of biology, chemistry, computer science, physics, and mathematics. Many scientists in biotechnology fields have diverse skills that bring together two or more science disciplines. Biotechnology is also practiced as a working relationship between two or more scientists who collaborate on projects by sharing their expertise and experiences. Certain types of biotechnology involve many specialized techniques that only a few people are capable of performing. Yet, other procedures and scientific instruments used in biotechnology are fairly simple. The biotechnology concepts and techniques taught only to graduate and postdoctoral students in the 1970s are now covered in high school science classes.

Unlike earlier scientific endeavors, biotechnology relies heavily on its ability to be commercialized into diverse of procedures and products that benefit humans. More and more scientists who enter biotechnology as a career are discovering that they need a strong business background. A great proportion of biotechnology is being applied in industrial settings. Academic biotechnology at most universities is not carried out solely for the pursuit of information. Many of the discoveries make their way into consumer and medical products through a process called technology transfer. Technology transfer is defined as the process of converting scientific findings from research laboratories into useful products by the commercial sector. The great potential for profits that biotechnological innovations can offer has changed the nature of scientific information over the past 30 years.

Scientific discoveries were once freely shared between scientists by publishing findings in professional journals. The journals were peer-reviewed, meaning that other scientists familiar with the field evaluated the accuracy and validity of the information before it was published. Information in the journals was then made available through professional scientific societies and university and industrial libraries. The advent of computer-to-computer communication systems and the Internet paved the way for inexpensive means to disseminate scientific information rapidly. Almost every new finding in biotechnology could be used to make huge profits for enterprising scientists. This started a trend in which biotechnology information is not shared freely anymore. Many scientists argue that this secrecy is stifling the progress of science and may restrict the growth of science to profit-making endeavors.

Most new biotechnology discoveries are patented or are protected by intellectual property rights. Patenting and intellectual property rights permit the scientists to protect their discoveries. This protection prohibits others from using the discoveries or ideas without permission or some type of payment. A patent is described as a set of exclusive rights approved by a government to a person for a fixed period. The patent does have a limitation in that the public has the right to know certain details of the discovery. Patents are awarded only to inventions or procedures. The person applying for a patent need not be the scientist who made the discovery. Many scientists working for biotechnology companies are required to let the owners of the company patent the discovery.

UNIT INTRODUCTION

Biotechnology, a dynamic and interdisciplinary field, harnesses biological systems, organisms, or derivatives to develop innovative solutions for a wide range of applications. Biotechnology involves manipulating biological materials and processes to address challenges in medicine, agriculture, environmental management, and industry. This rapidly evolving field has the potential to revolutionize various aspects of our lives by providing novel tools and techniques that enhance our understanding of living organisms and enable the development of groundbreaking technologies.

In the realm of medicine, biotechnology has played a pivotal role in the discovery and production of therapeutic drugs, vaccines, and diagnostic tools. Genetic engineering techniques, such as recombinant DNA technology, have facilitated the creation of genetically modified (GM) organisms capable of producing pharmaceuticals or serving as platforms for vaccine development. This has significantly accelerated the pace of drug discovery and revolutionized the treatment of various diseases.

In agriculture, biotechnology has led to the development of GM crops with enhanced traits such as resistance to pests, diseases, and environmental stress. This has not only increased crop yields but has also contributed to more sustainable and environmentally friendly farming practices. Biotechnological applications in agriculture also extend to improving livestock through selective breeding and genetic engineering.

Environmental biotechnology addresses pressing environmental challenges by utilizing microorganisms to remediate polluted sites, degrade waste, and produce biofuels. The field offers innovative solutions for waste management, pollution control, and the development of renewable energy sources, contributing to a more sustainable and eco-friendly future.

Industrial biotechnology focuses on utilizing biological systems to manufacture various products, ranging from chemicals and materials to biofuels and enzymes. The application of biocatalysts and fermentation processes in industrial settings has led to the development of more efficient and environmentally friendly manufacturing processes, reducing the reliance on traditional chemical methods.

1.1. OVERVIEW OF BIOTECHNOLOGY

→Learning Objectives

- Understand the various stages of biotechnology development.
- Identify different types of biotechnology
- Evaluate the advantages and disadvantages of utilizing biotechnology.

Biotechnology is referred to as any method, procedure, or technology that uses living things or systems for human benefit. When we dissect the term, we find that "bio" refers to life and "technology" refers to the use of science for a particular goal. In general, biotechnology modifies or produces goods for particular uses using living cells. Genetically engineered plants are a prime example. It is a field with five branches: animal, plant, industrial, environmental, and human. It is associated with genetic engineering. It helps us fight disease and hunger, cut greenhouse gas emissions, save energy, and improve production safety, cleanliness, and efficiency.

1.1.1. Biotechnology and Its Various Stages of Development

It is widely recognized that biotechnology refers to the technical application of biological material. It is crucial to consider the raw material or origin of biotechnology processes in order to comprehend how they operate. In general, biotechnology creates new products for use in a variety of pharmaceutical, environmental, medical, and agricultural applications by using living materials or biological products. The ultimate purpose of biotechnology is to advance humankind through the creation of resilient plants, vegetables, recombinant proteins, more intelligent milk-producing animals, etc.

Biotechnology has gone through several stages of development to satisfy the diverse demands of people over time. Its creation was primarily guided by observations and the application of those findings to real-world situations. The complexity of biotechnology has grown due to the development of new technologies and a deeper comprehension of numerous life science concepts. The key biotechnology historical occurrences are summarized in Table 1.1. The development of biotechnology can be divided into broad stages or categories, including:

- Ancient biotechnology (8000–4000 BC): Early history as related to food and shelter, including domestication of animals.

- Classical biotechnology (2000 BC; 1800–1900 AD): Built on ancient biotechnology, fermentation promotes food production and medicine.
- 1900–1953: Genetics.
- 1953–1976: DNA research, science explodes.
- Modern biotechnology (1977): Manipulates genetic information in organisms; genetic engineering; various technologies enable us to improve crop yield and food quality in agriculture and to produce a broader array of products in industries.

Keyword

Greenhouse Gas Emissions – These refer to the release of gases into the Earth's atmosphere that trap heat, contributing to the greenhouse effect and global warming. These gases include carbon dioxide (CO_2), methane (CH_4), nitrous oxide (N_2O), fluorinated gases, and others.

Did you know?

In 1982, the first synthetic human insulin was produced using GM bacteria, a landmark achievement in medical biotechnology.

Table 1.1. Historical Events in Biotechnology

Periods of Biotechnology History	Events
Pre-1800 (Early applications and speculation)	6000. BC
	Yeast was utilized to prepare beer (Sumerians and Babylonians).
	4000. BC
	In Egypt, a process was discovered to prepare leavened bread by means of yeast.
	420. BC
	Greek philosopher Socrates (470–399 BC) hypothesized similar characteristics between parents and their offspring.
	320. BC
	Greek philosopher Aristotle (384–322 BC) theorized that all inheritance originates from the father.
	1000. AD
	Hindus recognized that some illnesses may 'run in the family.' At the same time, the theory of abiogenesis, or spontaneous generation based on the idea that organisms arise from non-living matter, developed. According to this theory, maggots could develop from horse hair.
	1630
	William Harvey explained that plants and animals are similar in their reproduction, i.e., they reproduce sexually.
	1660–1675

	Marcello Malpighi (1628–1694) investigated blood circulation in capillaries using a microscope and found that the brain is connected to the spinal cord by bundles of fibers that form the nervous system.
	1673
	Antonie van Leeuwenhoek (1632–1723) was the first researcher to explain microorganisms such as protozoa and bacteria, and also identify that these microorganisms play an active role in fermentation.
	1701
	Giacomo Pylarini found that the deliberate administration of smallpox could prevent its occurrence later in life, especially in children. Later, this procedure was termed 'vaccination' and a process that uses cowpox instead of smallpox was established as the most reliable treatment.
1800–1900 (Significant advances in basic understanding)	1809
	Nicolas Appert invented a technique using heat to can and sterilize food.
	1827
	In the field of heredity, there had long been a hunt for the so-called mammalian egg. It had proved elusive; however, in 1827, the first report of canine eggs offered a basic clue to major breakthroughs in reproduction, at first in lower animals.
	1850
	Ignaz Semmelweis utilized epidemiological examinations to suggest the theory that puerperal fever could be transmitted from mother to mother by physicians. He also suggested that all physicians should wash their hands after investigating each patient. For this suggestion, he was criticized by medical professionals and ultimately lost his employment.
	1856
	Carl Ludwig discovered a procedure for keeping animal organs alive under *in vitro* conditions. This was done by supplying blood to them. In contrast to the concepts of Justus von Liebig, Pasteur (1822–1895) suggested that microbes are responsible for fermentation.
	1859

	Charles Darwin (1809–1882) speculated that animal populations adapt their forms to eventually best utilize the surroundings, a process he described as 'natural selection.' During his stay in the Galapagos Islands, he saw how the finches' beaks on each island were adapted to the environment, especially regarding food sources.
	1863
	Pasteur discovered the method of pasteurization. In this method, he heated wine enough to inactivate microbes (that would otherwise convert the 'vin' to 'vin aigre' or 'sour wine') and realized that this procedure did not affect the flavor of the wine.
	Heinrich Anton de Bary established that a fungus was responsible for potato blight. A major challenge for researchers during this period was to differentiate whether a microbe was responsible for this or whether it was the outcome of a disease.
	1865
	Mendel (1822–1884) suggested the laws of heredity to the National Scientific Society (Brunn, Austria). Mendel anticipated that imperceptible core units of information were responsible for noticeable characteristics. He called these 'factors,' which were later called genes (units that were inherited by one generation from its parents). The research done by Mendel was overlooked and not acknowledged due to Darwin's more sensational publication five years earlier, until 1900, when Hugo de Vries, Erich von Tschermak and Carl Correns supported Mendel's theory of heredity.
	1868
	Casimir Joseph Davaine cured plants suffering from bacterial infection by a novel heat treatment. While working in a hospital, Johannes Friedrich Miescher separated nuclein (a compound made of nucleic acid) from pus cells. These pus cells were derived from waste bandages.
	1870
	Walther Flemming discovered mitosis.
	1871
	During the period 1873–1876, interest in DNA research began. DNA was initially derived from the sperm of trout (found in the river Rhine). During this period, Koch investigated anthrax and explored certain techniques to identify, culture, and stain microorganisms. He also took their images which were later supported by Gram, Cohn, and Weigart.

	1880
	While working on fowl cholera, Louis Pasteur explored weakened (attenuated) strains of microorganisms that might not be virulent but could nevertheless potentially prevent healthy individuals against severe forms of a similar disease.
	1881
	Koch explained techniques for harvesting bacterial colonies on potato slices, gelatin, and agar medium. For the isolation of pure culture and for distinguishing the nutrients needed for genetic mutations, the agar technique was one of the most common methods. Thomas D Brock considered this breakthrough as the single most important discovery in the development of microbiology.
	During the same period, Pasteur explored the application of the attenuation process in the production of vaccines against certain bacterial pathogens, e.g., fowl cholera and anthrax; this was an early stage in immunology, which led to the exploration of areas such as preventive medicine.
	1884
	Koch established his 'claims' for assessing whether a microbe or another agent is responsible for the disease.
	During the same period, Pasteur established a rabies vaccine.
	Gram described the differential staining technique for cellular peptidoglycan-containing bacteria, now known as Gram staining.
	Mendel passed away after 41 years of predominantly investigating the heredity 'factors' of pea plants. He did not receive any technical support during his lifetime, but said before his death, 'My time will come.'
1900–1953 (Genetics: Converging on DNA)	1900: Mendel's work finally took on importance
	Mendel's work had given birth to genetic science. It was revived again by three researchers, de Vries, von Tschermak and Correns, who were working on the application of the original work done by Mendel.
	1902: Human genetics is born
	Sutton found that chromosomes (paired) contain certain elements which are transferred from one generation to another. During this transfer, traits are transported through carriers called chromosomes. He also advised that Mendel's 'factors' are located on chromosomes.
	1905: X and Y chromosomes related to gender
	Edmund Beecher Wilson and Nettie Stevens shared the same idea of separating X and Y chromosomes for the determination of sex. They also demonstrated that a single Y chromosome determines maleness, while two copies of the X chromosome decide femaleness.

	1905–1908
	William Bateson, R C Punnett, and other researchers, found that several genes alter or modify the action of other genes.
	1906
	Paul Erlich also investigated atoxyl compounds and discovered the important features of Salvarsan (the first chemotherapeutic agent).
	1907
	Thomas Hunt Morgan started his investigation into fruit flies that would reveal that chromosomes have a defined role in heredity; additionally, he discovered mutation theory. This resulted in an understanding of the basic concepts and mechanisms of heredity.
	1909: Mendel's laws to animals
	Wilhelm Johannsen used the word 'gene' to mean the carrier/transporter of heredity. He also coined the terms 'genotype' and 'phenotype'; the genotype is the genetic composition/establishment of an organism, whereas the phenotype describes the actual organism or its morphological characteristics, resulting from a blend of the genotype and a range of external/ environmental factors.
	1910: Basis of modern genetics
	Morgan also demonstrated that carriers of genetic information, called genes, are present on chromosomes, creating the basis for modern genetics. This work later assisted him in utilizing *Drosophi*la fruit flies to examine heredity.
	1911
	During the same period, Morgan established the separation of certain inherited features that are generally linked to the separation/breaking of chromosomes during the process of cell division. He also investigated the mapping of the genetic sites present on the chromosomes of the fruit fly.
	1912
	Crystallography era: William Lawrence Bragg discovered the application of X-rays in the determination of the molecular structure of crystalline substances.
	1918
	Herbert M Evans stated (mistakenly) that human genetic material is made up of 48 chromosomes.
	1924: Eugenics in the United States
	Several US diplomats, encouraged by the eugenics movement, accepted the US Immigration Act (1924), limiting the admission of illiterate refugees from Southern and Eastern Europe on the basis of their alleged genetic inferiority.
	1926

	Morgan published *The Theory of the Gene*.' This was based on Mendelian genetics (breeding investigations and optical microscopy).
	Hermann Joseph Muller discovered that X-rays are responsible for genetic mutations in fruit flies taking place 1500 times faster than under normal conditions. This innovation offered researchers and scientists a procedure to induce mutations. Later, various mutagens were explored to understand the complexity behind different genotypes.
	1928
	Frederick Griffiths observed the 'transforming principle' in which a rough type of bacterium is transformed to a smooth type when a mysterious 'transforming element' from the smooth type is present. After 16 years, Oswald Theodore Avery discovered that the 'transforming element' as DNA.
	Alexander Fleming studied an old culture of bacteria infected with fungal growth and found that it did not show any bacterial growth in a radius surrounding a piece of mold (fungi) in a petri dish. This breakthrough gave birth to the antibiotics era, or penicillin age, and penicillin was accessible to patients 15 years later for therapeutic use.
	1938
	Proteins and DNA were studied by means of X-rays. This was the dawn of a new age of crystallography in which large molecular weight complex proteins can be studied by X-rays.
	The term 'molecular biology' was coined.
	1941: One gene, one enzyme
	George Wells Beadle and Edward L Tatum examined *Neurospora crassa*, a mold that usually invades and grows on bread, and proposed 'one gene, one enzyme' theory: each gene encodes for or is translated into an enzyme to accomplish tasks within an organism.
	1943
	The Rockefeller Foundation (New York) collaborated with the Mexican government to start the Mexican Agricultural Program. This was the first step toward plant breeding at a global level.
	1943–1953
	Cortisone (17α,21-dihydroxypregn-4-ene-3,11,20-trione), a pregnane (21-carbon) steroid hormone, was first produced in great amounts. Cortisone is considered the first biotech product.
	1944
	Selman Abraham Waksman (a Ukrainian-American researcher) explored streptomycin, an active antibiotic against TB.

	1945
	The United Nations Food and Agriculture Organization was established in Quebec, Canada, with the objective of encouraging agricultural practices.
	1945–1950
	For the first time, animal cell cultures were harvested in laboratories, giving birth to the field of animal tissue culture.
	1947
	Barbara McClintock first demonstrated 'transposable elements' known as 'jumping genes' that can move (or jump) from one site on the genome to another site. Scientific society did not welcome the implications of her discovery at the time.
	1950
	Erwin Chargaff discovered that the same levels of adenine and thymine are present in DNA, as are the same levels of guanine and cytosine. These associations were later named 'Chargaff's rules.' Later, Chargaff's rules functioned as an important principle for James Watson and Francis Crick in measuring different models for the structure of DNA.
DNA research, science explodes (1953–1976)	1953–1976: Expanding the boundaries of DNA research
	The discovery of the structure of DNA finally resulted in an explosion of research into molecular biology and genetics, providing the resources for biotechnology development.
	1953
	The journal *Nature* published Watson and Crick's article based on unfolding the double-helix structure of DNA.
	1953
	Based on his technical exposure, George Otto Gey developed the HeLa human cell line. Cells taken from cancer patient Henrietta Lacks (who died in 1951) became the first immortal human cells and were cultured to develop a polio vaccine.
	1957: Central dogma of DNA—how DNA makes a protein
	Crick and Gamov studied 'central dogma,' demonstrating how DNA functions to construct a protein.
	1959
	François Jacob and Jacques Lucien Monod documented the veracity of gene-based regulation. They explained gene mapping with mappable control functions sited on the chromosome in the DNA sequence, which they later named the 'repressor' and 'operon.'

	1962
	Watson and Crick were awarded the Nobel Prize in Physiology or Medicine with Maurice Wilkins. Disappointingly, Rosalind Franklin, who actually contributed to the discovery of the double-helical structure of DNA, died before this date, and Nobel Prize conventions do not permit a prize to be awarded posthumously.
	1966: Genetic code cracked
	The genetic code was explored by several researchers. Marshall Warren Nirenberg, J Heinrich Matthaei, and S Ochoa reported that a genetic sequence of three nucleotide bases (called codons) decides each of the 20 amino acids.
	1967
	Arthur Kornberg reported a study using single-stranded natural viral DNA to assemble 5,300 nucleotide building blocks, and at the same time his Stanford group synthesized viral DNA.
	1970: Oncogenes
	Virologists Peter H Duesberg and Peter K Vogt identified the first oncogene in a virus. This gene can be utilized to study various human cancers.
	1972: First recombinant DNA molecule
	Paul Berg, a biochemist, utilized a restriction enzyme to cut DNA into fragments. He employed a ligase enzyme to join two DNA strands concurrently to form a hybrid circular molecule. This was the first synthesized recombinant DNA (rDNA) molecule.
	1972: National Institutes of Health (NIH) guidelines for rDNA
	Berg and other researchers at the National Institutes of Health (NIH) worked hard to establish guidelines to sanction the strategy for DNA splicing. Their concerns resulted in the Asilomar Conference (1975).
	1973: Ames test
	Bruce Nathan Ames, a biochemist at UC Berkeley, developed an investigation to distinguish chemicals that damage DNA. Later, the Ames test became extensively used to identify cancer-causing substances.
	1975: rDNA moratorium
	A global meeting was held in Asilomar, California, with the objective of approving guidelines regulating rDNA experimentation. All the scientists involved discussed the development of 'safe' bacteria and plasmids.
	1976: More about oncogenes

	J Michael Bishop and Harold Varmus at the University of California, San Francisco (UCSF) established that cancer-causing genes called oncogenes become visible on animal chromosomes, and modifications in their structure or expression can result in metastatic growth.
	1976: Release of NIH guidelines
	The NIH released the first set of guidelines for rDNA experimentation. Later, these guidelines restricted several types of trials.
1977–present (modern biotechnology)	1977–present: The dawn of biotech
	With the advent of genetic engineering, it was possible to produce human protein in bacteria for the first time. Biotech-based organizations started focusing more on the applications of genetic engineering. In 1978, Herbert W Boyer at UCSF synthesized synthetic human insulin by introducing the insulin gene into the bacterium *E,* coli. This breakthrough opened the gateway for further developments in DNA sequencing and cloning techniques.
	1977
	Genentech Inc. was the first organization to achieve the synthesis of a human protein (somatostatin) in a bacterium. Somatostatin is a human growth hormone (hGH)-releasing inhibitory factor. A synthetic, recombinant gene was, for the first time, employed to clone a protein. Several researchers believed that this was the beginning of the age of modern biotechnology.
	1978: Recombinant insulin
	Genentech Inc. announced that its laboratory had achieved the synthesis of human insulin using rDNA technology.
	1980: Patents allowed
	The US Supreme Court granted that genetically modified (GM) living organisms could be patented. According to a Supreme Court decision (1980) the Exxon oil company was allowed to patent an oil-eating micro-organism.
	Kary Mullis and other researchers at UC Berkeley, California, established a tool for multiplying DNA sequences *in vitro* using the polymerase chain reaction (PCR).
	1982: Site-directed mutagenesis
	Genentech Inc. signed an agreement with the US Food and Drug Administration (FDA) to further market genetically engineered human insulin. In 1982, the FDA allowed the first genetically engineered drug in the form of human insulin produced by bacteria.
	Michael Smith at the University of British Columbia, Vancouver, established a procedure for producing precise amino acid changes anywhere in a protein.
	1983: Site-directed mutagenesis

	Eli Lilly obtained a license to make and sell insulin.
	1985
	During this period, genetic fingerprinting stepped into the court room.
	Cal Bio produced a gene by a cloning method that encodes human lung surfactant protein, an important step toward reducing premature birth complications.
	For the first time, GM plants resistant to insects, viruses, and bacteria were examined.
	The NIH published guidelines for performing experiments in gene therapy on humans.
	1986
	Chiron Corp. obtained FDA approval for the production of the first recombinant vaccine for hepatitis.
	A GM crop (the tobacco plant) was allowed by the Environmental Protection Agency (EPA).
	1987
	Calgene Inc. obtained a patent for the tomato polygalacturonase DNA sequence, which was later used to synthesize an antisense RNA sequence that can further extend the shelf life of fruit.
	1988
	Harvard molecular geneticists Philip Leder and Timothy A Stewart were granted the first patent based on a GM animal (a mouse that is highly susceptible to breast cancer).
	1990
	UCSF and Stanford University achieved their 100th rDNA patent license. At the end of the 1991 financial year, both organizations had received $40 million from the patent.
	1990: Patents and money
	The first gene-based treatment was performed on a four-year-old girl suffering from an immunological disorder known as adenosine deaminase deficiency (ADA) deficiency. Gene therapy emerged, however ethical concerns surrounding gene therapy were highly debated.
	Commencement of the Human Genome Project, with the global objective to plot all of the genes in the human body. The expected cost was $13 billion.
	Michael Crichton's novel *Jurassic Park* was released, in which bioengineered dinosaurs wander in a paleontological theme park; the project goes wrong, with deadly outcomes.
	1992

	The US Army started taking blood and tissue samples from all new employees as part of a 'genetic dog-tag.' This course of action was intended for better identification of soldiers killed in battle.
	1993
	Researcher Kary Mullis won the Nobel Prize in Chemistry for inventing the PCR tool.
	1996
	A ground breakingly efficient diagnostic biosensor test allowed for the first time the instant detection of the toxic strain of *E. coli* (strain 0157:H7), the bacteria responsible for several food poisoning outbreaks. The possibility of its use against anthrax and other bioterrorism agents was also assessed.
	The discovery of a gene linked to Parkinson's disease offered researchers a significant new chance for the determination of the cause of, and potential treatments for, the incapacitating neurological disorder.
	Reports showed that there were public concerns about research into the human genome and gene therapy, with a combination of fear and mistrust.
	1997
	Researchers at the Roslin Institute in Scotland announced that they had cloned a sheep called Dolly from the cell of an adult ewe. Dolly was the first mammal cloned by a technique called nuclear transfer technology. Nuclear transfer allows the introduction of complete genetic material from one cell into another enucleated unfertilized egg cell.
	1998
	A group of researchers succeeded in culturing embryonic stem cells. A number of researchers at Japan's Kinki University cloned eight identical calves by means of cells taken from a single adult cow. A rough draft of the human genome map was created, presenting the sites of more than 30,000 genes.
	1999
	A fatal neurological disease called bovine spongiform encephalopathy (BSE), also known as mad cow disease, that spread from cattle to humans, was diagnosed by a new medical diagnostic examination that facilitated the quick detection of BSE/Creutzfeldt–Jakob disease (CJD).

Source: Saurabh, B., (2018). Introduction to Pharmaceutical Biotechnology, 1, IOP Publishing Ltd., doi: 10.1088/978-0-7503-1299-8ch1.

1.1.1.1. Old and New Biotechnology

Although the term "biotechnology" is relatively new, the field of study has a long history. As early as 5000 BC, people started using microorganisms to make wine, vinegar, curd, leavened bread, and other products. These procedures were widely used domestically and are now so much a part of standard food processing techniques that we might be reluctant to call them biotechnology. These methods, which rely on microorganisms' innate abilities, are frequently referred to as "old" biotechnology.

1.1.1.2. Ancient Biotechnology (Pre-1800)

Biotechnological developments can be defined as events that occurred prior to 1800 and were based on widely accepted natural observations. The three fundamental requirements of human civilization are clothing, food, and shelter. Due to food scarcity in the past, food products were domesticated—a process known as "agriculture." In order to domesticate various wild animals, improve their living conditions, and satiate their hunger, humans had to understand the importance of light, water, and other environmental factors in ancient times. The observation, comprehension, and practical applications of animal breeding began with the domestication of wild animals. During this early stage of farming's evolution, techniques for storing and preserving food evolved further. Food preservation for long-term storage started with the use of cold caves or pots (made of leather bags and clay jars). Following their discovery of the fundamentals underlying the domestication of wild animals and food crops, humans went on to create new inventions like cheese and curd. Since cheese was made by combining rennet, an enzyme that is present in calves' stomachs, with sour milk, it can be regarded as one of the earliest direct products (or by-products) of biotechnology. This is only possible (though at the time there was no understanding of this) when milk is exposed to microbes. Yeast is one of the oldest microbes that humans have used for their own purposes out of all the strains of microbes. For a very long time, this archaic microbe has been used to produce alcoholic beverages like whiskey, wine, beer, etc. Being one of the oldest preservatives, vinegar is important because of its low pH and ability to stop some microbes from growing, which makes it useful for preserving food. The importance of these discoveries enabled people to focus on further refining the involved processes. However, while processes such as the decomposition of debris or other materials, later called fermentation, were powerful tools to improve their living conditions, people were ignorant of the inherent principles.

Keyword

Food Preservation – It can be defined as the process of treating and handling food in such a way as to stop or greatly slow down spoilage and prevent foodborne illness while maintaining nutritional value, texture, and flavor.

Mules are among the earliest instances of crossbreeding for human benefit. Before tractors and trucks were invented, people used mules for farming, transportation, and load-carrying. Hinnies, the progeny of a male horse and a female donkey, are more difficult to come by than mules. Unlike the donkey (62) and the horse (64) both hinnies and mules have 63 chromosomes. Table 1.2 lists a few of the procedures and goods created during the prehistoric era.

Table 1.2. Biotechnological Processes and Products Developed in the Ancient Period

Process or Product Developed	Events That Contributed to its Development
Domestication	• Food supplies were often seasonal. In winter, food supplies could become quite low. • People came up with ways of capturing fish and small animals. • 15,000 years ago, large animals were difficult to catch. • People may only have had meat when they found a dead animal. • Domestication most likely began 11,000–12,000 years ago in the Middle East. • It involved the adaptation of organisms so they could be cultured. • Seen by scientists as the beginning of biotechnology. • Cattle, goats, and sheep were the earliest domesticated food animals.
Food preservation	• People knew that some foods rotted, while others changed form and continued to be good to eat. • Foods stored in a cool cave did not spoil as quickly. • Foods heated by fire also did not spoil as quickly. • Immersion in sour liquids prevented food decay. • Food was stored in bags of leather or jars of clay. • Fermentation occurs if certain microorganisms are present; it creates an acid condition that slows or prevents spoilage.
Cheese	• One of the first food products made through biotechnology. • Strains of bacteria were added to milk, resulting in sour milk. • An enzyme called 'rennet' was added. • Rennet comes from the lining of the stomachs of calves. It is genetically engineered today. • Not all cheese is made from rennet. • It may have been first developed by nomadic tribes in Asia some 4000 years ago.

Yeast	• Long used in food preparation and preservation. • Used in bread baking. • Produces a gas in the dough, causing it to rise. • Used in fermented products such as vinegar. • Ethanol production requires the use of yeast in at least one stage of production.
Vinegar	• Used in pickling. • Keeps foods from spoiling. • Juices and extracts from fruits and grains can be fermented. • Biblical references to wine indicate the use of fermentation some 3000 years ago. • Ancient products used to preserve food.
Fermentation	• Process in which yeast enzymes chemically change compounds into alcohol. • In making vinegar, the first product of fermentation is alcohol. • In ancient times, this likely happened by accident. • Advancements occurred in the 1800s and early 1900s.
Fermenters	• Allowed better control, especially with vinegar. • New products such as glycerol, acetone, and citric acid resulted.
Antibiotics	• The first drug produced by microbes. • Used in both human and veterinary medicine. • Use of fermentation hastened the development of antibiotics. • Penicillin was developed in the late 1920s. • Introduced in the 1940s as a drug used to combat bacterial infections. • Many kinds are available today. • Limitations in their use keep disease-producing organisms from developing immunity to antibiotics. • Some disease organisms are now resistant to certain antibiotics.

Source: Saurabh, B., (2018). Introduction to Pharmaceutical Biotechnology, 1, IOP Publishing Ltd., doi: 10.1088/978-0-7503-1299-8ch1.

1.1.1.3. Classical Biotechnology

The second stage of biotechnology development is known as classical biotechnology. This historical period spanned from 1800 until nearly the mid-1900s. Various observations began to be made during the classical era and were backed up by scientific data. The mysteries of biotechnology could now be resolved thanks to these observations. Every single observation has advanced the search for novel discoveries in its own unique way. The basic tenet of biotechnology is the transmission of genetic information from one generation to the next. Gregor John Mendel (1822–1884), an Augustinian monk from Austria, was the first to decipher information on the transmission of genetic information. Mendel addressed the Natural Science Society in Brunn, Austria, with his theories regarding heredity. His initial observation of genetic information transfer in a plant was in Pisum sativum, also referred to as the pea plant. Furthermore, Mendel postulated that observable traits were explained by an unseen internal unit of information. These "factors," subsequently identified as genes, were inherited by the following generation. However, the tragic aspect of his tale is that, despite other scientists like Hugo de Vries, Erich von Tschermak, and Carl Correns validating his work in 1900, Mendel did not receive proper credit for his invention for nearly 34 years after his passing. Mendel's discovery was largely disregarded for a very long time because Charles Darwin's theory of evolution was so dominant at the time and eclipsed the significance of Mendel's work. Around this period, the discovery of the cell nucleus occurred, and Swiss biologist Fredrich Miescher announced the discovery of nuclein—a substance made of nucleic acid that he had isolated from pus cells—i.e., white blood cells. The identification of DNA as a genetic material and its function in the transmission of genetic information gave rise to the DNA era, which is now the foundation of modern molecular biology. Meanwhile, German physician Robert Koch (1881) first proposed the bacterial propagation method by describing bacterial colonies growing on potato slices (the first solid medium). Walter Hesse, a colleague in Koch's laboratory, found that nutrient agar is the most appropriate and practical medium for obtaining pure microbial cultures and for their identification while investigating the reason behind the solidification of jelly. When he asked his wife what kept the jelly solid even in the sweltering summer heat, she told him about agar. According to her, the nutrient was agar, and microbial cultures have since used it. The term "chromosome" was first used in the 19th century by the German scientist Heinrich Wilhelm Gottfried von Waldeyer-Hartz to describe an organized structure of protein and DNA found in cells or a single coiled DNA strand that contains numerous genes, regulatory elements, and other nucleotide sequences. Among various other prominent discoveries during this period, vaccinations against smallpox and rabies were developed by Edward Jenner, a British physician, and Louis Pasteur, a French biologist, respectively. The development of biological sciences seemed to be reaching an exponential phase. During this period, the principle of genetics in inheritance was redefined by T H Morgan. He showed inheritance and the role of chromosomes in inheritance using fruit flies (*Drosophila melanogaster*). Later, in 1926, this work was published in Morgan's book *The Theory of the Gene.* Prior to Morgan's 1909 work, the term 'gene' had already been coined by Wilhelm Johannsen (1857–1927). He described the gene as the carrier of heredity. Afterwards, Johannsen coined the terms 'genotype' to describe the genetic constitution of an organism, and 'phenotype' to describe the actual organism. After this exploration, genetics started gaining importance. This led to the beginning of the eugenics movement in the USA (1924). At the same time, Alexander Fleming, a British physician, discovered antibiotics when he observed that one micro-organism can be used to kill another micro-organism. The basic idea behind it was a true representation of the 'divide and rule' policy of humans. He noticed that all bacteria (*Staphylococci*) died when a mold was growing in a petri dish. Afterwards he discovered penicillin, the antibacterial toxin from the mold *Penicillium notatum*, which could be used against many infectious diseases. He wrote, 'When I woke up just after dawn on September 28, 1928, I certainly didn't plan to revolutionize all medicine by discovering the world's first antibiotic, or bacteria killer.' He also concluded that vaccines and antibiotics would turn out to be the best saviors of humanity: 'Can we attribute these two discoveries for the ever-increasing population as well the ever-aging population of the world?'

1.1.1.4. Modern Biotechnology

The Second World War constituted a significant barrier to scientific advancements. Several important discoveries were investigated after the war. These discoveries have led to the current state of biotechnology and serve as its foundation. Table 1.3 highlights some of the major biotechnology-related events of the modern era.

Table 1.3. Historical and Current Events That Form the Basis of Modern Biotechnology

Year	Discoveries
	The 1950s
1952	• George Otto Gey created a continuous cell line taken from a human cervical carcinoma. This cell line, known as HeLa, is still used in therapeutic research.
1953	• Watson and Crick explored DNA as a genetic material and discovered its structure, called the double-helix.
1954	• Joseph Murray carried out the first kidney transplant between identical twins.
1957	• Scientists revealed that sickle-cell anemia occurs due to an alteration in a single amino acid in hemoglobin cells.
1958	• Arthur Kornberg created DNA in a test tube for the first time. The first mechanical protein sequencer, the Moore–Stein amino acid analyzer, is developed.
	The 1960s
1960	• A French researcher discovered messenger RNA (mRNA).
1961	• François Jacob and Jacques Monad demonstrated the concept of Operon.
1962	• Osamu Shimomura explored the green fluorescent protein in the jellyfish *Aequorea victoria*. He afterward developed it into a technique for examining formerly invisible cellular processes.
1963	• Samuel Katz and John F Enders developed the first vaccine for measles.

1963	• Autonomous groups in the USA, Germany, and China produced insulin, a pancreatic hormone.
1964	• The existence of reverse transcriptase was predicted.
1967–1971	• Maurice Hilleman made the first American vaccine for mumps. • The first vaccine for rubella was developed. • Rubella was combined with the measles and mumps vaccines to yield the measles/mumps/rubella (MMR) vaccine.
1968	• How the arrangement of nucleotides in nucleic acids regulates the cell's synthesis of proteins was discovered.
The 1970s	
1970	• Restriction enzymes were discovered.
1972	• DNA ligases, which join DNA fragments together, were used for the first time. • The DNA composition of humans was discovered to be 99% similar to that of chimpanzees and gorillas. • The purified enzyme reverse transcriptase was first employed to prepare complementary DNA from purified messenger RNA in a test tube.
1973	• Stanley Cohen and Herbert Boyer used bacterial genes to perform the first successful rDNA experiment. • Sir Edwin Mellor Southern developed a blotting technique for DNA called the Southern blot.
1974	• The NIH formed a Recombinant DNA Advisory Committee to supervise recombinant genetic research. • The first vaccine for chicken pox was developed in Japan.
1975	• Colony hybridization and Southern blotting were explored to identify specific DNA sequences. • The first monoclonal antibodies were prepared. • César Milstein, Georges Jean Franz Kohler, and Niels Kaj Jerne explored the monoclonal antibody technique by fusing immortal tumor cells with antibody-producing B-lymphocyte cells to generate hybridomas that constantly produce identical antibodies.
1975	• The theory of cytoplasmic hybridization was proposed, and the first ever monoclonal antibodies were synthesized.
1976	• The NIH published the first guidelines for rDNA research. • Molecular hybridization was employed for the prenatal diagnosis of alpha thalassemia. Yeast genes were expressed in *E. coli* bacteria.

1977	•	Procedures were developed to swiftly sequence long sections of DNA. Genetically engineered bacteria were employed to manufacture the human growth protein somatostatin, marking the first time a synthetic recombinant gene was employed to clone a protein. Several believed this to be the arrival of the 'age of biotechnology.'
	•	Austrian et al. at the University of Pennsylvania developed the first vaccine for pneumonia.
1978	•	Boyer synthesized the human insulin gene (i.e., a synthetic version of it) and inserted it into the bacterium *E. coli*, allowing the bacterium to produce human insulin.
	•	Louise Brown, the first test-tube baby, was born in the UK. The first vaccine for meningococcal meningitis was developed.
		The 1980s
1980	•	According to the US Supreme Court, genetically altered life forms could be patented, creating vast possibilities for commercially exploiting genetic engineering.
	•	The first patent of this nature was awarded to the Exxon oil company to patent an oil-eating micro-organism, which would afterward be employed in the 1989 cleanup of the Exxon oil spill at Prince William Sound, Alaska. S Cohen and D H Boyer received a US patent for gene cloning.
	•	The first automatic gene machine was developed in California.
	•	Launch of Amgen, which would grow to become the world's largest biotechnology medicines company.
1981	•	Baruch Blumberg and Irving Millman developed the first vaccine for hepatitis B.
	•	Researchers in Switzerland cloned mice.
	•	The first transgenic animals were produced by transforming genes from other animals into mice.
1982	•	The FDA supported the first recombinant protein.
1983	•	Luc Montagnier of the Pasteur Institute in Paris isolated the AIDS virus.
	•	Kary Mullis discovered the polymerase chain reaction (PCR), a technique for multiplying DNA sequences.
	•	PCR was identified as the most innovative molecular biology technique. The FDA sanctioned a monoclonal antibody-based diagnostic analysis to identify *Chlamydia trachomatis*.
	•	The first artificial chromosome was produced and the first genetic markers for specific inherited diseases were discovered.

1984	• The DNA fingerprinting technique was discovered. When a restrictive enzyme is applied to DNA from various individuals, the ensuing sets of fragments sometimes vary noticeably from one person to the next. Such differences in DNA are called restriction fragment length polymorphisms and are particularly helpful in genetic investigations. • The first genetically engineered vaccine was discovered for hepatitis B. The whole genome of the human immunodeficiency virus (HIV) virus was cloned and sequenced.
1985	• Genetic fingerprinting stepped into the courtroom. • Genentech became the first biotechnology organization to launch its own biopharmaceutical product. • Genetically engineered plants resistant to viruses, insects, and bacteria were field-tested for the first time. • Cloning of the gene that encodes human lung surfactant protein was achieved. This was a major step toward reducing premature birth problems. • The NIH sanctioned guidelines for executing trials of gene therapy on humans.
1986	• Peter G Schultz from UC Berkeley explained how to conjugate antibodies and enzymes (abzymes) to create therapeutics. • The automated DNA sequencer was developed in California. The FDA sanctioned the first monoclonal antibody treatment to fight kidney transplant rejection. • The FDA sanctioned the first biotech-derived interferon drugs to treat cancer. Drugs to treat Kaposi's sarcoma, a complication of AIDS, were discovered. • The FDA sanctioned the first genetically engineered human vaccine to avert hepatitis B.
1987	• The FDA sanctioned a genetically engineered tissue plasminogen activator to treat heart attacks. • Maynard Olson and colleagues at Washington University discovered yeast artificial chromosomes, which are expression vectors for large proteins. • Reverse transcription and the PCR were linked to augment messenger RNA sequences. DNA microarray technology, the use of a set of various DNAs in arrays for expression outline, was first explained. • The collection of DNA was used to recognize genes whose expression is altered by interferon. • The FDA sanctioned a diagnostic serum tumor marker test for ovarian cancer.

1988	• Congress financed the Human Genome Project, a huge international attempt to map and sequence the human genetic code as well as the genomes of other species. • The first contract between two organizations with parallel patents for cross-licensing of biotech products occurred and became the example.
1989	• The FDA sanctioned Amgen's first biologically derived human therapeutic. • Oil-eating bacteria were employed to clear up the Exxon Valdez oil spill. • A gene responsible for cystic fibrosis was explored.
The 1990s	
1990	• The first nationally sanctioned gene therapy treatment was executed effectively on a four-year-old girl suffering from an immune disorder called adenosine deaminase deficiency. • The Human Genome Project was launched. • The FDA approved the first hepatitis C antibody test, which helped to guarantee the purity of blood bank products. • The FDA sanctioned a bioengineered form of the protein interferon gamma to treat chronic granulomatous disease. • The FDA sanctioned a modified enzyme for enzyme replacement therapy to treat severe combined immunodeficiency disease. It was the first successful application of enzyme replacement therapy for an inherited disease.
1992	• The US Army accumulated blood and tissue tests from all new recruits as part of a genetic dog-tag plan intended to achieve better recognition of soldiers killed in combat. • The FDA sanctioned the first genetically engineered blood-clotting factor—a recombinant protein used to treat hemophilia A. • The FDA sanctioned a recombinant protein to treat renal cell cancer. American and British researchers revealed a technique for analyzing embryos *in vitro* for genetic abnormalities, e.g., cystic fibrosis and hemophilia.
1993	• The FDA approved a recombinant protein to treat multiple sclerosis—marking the first new multiple sclerosis treatment in 20 years. • A global research team led by Daniel Cohen from the Center for the Study of Human Polymorphisms, Paris, created a rough map of all 23 pairs of human chromosomes. • Two minor trade organizations merged together to form the Biotechnology Industry Organization, an international biotechnology support group.

1994	• The FDA sanctioned a recombinant protein to deal with growth hormone (GH) deficiency. • Mary-Claire King at UC Berkeley explored the first breast cancer gene, BRCA1. The FDA sanctioned a modified enzyme to deal with Gaucher's disease. • A number of genes, human and otherwise, were identified and their functions explained. These comprised: Ob, a gene inclining to obesity; BCR, a breast cancer receptiveness gene; BCL-2, a gene linked to apoptosis (programmed cell death); Hedgehog genes (named because of their shape), which synthesize proteins that direct cell differentiation in complex organisms; and Vpr, a gene regulating the reproduction of the HIV virus. • Genetic linkage studies recognized the role of genes in a variety of disorders, including bipolar disorder, cerulean cataracts, melanoma, dyslexia, prostate cancer, thyroid cancer, hearing loss, sudden infant death syndrome, and dwarfism. • The FDA sanctioned a genetically engineered description of human DNase, which breaks down protein accretion in the lungs of cystic fibrosis patients. This corresponded to the first new therapeutic drug for treating cystic fibrosis.
1995	• The first baboon-to-human bone marrow transplant was executed on an AIDS patient. • The first vaccine for hepatitis A was explored. • The NIH, the US Army, and the Centers for Disease Control and Prevention were considerably involved in the growth and clinical testing of the vaccine. • Researchers at the Institute for Genomic Research completed the first full gene sequence of a living organism for the bacterium *Haemophilus influenzae*. • A European study group determined a genetic defect that turned out to be the most frequent cause of deafness.
1996	• Researchers at the Department of Biochemistry at Stanford University and Affymetrix developed the gene chip, a small glass or silica microchip that contains thousands of individual genes that can be examined simultaneously. This symbolized a scientific breakthrough in gene expression and DNA sequencing technology. • Research groups sequenced the complete genome of a complex organism, *Saccharomyces cerevisiae*, otherwise known as baker's yeast. The accomplishment symbolizes the entire sequencing of the largest genome to date. A novel, economic diagnostic biosensor test was developed to hasten the detection of a toxic strain of *E. coli*, the bacteria responsible for several food-poisoning outbreaks.

1997	•	The first human artificial chromosome was discovered.
	•	A mixture of natural and synthetic DNA was used to synthesize a genetic cassette that could possibly be adapted and employed in gene therapy.
	•	The FDA sanctioned a recombinant follicle-stimulating hormone to deal with infertility.
	•	The FDA permitted the first bloodless HIV-antibody analysis, which used cells from patients' gums. Researchers at the Institute for Genomic Research sequenced the entire genome of the Lyme disease pathogen, *Borrelia burgdorferi*, along with the genome for the organism associated with stomach ulcers, *Helicobacter pylori*.
	•	Researchers at the University of Wisconsin–Madison sequenced the *E. coli* genome.
	•	The FDA permitted the first therapeutic antibody to treat cancer in the USA. It was employed for patients with non-Hodgkin's lymphoma.
1998	•	Human skin was created in the laboratory for the first time.
	•	Two research groups cultured embryonic stem cells.
	•	Embryonic stem cells were employed to regenerate tissue and produce disorders mimicking diseases.
	•	Researchers at the Sanger Institute in the UK and at the Washington University School of Medicine in St Louis, USA, sequenced the first whole animal genome for the *Caenorhabditis elegans* worm.
	•	A rough draft of the human genome map was created, displaying the sites of more than 30,000 genes.
	•	The first vaccine for Lyme disease was discovered.
	•	The FDA sanctioned a novel monoclonal antibody to treat Crohn's disease.
	•	A monoclonal antibody therapy employed against breast cancer had positive outcomes, indicating a new age of management based on the molecular targeting of tumor cells.
	•	Support for using the HER2 inhibitor for the management of breast cancer in patients who had tested positive for the HER2 mutation brought personalized medicine to oncology.
The 2000s		
2000	•	Har Gobind Khorana synthesized DNA in a test tube.
	•	Kary Mullis added value to Har Gobind Khorana's findings by amplifying DNA in a test tube to create 1,000 times more than the original amount of DNA.
	•	Sir Ian Wilmut cloned an adult sheep and called the cloned sheep 'Dolly.'

	•	Craig Venter sequenced the human genome; the first publicly accessible genomes would later be those of James Watson and Ventor.
	•	Researchers at Celera Genomics and the Human Genome Project successfully completed a rough draft of the human genome.
2001	•	The journals *Science* and *Nature* reported the human genome sequence, making it feasible for researchers all over the world to start investigating innovative treatments for diseases that have genetic origins, e.g., heart disease, cancer, Parkinson's, and Alzheimer's.
2002	•	An era of very rapid shotgun sequencing of major genomes was completed. Included were the mouse, chimpanzee, dog, and hundreds of other species.
2003	•	Celera and the NIH successfully finished the sequencing of the human genome.
2004	•	The FDA supported the first monoclonal antibody (i.e., antiangiogenic, inhibiting blood vessel formation or angiogenesis) for cancer therapy.
	•	The FDA approved a DNA microarray analysis system, which helped in selecting medications for different conditions. This was a significant step toward modified medicine.
2006	•	The FDA sanctioned a recombinant vaccine against human papillomavirus, which causes genital warts and cervical cancer.
	•	Researchers established the three-dimensional (3D) structure of HIV, which causes AIDS.
2007	•	Researchers discovered how to use human skin cells to produce embryonic stem cells.
2008	•	Venter replicated a bacterium's genetic structure completely from laboratory chemicals, taking a step nearer to generating the world's first living artificial organism.
2008	•	Japanese chemists developed the first DNA molecule made nearly entirely of artificial parts. The finding could be used in areas of gene therapy.
2009	•	Former US President Barack Obama signed an administrative order releasing national funding for broader research on embryonic stem cells.
	•	Scientists identified three new genes connected with Alzheimer's disease, paving the way for possible new diagnostics and therapeutics.
	•	Geron commenced the first FDA-approved clinical trial employing embryonic stem cells.
2010	•	The FDA sanctioned a modified prostate cancer medicine that improves a patient's immune cells to distinguish and attack cancer cells.
	•	The FDA sanctioned an osteoporosis treatment that was one of the first medicines based on genomic investigations.
	•	Craig Venter showed that a synthetic genome could duplicate alone.
2011	•	A trachea developed from stem cells was transplanted into a human recipient. Progress in 3D printing technology resulted in 'skin-printing.'

	• The FDA sanctioned the first cord blood therapy to be employed in hematopoietic stem cell transplantation protocols in patients with disorders influencing the hematopoietic system.
2012	• The FDA issued draft regulations for biosimilar drugs.

Source: Saurabh, B., (2018). Introduction to Pharmaceutical Biotechnology, 1, IOP Publishing Ltd., doi: 10.1088/978-0-7503-1299-8ch1.

1.1.2. Types of Biotechnology

The science of biotechnology is further categorized into subdisciplines that are color-coded based on common uses and applications.

1. **Red Biotechnology:** It involves medical processes, such as using organisms to produce new drugs and stem cells to regenerate damaged human tissues and grow and regrow entire organs.
2. **White or Gray:** It refers to industrial processes, such as the development of new chemicals or new biofuels for vehicles.
3. **Green:** It covers agricultural processes, such as producing pest-resistant crops, disease-resistant animals and environmentally friendly agricultural practices.
4. **Gold:** Also known as bioinformatics, is a cross between biological processes and informatics. It refers to the methods healthcare workers use to gather, store, and analyze biological data to treat patients.
5. **Blue:** It encompasses processes in marine and aquatic environments, such as converting aquatic biomass into fuels and pharmaceuticals.
6. **Yellow:** It refers to processes that aid food production, the most popular application being the fermentation of alcohol and cheese.
7. **Violet:** It ensures the practice of biotechnology is in compliance with laws and ethical standards governing each field.
8. **Dark:** It is the use of biotechnology for weapons or warfare.

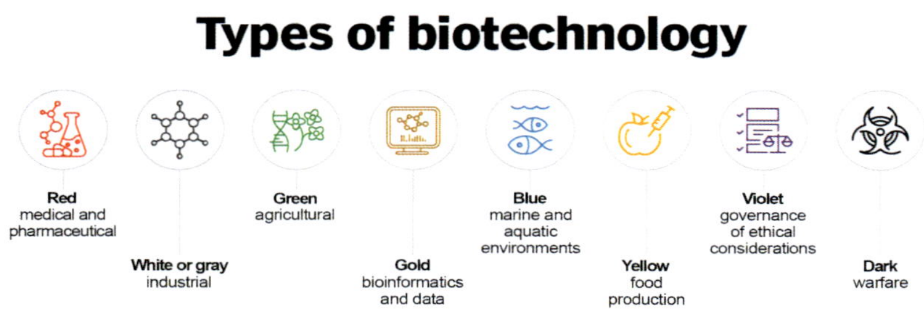

Figure 1.1. Types of biotechnology.

Source: https://cdn.ttgtmedia.com/rms/onlineimages/types_of_biotechnology-f.png.

1.1.3. Advantages and Disadvantages of Biotechnology

1.1.3.1. Advantages of Biotechnology

1. **Medical Advancements:** Biotechnology has revolutionized medicine by enabling the development of novel drugs, vaccines, and therapies. It has played a crucial role in the treatment of various diseases, including genetic disorders and certain types of cancer.
2. **Increased Agricultural Productivity:** GM crops have been designed to resist pests, diseases, and harsh environmental conditions, leading to higher crop yields. This helps address food security concerns and reduce the need for chemical pesticides.
3. **Environmental Sustainability:** Biotechnology offers solutions for environmental issues, such as using microorganisms to clean up pollutants and the development of biofuels as alternatives to traditional fossil fuels. These approaches contribute to more sustainable practices.
4. **Improved Industrial Processes:** Industrial biotechnology has led to more efficient and eco-friendly manufacturing processes. The use of biocatalysts and fermentation in industries reduces the reliance on harsh chemicals and minimizes environmental impact.
5. **Bioremediation:** Biotechnology can be employed for the cleanup of contaminated environments. Microorganisms are utilized to break down pollutants, helping to restore ecosystems affected by pollution.

Keyword

Bioinformatics – It is a subdiscipline of biology and computer science concerned with the acquisition, storage, analysis, and dissemination of biological data (Figure 1.1).

1.1.3.2. Disadvantages of Biotechnology

1. **Ethical Concerns:** Manipulating genes and organisms raises ethical questions, particularly in the areas of human genetics and cloning. The ethical implications of certain biotechnological applications, such as genetic modification, can be complex and divisive.
2. **Environmental Risks:** The release of GM organisms into the environment may pose potential risks, such as unintended ecological consequences or the development of resistant pests and diseases. Assessing and managing these risks is a significant challenge.
3. **Social and Economic Inequities:** The accessibility and affordability of biotechnological advancements may create

disparities, with certain populations or countries benefiting more than others. This can lead to social and economic imbalances on a global scale.

4. **Unknown Long-Term Effects:** The long-term effects of some biotechnological interventions, especially in the field of genetic engineering, may not be fully understood. Unintended consequences could emerge over time, raising concerns about the safety of these technologies.
5. **Loss of Biodiversity:** The widespread adoption of GM crops may lead to the displacement of traditional crop varieties, potentially resulting in a loss of biodiversity. This can have negative implications for ecosystem resilience and adaptability.

Did you get it?

1. Discuss the advantages associated with the use of biotechnology.
2. Explore potential drawbacks by outlining the disadvantages of biotechnology.

1.2. BASIC APPLICATIONS OF BIOTECHNOLOGY

→Learning Objectives

- Understand the most common applications of biotechnology;
- Define cloning and DNA fingerprinting.

As Table 1.4 illustrates, there are many well-established biotechnology applications that are divided into their own distinct categories.

Table 1.4. Applications of Biotechnology in Different Areas

Area	**Applications**
Plant biotechnology	Transgenic plants, production of secondary metabolite, production of pathogen-free plants or crop improvement, production of herbicide-resistant crops, pest-resistant ('Bt concept' pest-resistant transgenic) plants, drought resistance, flood resistance, salt tolerance, high-yielding GM crops, nitrogen fixing ability, acidity and salinity tolerance, *in vitro* germplasm conservation, genetic variability, *in vitro* pollination, induction of haploidy, somatic hybridization, genetic transformation, molecular pharming, somatic embryogenesis, organogenesis, phytoremediation, in vitro plant germplasm conservation, mutant selection, somaclonal variation, plant genome analysis, hybrid seeds, artificial seeds.
Animal biotechnology	Biopharmaceuticals: Production of hormones, growth factors, interferons, enzymes, recombinant proteins, vaccines, blood components, oligonucleotides, transcription factor-based drugs, oligonucleotides.
	Antibiotics
	Replacement therapies: Lack of production of normal substances (factor VIII—missing in hemophilia, insulin).
	Diagnostics: antibodies, biosensors, PCR, therapeutics, vaccines, medical research tools, human genome research, development of biosensors.
	In Vitro Fertilization, Embryo Transfer
	Gene therapy
	Stem cell therapy
	Animal tissue culture: Cell, tissue, and organ culture.
	Gene cloning: rDNA technology, genetic engineering, transgenic animals, antibiotics, DNA markers, animal husbandry, xenotransplantation, medical biotechnology.

	Therapeutics: Natural products such as from the foxglove (*Digitalis*, heart conditions) and yew tree (cancer agent, taxol) for breast and ovarian cancers, endogenous therapeutic agents, i.e., proteins produced by the body that can be replicated by genetically engineered, tPA—tissue plasminogen factor (dissolves blood clots), biopharmaceuticals (drug or vaccine developed through biotechnology), therapeutics, i.e., products used to maintain health or prevent disease, biopharming, i.e., production of pharmaceuticals in cultured organisms, certain blood-derived products needed in human medicine can be produced in the milk of goats.
	Biopolymers and medical devices: natural substances useful as medical devices: hyaluronate, an elastic, plastic-like substance used to treat arthritis, prevent post-surgical scarring in cataract surgery, used for drug delivery, adhesive substances to replace stitches.
	Designer drugs: Using computer modeling to design drugs without the lab-protein structure.
	Evolutionary and ecological genomics: Finding genes associated with ecological traits and evolutionary diversification. Common goals are health and productivity.
Agricultural biotechnology	The applications of animal biotechnology, crop biotechnology, horticultural biotechnology, tree biotechnology, food processing, plant biotechnology (photosynthesis improvers, bio-fertilizers, stress-resistant crops and plants, bio-insecticides, and biopesticides), food biotechnology.
	Food: Increased milk production, leaner meat in pork, growth hormones in farm-raised fish that result in earlier market-ready fish.
	Pharmaceuticals: Animals engineered to produce human proteins for drugs, including insulin and vaccines.
	Breeding disease tolerance, exact copies of desired stock, increased yields.
	Health: Microorganisms introduced into feed for beneficial purposes, diagnostics for disease and pregnancy detection, animals engineered to produce organs suitable for transplantation into humans.
Environmental biotechnology	Environmental monitoring: Diagnosis of environmental problems via biotechnology.
	Waste management: Bioremediation is the use of microbes to break down organic molecules or environmental pollutants.
	Pollution prevention: Renewable resources, biodegradable products, alternative energy sources.
Fuel and fodder	Provides a clean and renewable alternative to traditional fossil fuels, the burning of which contributes to global warming.
	Tissue culture technique offers rapid afforestation of degraded forests and regeneration of green cover.
	Biotechnology could play an important role in three ways in the productivity of biomass.
	Can be used to generate methane.

Industrial biotechnology	Metabolite production (acetone, butanol, alcohol, antibiotics, enzymes, vitamins, organic acids), anaerobic digestion (for methane production), waste treatment (both organic and industrial), production of bio-control agents, fermentation of food products, bio-based fuel and energy, industrial microbiology, biotechnology in the galvanizing industry, recovery of metals and minerals, bioethanol, bioconversion of synthesized gas to liquid fuels such as methanol, using bacteria to remove byproducts, pulp, and paper, sugars from starches, animal feed, food, textiles, and leather, pharmaceuticals, an enzymatic process for producing antibiotics.
Aquatic biotechnology	Aquaculture, restoring, and protecting marine ecosystems, improving seafood quality, environmental remediation, marine byproducts for human health, biomaterial and bioprocessing, marine molecular biotechnology.

Source: https://iopscience.iop.org/book/mono/978-0-7503-1299-8/chapter/bk978-0-7503-1299-8ch1.

1.2.1. Most Common Applications

Some of the most common applications of biotechnology are highlighted in subsections.

1.2.1.1. Cloning

The word "cloning" refers to a variety of techniques that can be applied to create genetically identical copies of living things. A clone is a copy of material that shares the same genetic makeup as the original. Table 1.5 lists a number of historical occurrences that influenced the development of cloning.

Table 1.5. The Timeline of Cloning

Year	Cloning Event
1885	First ever display of artificial embryo twinning.
1901	Cloning hypothesized: Hans Spemann splits a two-celled newt embryo into two parts, effectively producing two larvae. (Later, in 1938, Spemann hypothesized that animals could be cloned by fusing an embryo with an egg cell.)
1902	Artificial embryo twinning in a vertebrate achieved.
1928	The first discovery that cell nucleus regulates embryonic development.
1952	First thriving nuclear transfer (frog).
1958	Nuclear transfer from a differentiated cell (frog).
1963	Term 'clone' coined: John Burdon Sanderson Haldane used 'clone' in his speech on the biological potential for the human species over the next 10,000 years.
1963	First cloned fish: Tong Dizhou, an embryologist from China, developed the world's first cloned fish by incorporating the DNA from a cell of a male carp into an egg from a female carp.
1973	Stanley Norman Cohen and Herbert Boyer discovered the tool of DNA cloning, which copies genes to facilitate their transplantation between various biological species.
1975	First mammalian embryo produced by nuclear transfer (rabbit).
1980	First transgenic (genetically modified) mouse.
1982	Giant mouse created by transferring GH genes from a rat.
1984	First mammal produced by nuclear transfer (sheep).

1985	First transgenic domestic animal, a pig was developed.
1987	A sequence of transgenic mice developed carrying human genes.
1987	Nuclear transfer from an embryonic cell (cow).
1995	Transgenic pig hearts made: researchers discover transgenic pig hearts that endure up to 30 h when transplanted to baboons. The FDA sanctions the use of transgenic pig livers as bridge organs for transplant candidates awaiting organs.
1996	Nuclear transfer from laboratory cells (sheep).
1996	The birth of the first cloned animal, Dolly the sheep, was proclaimed.
1997	First primate engineered by embryonic cell nuclear transfer (rhesus monkey).
1997	Cloning of a transgenic lamb (Polly) from cells engineered with a marker gene and a human gene was announced. In this fashion, the genetic alteration of a lamb was combined with the techniques of cloning, thereby generating animals that produce a new protein.
1997	Nuclear transfer from genetically engineered laboratory cells (sheep).
1998–1999	Additional mammals cloned by somatic cell nuclear transfer (mice, cows, and goats).
2000	First pigs cloned: PPL Therapeutics clones the first pigs, engineered to create organs for human transplant.
2000	First transgenic pigs cloned: Infigen clones the first transgenic pigs, as a potential source of organs and tissues for transplant for humans.
2000	The Raelian sect claimed it would clone a human within the year.
2001	Endangered animals cloned by somatic cell nuclear transfer.
2001	A patent from the University of Missouri was approved for a technology for cloning mammals; the University then licensed the patent to Massachusetts company Biotransplant. Critics were troubled since the patent application did not prohibit humans and specifically mentioned human eggs. Others challenged that the patent was for just the process and not the product, although the patent says it covers 'cloned products.'
2001	Human clones banned in England: The Human Reproductive Cloning Act was passed, barring the implanting of cloned embryos in the womb.
2001	Researchers produced the first clone of an endangered species: a type of Asian ox known as a guar. Sadly, the baby guar, which had developed inside a surrogate cow mother, died just a few days after its birth.
2002	US President Bush set up the Council on Bioethics to advise him on issues such as stem cell research and cloning.
2002	Cloning to make body parts: Advanced Cell Technology proclaimed that cells from cloned cow embryos were employed to cultivate kidney-like organs.
2002	CC (Carbon Copy), the first cloned pet: CC is not a phenotypic copy of the animal she was cloned from. A number of people have expressed interest in having their deceased pets cloned in the hope of getting a similar animal to replace the dead one. But as demonstrated by CC, the cloned cat, a clone may not turn out precisely like the original pet whose DNA was employed to create the clone.
2003	Cloning approved for research purposes: the Kentucky House Judiciary Committee banned reproductive cloning but permitted cloning for research purposes. The bill would make the shipping or use of cloned embryos for reproductive purposes a crime punishable by 10–20 years in prison, as well as requiring those conducting cloning research to record it with the state Cabinet for Health Services 30 days before the commencement of research.
2003	An endangered type of ox called the Banteg, was successfully cloned.
2006	A pig was developed to generate omega-3 fatty acids via the incorporation of a roundworm gene.
2007	Primate (rhesus monkey) embryonic stem cells were developed by somatic cell nuclear transfer.
2013	Human embryonic stem cells were developed by somatic cell nuclear transfer.

Source: https://iopscience.iop.org/book/mono/978-0-7503-1299-8/chapter/bk978-0-7503-1299-8ch1.

1. **Reproductive Cloning:** In order to create an egg, genetic material from another source must be used. After that, the egg becomes an embryo, which is then placed in the uterus of a female host to continue developing. Through reproductive cloning, entire animals are duplicated.
2. **DNA/Gene Cloning:** A less complicated technique is DNA cloning, which involves removing DNA from a host and using plasmids to replicate it. This technique can even be used to clone individual genes. Gene cloning creates duplicates of specific genes or DNA segments. Figures 1.2 and 1.3 provide a basic illustration of gene cloning.

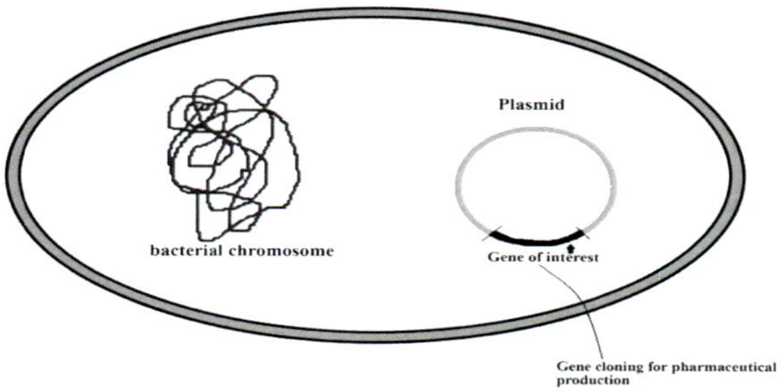

Figure 1.2. Plasmids are transformed into cells during the process of gene cloning for the production of biomolecules.

Source: Saurabh, B., (2018). Introduction to Pharmaceutical Biotechnology, 1, IOP Publishing Ltd., doi: 10.1088/978-0-7503-1299-8ch1.

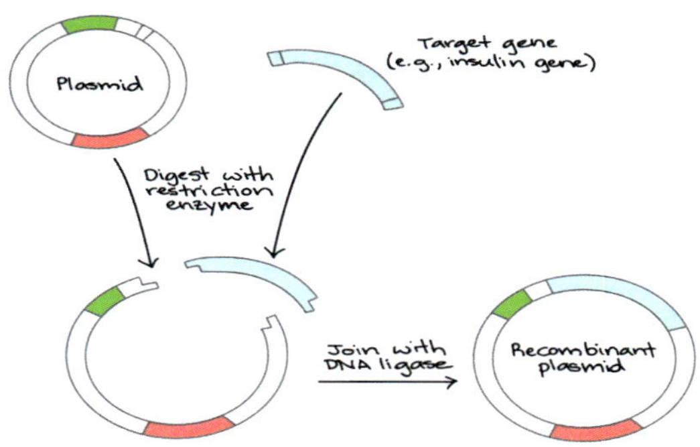

Figure 1.3. Shows the Process of gene cloning.

Source: Overview: DNA cloning (article) | Khan Academy. (n.d.). Khan Academy. https://www.khanacademy.org/science/ap-biology/gene-expression-and-regulation/biotechnology/a/overview-dna-cloning.

Keyword

Inner Cell Mass (ICM) – Or embryoblast (known as the pluriblast in marsupials) is a structure in the early development of an embryo.

3. **Therapeutic Cloning:** Comparable to reproductive cloning, this kind of cloning involves the extraction of embryonic

stem cells for application in host treatment. The destruction of the embryo after stem cell extraction makes this type of cloning controversial despite its many medical benefits for treating a wide range of diseases. Figure 1.4 depicts a basic demonstration of therapeutic cloning. Embryonic stem cells are created through therapeutic cloning and are used in research to create tissues to replace damaged or diseased tissues. The embryonic stem cells come from the inner cell mass (ICM). By dissecting the embryo into individual cells for the ICM cell collection, the embryo is destroyed. Although adult stem cells exist, embryonic stem cells offer the greatest therapeutic potential.

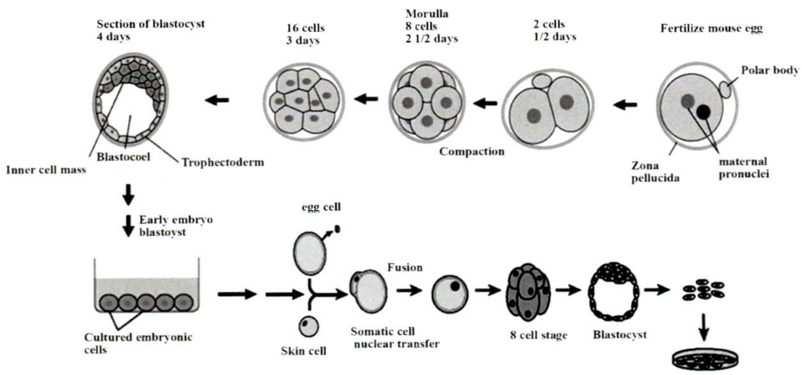

Figure 1.4. Therapeutic cloning from embryonic cells.

Source: https://www.researchgate.net/figure/Therapeutic-cloning-from-embryonic-cells_fig4_326560098.

According to *in vivo* research, therapeutic cloning has enormous promise for de novo organogenesis and the long-term treatment of incurable conditions like diabetes mellitus, Parkinson's disease, and Duchenne muscular dystrophy in the context of cell replacement therapy. Tumorigenicity, epigenetic reprogramming, mitochondrial heteroplasmy, interspecies pathogen transfer, and limited oocyte availability are the main barriers to therapeutic cloning progress. Furthermore, based on the theory of potential, therapeutic cloning is frequently linked to ethical questions about the origin, disposal, and moral standing of IVF embryos. Funding and legislative concerns must also be resolved. Future legislative formulations would take into account distinguishing between therapeutic and reproductive cloning.

1.2.1.1.1. Applications

Researchers may be able to replicate animals through reproductive cloning, which could have advantages for the medical and agricultural domains. Following the discovery of Dolly, the sheep by Scottish scientists, other sheep have undergone genetic modification to yield milk containing a human protein necessary for blood clotting. This study was carried out to determine how to extract milk protein for people whose blood clots improperly. This study, among numerous others, illustrates the potential of cloned animals for testing novel medications and therapeutic approaches. Since cloned animals share the same genetic makeup, using them for drug testing is primarily motivated by this fact. This implies that, unlike animals with diverse genetic makeups, their reactions to the medications ought to be consistent rather than varied.

The FDA ruled in 2008 that milk and meat from cloned animals, including goats, pigs, and cattle, are just as safe as products from non-cloned animals. This ruling permits researchers to employ cloning techniques without restriction. Animals with desired agricultural traits, like high milk production or lean meat, can be replicated using these techniques. However, it will probably take years before food products derived from cloned animals actually show up in supermarkets because cloning is still incredibly expensive. Furthermore, cloning can be used to produce clones in order to restore populations of animal species that are at risk of going extinct. In the era of genetic engineering, transgenesis using rDNA and cloning as a fundamental tool for the acquisition and optimal use of elite traits have dominated research efforts.

1.2.1.2. DNA Fingerprinting

Different alleles are carried by different individuals. The majority of alleles that are helpful for DNA fingerprinting vary in the quantity of repetitive DNA sequences that they contain. DNA fragments of various sizes will result from cutting DNA with a restriction enzyme that can identify sites on either side of the variable region. The process of creating a DNA fingerprint involves examining the length variations of DNA fragments that are generated from various genomic locations. The fingerprint becomes more informative the more often the length variation occurs at a specific site and the more sites are examined. Figure 1.5 shows a basic demonstration of DNA fingerprinting.

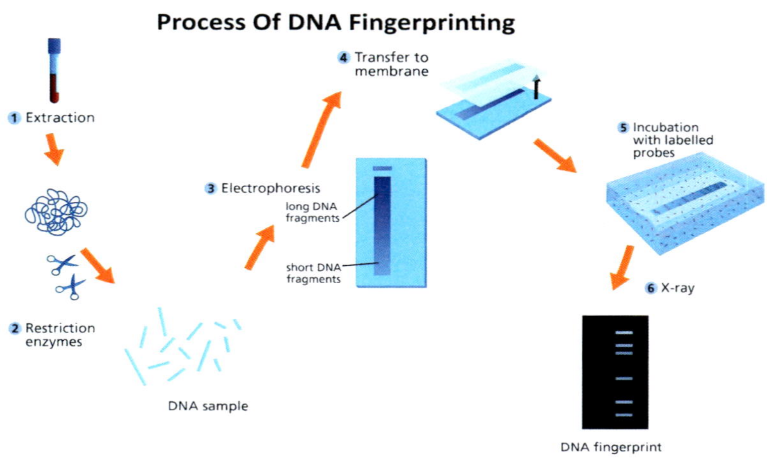

Figure 1.5.Steps in DNA fingerprinting.

Source: Staff, C. (2022, December 17). What is DNA Fingerprinting? - Civilsdaily. CivilsDaily. https://www.civilsdaily.com/news/dna-fingerprinting/ .

Applications for DNA fingerprinting can be found in the disciplines of biology, archaeology, paleontology, and medicine. Its application in biological classification is advantageous when only very small samples, like microscopic fragments of preserved tissue from extinct animals, are available. It can aid in demonstrating evolutionary change and relationships at the molecular level. The DNA fingerprint of a suspect's blood or other bodily material is compared to that of the evidence from the crime scene in criminal investigations to determine how closely the two match. It is also possible to establish paternity using this method. When done correctly, DNA fingerprinting is widely recognized as a trustworthy forensic technique. However, some scientists have advocated for larger human DNA samples to guarantee that the segments under analysis are, in fact, highly variable for all racial and ethnic groups. Although it is feasible to fabricate genetic samples and use them to mislead forensic investigators, samples generated through gene amplification methods can be identified from authentic DNA evidence.

1.2.1.2.1. Applications

1. **Personal Identification:** This is the concept of storing each person's DNA as a bar code on a computer. After discussion, it was determined that this idea would be extremely costly and impractical. It is extremely unlikely to be adopted as a standard system. For example, social security numbers and photo ID cards are far more reliable forms of identification that are unlikely to change.
2. **Paternity and Maternity:** Another well-known use of DNA fingerprinting is this. This test is used to identify the father of a newborn or child. Each person has an inherited variable number tandem repeat (VNTR) pattern from both parents. Although each person's pattern is unique, they are all similar enough to reconstruct the parents' VNTR. This technique can also be used to establish legal nationality or identify the true biological parents of an adopted child. When using a test like this, people should exercise caution because the unexpected results could be upsetting.
3. **Criminal Identification and Forensics**: This area of DNA fingerprinting is highly well-known. Because of the popular TV series CSI: Crime Scene Investigation, it has gained widespread recognition. Given that DNA fingerprinting can establish someone's guilt or innocence in relation to a crime, it is a highly significant application of the technology. A DNA sample from the crime scene must be collected and matched to the suspected person in order for it to be utilized. Next, VNTR patterns are used to compare the two pieces of DNA.
4. **Diagnosis and Cures for Inherited Diseases:** Genetically inherited diseases can also be identified and treated with DNA fingerprinting. Genetic disorders like hemophilia, Huntington's disease, cystic fibrosis, and many more can be identified through DNA fingerprinting. Early detection of the disease increases the likelihood of its defeat and allows for treatment. Some carriers of a disease look for genetic counselors who can use a DNA fingerprint to provide information

and support while also helping them understand the risks of having an affected child. Researchers can use the fingerprints to search for patterns associated with particular diseases and attempt to determine potential treatments.

1.2.1.3. rDNA Technology

Enzymes known as restriction enzymes only cleave DNA at specific sequences (*see* Figures 1.6 and 1.7). The recognition sequences of various restriction enzymes vary. This enables the creation of a large range of distinct gene fragments. The plasmids that produce human insulin are used to replicate rDNA in order to fully utilize the potential of rDNA technology. Bacteria contain tiny DNA circles called plasmids that replicate separately from the bacterial chromosome. A recombinant plasmid can be prepared by incorporating fragments of foreign DNA into a plasmid. Each recombinant plasmid replicates 50–100 times during replication in most cells. Figure 1.7 illustrates the path taken by bacteria to produce human insulin.

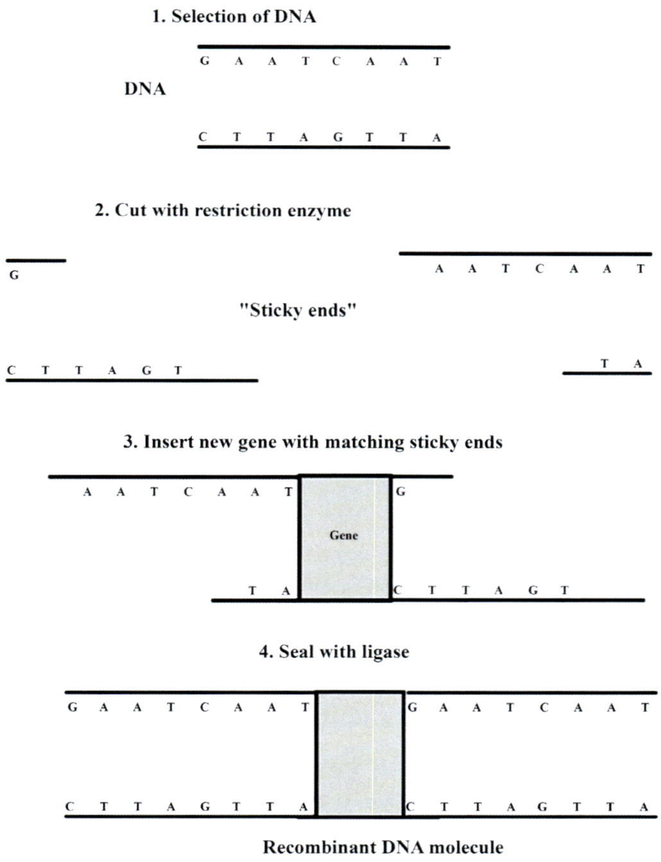

Figure 1.6. Role of restriction enzyme in rDNA technology.

Source: Saurabh, B., (2018). Introduction to Pharmaceutical Biotechnology, 1, IOP Publishing Ltd., doi: 10.1088/978-0-7503-1299-8ch1.

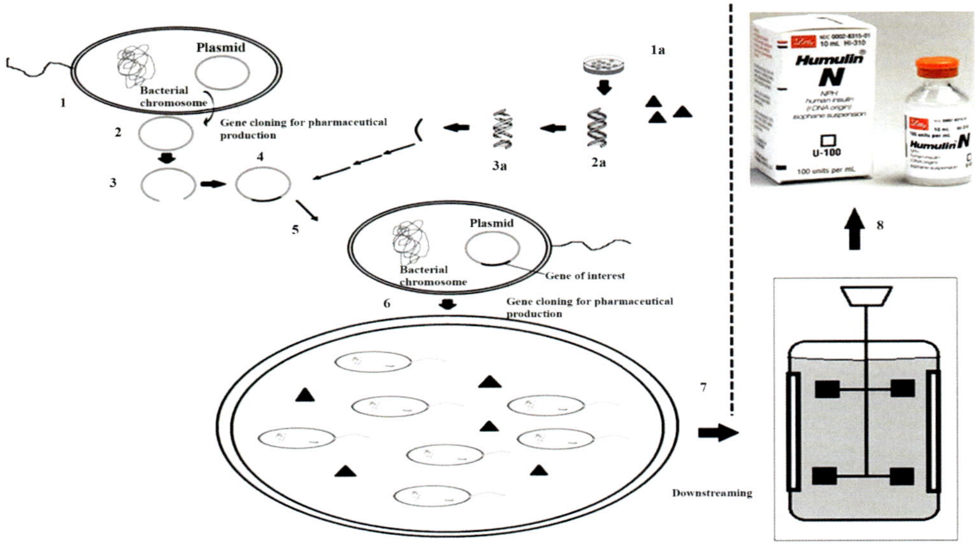

Figure 1.7. rDNA technology. Step 1: Select specific strains of bacteria. Step 2: Isolate the plasmid from the bacteria. Step 3: Cut with the restriction enzyme. Step 1a: Isolate human cells containing the gene of interest (GOI) that codes for the synthesis of protein of interest and grow in tissue culture. Step 2a: Isolate DNA from human cells, and cut with the same restriction enzyme. Step 3a: Allow the insertion of GOI into a plasmid to form a recombinant plasmid. Step 4: Transform the plasmid in bacteria. Step 5: Allow to them to grow on culture medium. Step 6: Multiple bacterial clones are screened for their GOI expression to produce protein. Step 7: Allow mass propagation of the new bacteria. Step 8: Isolate and purify human proteins such as insulin and develop formulations such as Humulin.

Source: Saurabh, B., (2018). Introduction to Pharmaceutical Biotechnology, 1, IOP Publishing Ltd., doi: 10.1088/978-0-7503-1299-8ch1.

In recent years, and even now, there has been a lot of interest in GM plants. Despite this, most people are still unaware of the true nature of GM plants and the benefits and drawbacks of the technology, especially when it comes to the variety of uses for which it can be put to use. Two main areas of concern have emerged since the first generation of GM crops: risks to human health and environmental dangers. The European Union is gradually allowing GM plants to proliferate, which may raise public concerns about possible health risks. Even though the media now regularly supports "health campaigns," the data they disseminate is frequently erratic and untrustworthy in comparison to the body of available scientific evidence. Understanding how transgenic plants are produced using rDNA technology is crucial (Figure 1.8).

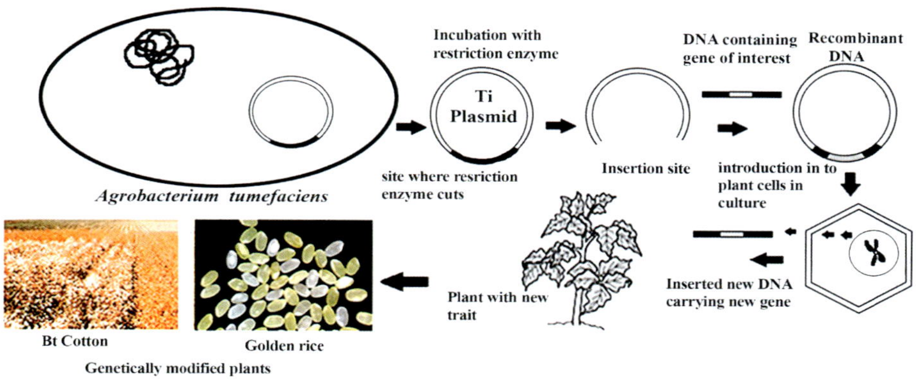

Figure 1.8. rDNA technology-based production of transgenic plant.

Source: https://www.researchgate.net/figure/Therapeutic-cloning-from-embryonic-cells_fig4_326560098.

1.2.1.4. Stem Cell Therapy

A stem cell is an undifferentiated, dividing cell that gives rise to a daughter cell like itself and a daughter cell that becomes a specialized cell type (Figure 1.9).

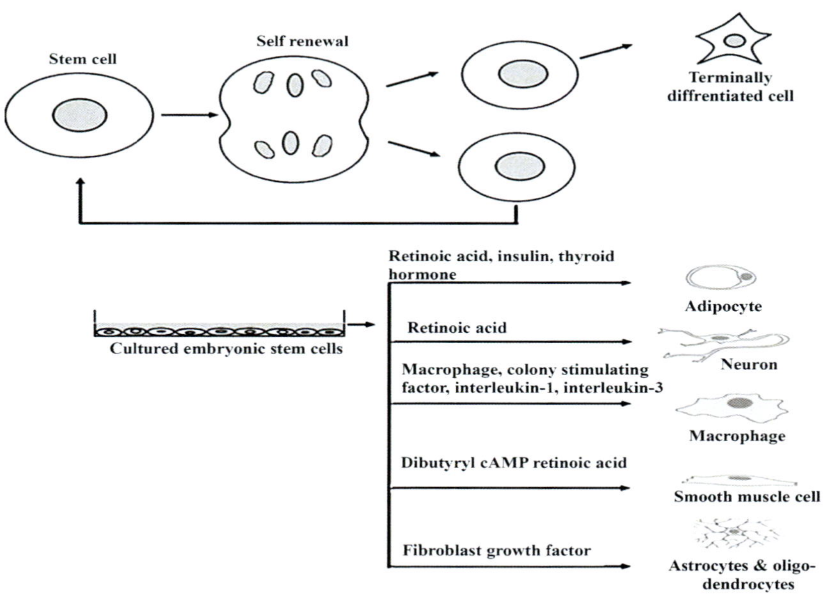

Figure 1.9. Stem cell and its development.

Source: Saurabh, B., (2018). Introduction to Pharmaceutical Biotechnology, 1, IOP Publishing Ltd., doi: 10.1088/978-0-7503-1299-8ch1.

1.2.1.4.1. Applications

Development can be studied using stem cells, i.e., they might clarify for us how an intricate creature grows from a fertilized egg. Scientists can track stem cells in the lab as they divide and diversify into different cell types, such as the skin, bone, and brain. We can learn more about what regulates normal development by identifying the signals and mechanisms that decide which specialized cell type a stem cell differentiates into and whether to continue replicating itself. Aberrant cell division and differentiation are the root cause of some of the most dangerous medical diseases, including cancer and birth defects. Gaining more knowledge about the genetic and molecular mechanisms governing these processes could provide insight into the etiology of these disorders and propose novel treatment approaches. This is a crucial objective for stem cell research.

Stem cells are able to heal illness and replace damaged cells. Patients with leukemia and other blood disorders can benefit from this property as it is already used to treat extensive burns and restore their blood systems. Additionally, stem cells might be the key to replacing cells lost to numerous other debilitating diseases for which there are no long-term treatments at this time. Although damaged tissue is now frequently replaced with donated tissues and organs, the demand for transplantable tissues and organs far exceeds the supply. If stem cells can be trained to differentiate into particular cell types, they may provide a replenishable supply of tissues and cells to treat conditions like diabetes, heart disease, Parkinson's disease, and stroke. Although this is an exciting prospect, there are still significant technical obstacles that will require years of intensive research to overcome.

One potential resource for testing novel medical interventions is stem cells. The need for animal testing could be minimized by testing new medications on specialized cells produced in large quantities from stem cell lines. This is already the case with a variety of cell line types. For example, cancer cell lines are used to screen possible anti-tumor medications.

Did you get it?

1. Enumerate three common applications of biotechnology.
2. Provide a concise definition of cloning in the context of biotechnology.

ROLE MODEL

MAURICE WILKINS: BRITISH BIOPHYSICIST

Maurice Wilkins (born December 15, 1916, Pongaroa, New Zealand—died October 6, 2004, London, England) New Zealand-born British biophysicist whose X-ray diffraction studies of deoxyribonucleic acid (DNA) proved crucial to the determination of the molecular structure of the DNA by James D. Watson and Francis Crick. For this work the three scientists were jointly awarded the 1962 Nobel Prize for Physiology or Medicine.

Wilkins, the son of a physician (who was originally from Dublin), was educated at King Edward's School in Birmingham, England, and St. John's College, Cambridge. His doctoral thesis, completed for the University of Birmingham in 1940, contained his original formulation of the electron-trap theory of phosphorescence and thermoluminescence. He participated for two years during World War II in the Manhattan Project at the University of California, Berkeley, working on mass spectrograph separation of uranium isotopes for use in the atomic bomb.

Upon his return to Great Britain, Wilkins lectured at the University of St. Andrews in Scotland. In 1946, he joined the Biophysics Unit of the Medical Research Council at King's College in London. In 1955 he became its deputy director, and from 1970 to 1980, he served as the unit's director. There he began the series of investigations that led ultimately to his X-ray diffraction studies of DNA. Wilkins headed a group that included Rosalind Franklin, a crystallographer who produced DNA pictures that also aided the work of Crick and Watson. Wilkins later applied X-ray diffraction techniques to the study of ribonucleic acid.

At King's College proper, Wilkins was professor of molecular biology (1963–70) and of biophysics (1970–81) and emeritus professor thereafter. There, he published literature on light microscopy techniques for cytochemical research. His autobiography, *The Third Man of the Double Helix*, was published in 2003.

SUMMARY

- Biotechnology refers to all practices, processes, or technology that harness or modify living organisms or systems for human use. When we break down the term, it will show 'bio' which is life and technology referring to the application of science for a certain purpose.
- In agriculture, biotechnology has led to the development of GM crops with enhanced traits such as resistance to pests, diseases, and environmental stress. This has not only increased crop yields but has also contributed to more sustainable and environmentally friendly farming practices.
- Environmental biotechnology addresses pressing environmental challenges by utilizing microorganisms to remediate polluted sites, degrade waste, and produce biofuels. The field offers innovative solutions for waste management, pollution control, and the development of renewable energy sources, contributing to a more sustainable and eco-friendly future.
- The term cloning describes a number of different processes that can be used to produce genetically identical copies of a biological entity. The copied material, which has the same genetic makeup as the original, is referred to as a clone.
- DNA cloning is a simpler method, whereby DNA is extracted from a host and then replicated using plasmids. Even individual genes can be cloned in this manner. Gene cloning produces copies of genes or segments of DNA.
- The inner cell mass (ICM) is the source of embryonic stem cells. The embryo is destroyed by separating it into individual cells to collect ICM cells. Stem cells are found in adults, but the most promising types of stem cells for therapy are embryonic stem cells.
- Restriction enzymes are enzymes that cut DNA only at particular sequences. Different restriction enzymes have different recognition sequences. This makes it possible to create a wide variety of different gene fragments.

REVIEW QUESTIONS

1. What are the key stages in the development of biotechnology?
2. Name and explain three types of biotechnology applications.
3. How do different stages of biotechnology contribute to its overall development?
4. Explain the concept of DNA fingerprinting and its relevance in biotechnology.
5. How are common applications of biotechnology integrated into various industries?
6. Discuss the significance of cloning in biotechnological advancements.

MULTIPLE CHOICE QUESTIONS

1. **What is biotechnology?**
   a. The study of living organisms
   b. The use of living organisms to develop new products
   c. The study of rocks and minerals
   d. The study of stars and galaxies
2. **What is recombinant DNA technology used for?**
   a. Cloning animals
   b. Modifying genes
   c. Producing genetically modified organisms
   d. All of the above
3. **Who is considered the father of biotechnology?**
   a. Gregor Mendel
   b. James Watson
   c. Paul Berg
   d. Herbert Boyer
4. **In 1973, which landmark event marked the beginning of modern genetic engineering?**
   a. Discovery of DNA
   b. Invention of PCR
   c. Recombinant DNA technology
   d. Human Genome Project
5. **The first genetically modified organism (GMO) was created in 1973. What was it?**
   a. *E. coli* bacteria
   b. Mouse
   c. Tobacco plant
   d. Yeast
6. **What does PCR stand for?**
   a. Polymerase Chain Reaction
   b. Protein Coding Region
   c. Primary Chromosome Replication
   d. Plasmid Cloning Reaction
7. **What is the purpose of gene cloning?**
   a. To create identical twins
   b. To produce multiple copies of a gene

c. To eliminate specific genes

d. To transfer genes between species

8. Which biotechnological technique is used to identify individuals based on their unique DNA patterns?

a. DNA sequencing

b. Gel electrophoresis

c. Southern blotting

d. DNA fingerprinting

9. Which biotechnological application is associated with the production of biofuels?

a. Recombinant DNA technology

b. Fermentation

c. Polymerase Chain Reaction (PCR)

d. Hybridization

10. What is the primary application of CRISPR-Cas9 technology?

a. DNA fingerprinting

b. Genome editing

c. Vaccine development

d. Protein synthesis

Answer to Multiple Choice Questions

1. (b); 2. (d); 3. (d); 4. (c); 5. (a); 6. (a); 7. (b); 8. (d); 9. (b); 10. (b)

REFERENCES

1. Arnold, J. P., (2005). *Origin and History of Beer and Brewing: From Prehistoric Times to the Beginning of Brewing Science and Technology* (p. 34). Cleveland, Ohio: BeerBooks.
2. Boscha, F., & Rosicha, L., (2008). The contributions of Paul Ehrlich to pharmacology: A tribute on the occasion of the centenary of his Nobel Prize. *Pharmacology, 82*, 171–179.
3. Kampowski, S., (2013). *A Greater Freedom: Biotechnology, Love, and Human Destiny*. Eugene, OR: Pickwick Publications.
4. Kilner, J. F., Christopher, H. C., & Diann, B. U., (2002). *Cutting-Edge Bioethics: A Christian Exploration of Technologies and Trends*. Grand Rapids: Eerdmans.
5. Kurzweil, R., (1999). *The Age of Spiritual Machines.* New York: Penguin.
6. Lake, C. B., (2013). *Prophets of the Posthuman: American Fiction, Biotechnology, and the Ethics of Personhood*. University of Notre Dame Press.

7. Langlois, A., (2013). *Negotiating Bioethics: The Governance of UNESCO's Bioethics Program*. New York: Routledge.
8. Lee, S. C., (2007). *The Family, Medical Decision-Making, and Biotechnology: Critical Reflections on Asian Moral Perspectives*. Dordrecht, Netherlands: Springer.
9. Lustig, B. A., Baruch, A. B., & McKenny, G. P., (2010). *Altering Nature: Religion, Biotechnology, and Public Policy* (Vol. II). New York: Springer.
10. Rajan, K. S., (2012). *Lively Capital: Biotechnologies, Ethics, and Governance in Global Markets*. Duke University Press.
11. Richter, F. C., (2015). Remembering Johann Gregor Mendel: A human, a catholic priest, an Augustinian monk, and abbot. *Mol. Gen. Genom. Med., 3*, 483–485.
12. Sharma, S. D., (2006). In: Stanley, W., (ed.), *"Biotechnology Revolution"; Encyclopedia of India* (Vol. 1, pp. 154–157). Thomson Gale.
13. Sutton, S. D., (2009). *Biotechnology: Our Future as Human Beings and Citizens.* State University of New York Press.

CHAPTER

2 Molecular Biology

LEARNING OBJECTIVES

After studying this chapter, you will be able to:

- Understand the DNA structure and function;
- Comprehend the RNA structure and function;
- Familiarize yourself with the genetic code and the concept of a gene;
- Learn the process of translation of RNA into proteins.
- Explain the Polymerase Chain Reaction (PCR)

INTRODUCTORY EXAMPLE

Ø Unraveling the Mysteries of Cellular Communication: Signal Transduction Pathways in Molecular Biology:

Molecular biology, a field delving into the intricate mechanisms governing life at the cellular level, has made significant strides in deciphering signal transduction pathways. These pathways, crucial for cellular communication, play a pivotal role in coordinating various physiological processes, responding to external stimuli, and maintaining cellular homeostasis.

1. **The Intricacies of Signal Transduction:** At the heart of molecular biology lies the fascinating world of signal transduction, where cells interpret and respond to signals from their environment. These signals, often in the form of hormones, growth factors, or environmental cues, set off a cascade of molecular events within the cell, orchestrating a tailored response.
2. **Key Components:** Signal transduction pathways involve a series of molecular players, including receptors, signaling molecules, and intracellular effectors. Receptors, typically proteins on the cell membrane or within the cell, recognize specific signals and initiate the transmission of information. Signaling molecules, such as second messengers, relay the signal inside the cell, triggering a cascade of events. Intracellular effectors then execute the cellular response, which may involve changes in gene expression, cell metabolism, or cell behavior.
3. **Examples in Cellular Regulation:** One prominent example of signal transduction in molecular biology is the cAMP (cyclic adenosine monophosphate) pathway. In response to extracellular signals, receptors activate an enzyme that produces cAMP. This second messenger, in turn, activates protein kinases, leading to the phosphorylation of target proteins and ultimately influencing cellular processes such as gene expression and metabolism.

   Another well-studied pathway is the mitogen-activated protein kinase (MAPK) pathway. This pathway is crucial in transmitting signals from the cell surface to the nucleus, regulating cell growth, differentiation, and apoptosis. Aberrations in these pathways can contribute to various diseases, including cancer.
4. **Disease Implications and Therapeutic Ta**rgets: Understanding signal transduction pathways at the molecular level has provided insights into the molecular basis of numerous diseases. Dysregulation of these pathways is often implicated in cancer, neurodegenerative disorders, and immune system malfunctions. Consequently, researchers are exploring targeted therapies that aim to modulate specific components of these pathways to treat diseases with greater precision and fewer side effects.

   Technological Advances: Technological advances, such as CRISPR-Cas9 gene editing and high-throughput omics technologies, have accelerated our ability to study signal transduction pathways. These tools allow researchers to manipulate genes and analyze cellular responses on a large scale, providing unprecedented insights into the complexity and interconnectedness of molecular networks.

UNIT INTRODUCTION

Molecular biology is a field of biology that explores the intricate mechanisms underlying the structure, function, and interactions of biological molecules within cells. Molecular biology delves into the molecular processes that drive life, examining the fundamental building blocks of living organisms at the smallest scales. One of its central focuses is the study of nucleic acids, particularly DNA and RNA, which carry the genetic information responsible for an organism's development, function, and inheritance.

The discovery of the DNA double helix by James Watson and Francis Crick in 1953 marked a watershed moment in molecular biology, providing a foundational framework for understanding genetic information storage and transfer. DNA, deoxyribonucleic acid, is a long, double-stranded molecule composed of nucleotide units. These nucleotides consist of a sugar-phosphate backbone and nitrogenous bases—adenine (A), thymine (T), cytosine (C), and guanine (G)—that form complementary base pairs, maintaining the structural integrity of the DNA helix.

RNA, or ribonucleic acid, plays a crucial role in molecular biology by serving as a messenger that translates the genetic instructions encoded in DNA into functional proteins. Transcription, the process by which RNA is synthesized from a DNA template, is a key step in this information transfer. The resulting messenger RNA (mRNA) carries the genetic code from the nucleus to the ribosomes, where translation occurs. During translation, transfer RNA (tRNA) molecules read the mRNA code and facilitate the assembly of amino acids into proteins.

Molecular biology extends beyond nucleic acids to encompass the study of proteins and other macromolecules. Proteins, composed of amino acids, play diverse roles in cellular structure, function, and signaling. Understanding the intricate interactions between DNA, RNA, and proteins is essential for unraveling the complexities of cellular processes, including cell division, metabolism, and response to environmental stimuli.

2.1. DNA STRUCTURE AND FUNCTION

→Learning Objectives

- Identify key properties of DNA;
- Explain types of DNA;
- Understand the DNA structure;
- Define functions of DNA;
- Understand the DNA translation.

Deoxyribonucleic acid, or DNA, is an organic compound that contains information needed for protein synthesis as well as genetic instructions. The majority of cells in all living things contain DNA. Genetic heredity takes place there, and it is an essential aspect of reproduction. Transferring the inheritance from the older generation to the future generations is, therefore, the primary goal.

Additionally, viruses experience the same issue. The majority of these organisms have genetic material in the form of DNA or RNA. For instance, the genetic material of many viruses is DNA, but RNA may be present in the genetic material of other viruses. The Human Immunodeficiency Virus (HIV) is RNA-based, but once it binds to a host cell, it changes into DNA. Furthermore, DNA is essential for the synthesis of proteins. The DNA found in the nucleus of every cell is known as nuclear DNA. Eukaryotic organisms are the name of the organisms that contain these cells. The majority of the genomes of the organisms can be coded by it. The DNA of the plastid and mitochondria handles the remainder in the interim. In addition, the DNA found in the cell's mitochondria is known as mitochondrial DNA, and it is passed down from mothers to their offspring. If we take human subjects, we can find about 16,000 base points in their mitochondrial DNA. Similarly, plastids have the ability to contain DNA. They are also essential to photosynthesis.

2.1.1. Properties of DNA

DNA (deoxyribonucleic acid) possesses several key properties that are fundamental to its role as the carrier of genetic information. Here are some important properties of DNA:

1. **Double Helix Structure:** DNA has a double-stranded helical structure, resembling a twisted ladder or spiral staircase. The two strands are held together by

hydrogen bonds between complementary base pairs: A pairs with T, and C pairs with G.

2. **Antiparallel Orientation:** The two strands of DNA run in opposite directions, meaning they are antiparallel. One strand runs from 5' to 3,' while the complementary strand runs from 3' to 5.' This antiparallel arrangement is crucial for DNA replication and other cellular processes.
3. **Base Pairing Specificity:** Adenine pairs specifically with thymine, forming two hydrogen bonds, while cytosine pairs specifically with guanine, forming three hydrogen bonds. This base-pairing specificity ensures the accurate transmission of genetic information during DNA replication and transcription.
4. **Complementary Base Pairing:** The sequence of bases on one strand of DNA determines the sequence on the complementary strand. The principle of complementary base pairing allows for the faithful replication and transcription of genetic information.
5. **Genetic Code:** The sequence of nucleotide bases in DNA encodes genetic information. The genetic code specifies the order of amino acids in proteins, and variations in the DNA sequence contribute to genetic diversity.
6. **Semiconservative Replication:** During DNA replication, each parental DNA strand serves as a template for the synthesis of a new complementary strand. The resulting DNA molecule consists of one original (parental) strand and one newly synthesized (daughter) strand. This semiconservative replication ensures the preservation of genetic information.
7. **High Stability:** The hydrogen bonds between complementary base pairs, along with the helical structure and interactions with proteins, contribute to the high stability of the DNA molecule. This stability is essential for the long-term storage of genetic information.
8. **Compact Packaging:** DNA is organized into structures called chromosomes. DNA molecules are tightly wound around histone proteins, forming nucleosomes. This compact packaging helps fit the extensive DNA strands into the cell's nucleus.
9. **Mutability:** While DNA is generally stable, it can undergo mutations—changes in the sequence of nucleotide bases. Mutations can occur due to errors during replication, exposure to mutagenic agents, or other factors. Mutations contribute to genetic diversity and evolution.
10. **Replication and Repair Mechanisms:** DNA has sophisticated

Keyword

Human Immunodeficiency Virus (HIV) – It is a retrovirus that attacks the immune system, specifically the CD4 cells (T cells), which play a crucial role in the body's defense against infections.

mechanisms for replication and repair. DNA replication ensures accurate transmission of genetic information during cell division, while repair mechanisms fix errors and damage to maintain DNA integrity.

2.1.2. Types of DNA

There are several types of DNA based on their structural and functional characteristics. Here are some of the main types of DNA:

1. **Genomic DNA:** This is the most well-known type of DNA, and it is found in the cell nucleus of eukaryotic organisms. Genomic DNA contains the complete set of genetic instructions for an organism.
2. **Mitochondrial DNA (mtDNA):** Found in the mitochondria of eukaryotic cells, mtDNA is separate from genomic DNA and is inherited maternally. It carries genetic information essential for mitochondrial function, which is crucial for energy production within the cell.
3. **Plasmid DNA:** Plasmids are small, circular DNA molecules often found in bacteria. They can carry extra genes that may provide advantages, such as antibiotic resistance. Plasmids are also widely used in genetic engineering and biotechnology.
4. **Chloroplast DNA (cpDNA):** Found in the chloroplasts of plant cells, cpDNA carries genetic information essential for photosynthesis. Like mitochondrial DNA, chloroplast DNA is inherited maternally.

6. **cDNA (Complementary DNA):** cDNA is synthesized from messenger RNA (mRNA) through a process called reverse transcription. It represents the coding sequences of genes and is often used in molecular biology research and genetic engineering.

2.1.3. DNA Structure

These bases are adenine, guanine, cytosine, and thymine in the case of DNA. Take note of how adenine and guanine, as well as cytosine and thymine, have similar shapes. C and T are categorized as pyrimidines, whereas A and G are purines. While naming things, take note of the terms "ribose" and "deoxyribose." Deoxyribose is simply a ribose without an oxygen atom, as the name suggests. More precisely, the 2-carbon of deoxyribose only has a hydrogen atom attached to it, whereas the 2-carbon of ribose has a hydroxyl group attached to it. There isn't any other distinction between the two sugars.

There are no specific guidelines for the arrangement of nucleotides in relation to their bases when creating a single strand of nucleic acid in vitro. Because phosphodiester bonds are formed between the phosphate group and the pentose, the identities of their nitrogenous bases are inconsequential. For this reason, it's frequently called the sugar-phosphate backbone. When we dissect the term "phosphodiester," we find that it accurately characterizes the relationship: two ester bonds (—O—) with phosphorous positioned between the sugars hold them together. The directionality of this bond and, by extension, of nucleic acids in general, is one of the concepts that frequently stumps students. For instance, we say that DNA polymerase functions in a 5-prime (5') to 3-prime (3') direction when discussing the enzyme that catalyzes the addition of nucleotides in living cells. Though it may sound like obscure molecular biology jargon, this is actually very straightforward. Examine two nucleotides that are connected by a phosphodiester bond once more (Figure 2.1 – bottom left). A cytosine and an adenine nucleotide are joined. The 5-carbon of one deoxyribose (or ribose in RNA) will always be joined to the 3-carbon of the subsequent sugar by a phosphodiester bond. This also implies that a chain of linked nucleotides will have a free 3' hydroxyl (-OH) group at one end and a free 5' phosphate (-PO4) group at the other. These specify an RNA or DNA strand's directionality.

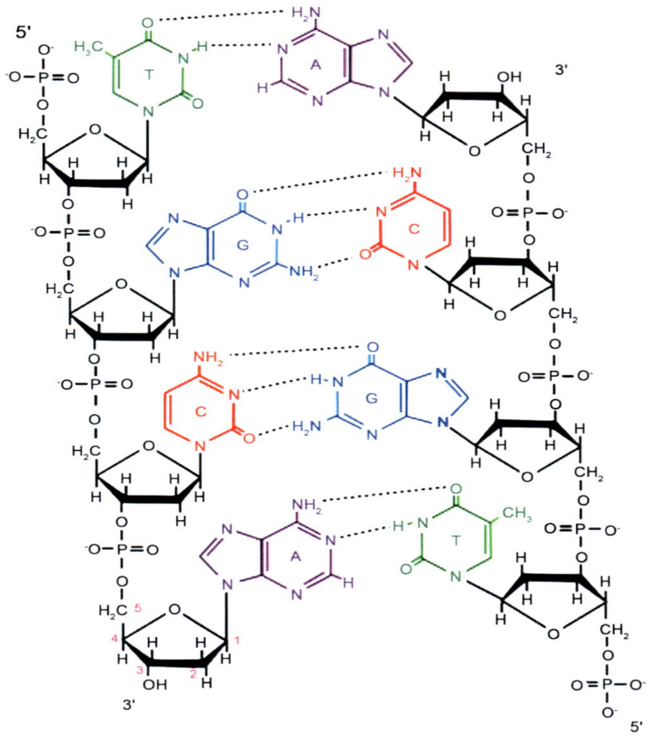

Figure 2.1. Deoxyribonucleic acid (DNA) is a polymer chain of nucleotides connected by 5' to 3' phosphodiester bonds. DNA normally exists as two antiparallel complementary strands held together by hydrogen bonds between adenines (A) and thymines (T), and between guanines (G), and cytosines (C).

Source: E. V. Wong, DNA Structure, bio.libretexts.org.

While RNA is primarily single-stranded, DNA is typically found in cells as double-stranded molecules. However, it is crucial to realize; that RNA and DNA can both become single-stranded or double-stranded under the right circumstances. Indeed, the molecules are so similar that one DNA strand and one RNA strand can be combined to form double-stranded hybrid molecules. Remarkably, compared to the more common **DNA-DNA double helix**, RNA-RNA, and RNA-DNA double helices are marginally more stable.

Base pairing is the fundamental mechanism underlying the double-stranded structure of DNA and, more broadly, the function of nucleic acids as a genetic information storage and transfer medium. Base-pairing is the process by which guanines and cytosines as well as adenines and thymines form hydrogen bonds. Compared to any association formed with the other potential bases, these pairs are noticeably more stable. Moreover, these base-pair associations exhibit uniform and highly stable spacing when they arise within the framework of two nucleic acid strands. As you may remember, hydrogen bonds are not very strong. But when it comes to DNA, hydrogen bonds give DNA its remarkable stability and suitability as a long-term genetic information storage medium. DNA double helices are at least thousands of nucleotides long, even in simple prokaryotes, which indicates that several 1,000 hydrogen bonds are holding the two strands together. Even though a tiny temperature increase or change in the ionic strength of the solution could easily temporarily disrupt any individual nucleotide-to-nucleotide hydrogen bonding interaction, a complete double-helix of DNA requires very high temperatures (usually over 90°C) to completely denature the double helix into individual strands.

The two strands are essentially backup copies of each other because the nucleotides are paired exactly one-to-one, providing a safety net in case nucleotides are lost from one strand. As long as the other strand remains intact in the damaged area, the essential information can still be written into the complementary sequence of the other strand, even if parts of both strands are damaged. However, bear in mind that although one DNA strand can serve as a "backup" for the other, the two strands are complementary rather than identical.

A DNA double helix has antiparallel strands. This indicates that one strand of a DNA double helix would be built in the 5' to 3' direction, while the complementary strand would be built in the 3' to 5' direction, if we were to view the molecule from left to right. We will talk about the role of these enzymes in creating and repairing DNA shortly. The left strand in Figure 2.1 is 5' to 3' from top to bottom, while the other strand is 5' to 3' from bottom to top. DNA molecules are negatively-charged (all those phosphates), and normally a double-helix with a right-handed twist. In this normal (also called the "B" conformation) state, one full twist of the molecule encompasses 11 base pairs, with 0.34 nm between each nucleotide base. Each of the nitrogenous base is planar, and when paired with the complementary base, forms a planar "rung" on the "ladder" of DNA. These are perpendicular to the longitudinal axis of the DNA. Most of the free-floating DNA in a cell, and most DNA in any aqueous solution of near-physiological osmolarity and pH, is found in this B conformation.

Did you know?

In 1958, Francis Crick proposed the Central Dogma, a conceptual framework outlining the unidirectional flow of genetic information from DNA to RNA to proteins. This conceptualization provided a roadmap for molecular processes within living organisms.

However, other conformations have been found, usually under very specific environmental circumstances. A compressed conformation, A-DNA, was observed as an artifact of in vitro crystallization, with slightly more bases per turn, shorter turn length, and base-pairs that are not perpendicular to the longitudinal axis. Another, Z-DNA, appears to form transiently in GC-rich stretches of DNA in which, interestingly, the DNA twists in the opposite direction. It has been proposed that DNA's A and Z forms are both physiologically significant. The A form may be present in RNA-DNA hybrid double helices and when DNA is complexed with specific enzymes, according to available data. DNA methylation may cause the Z conformation to happen. Moreover, the "normal" conformation of B-DNA is essentially an idealized structure predicated on being completely hydrated, which is most likely the case inside a cell. However, because of the minute fluctuations in that hydration state, the DNA conformation frequently deviates slightly from the B-conformation parameters shown in Figure 2.2.

Keyword

DNA-DNA double helix – It is a structure formed by two complementary strands of DNA coiled around each other in a spiral shape. This configuration, stabilized by hydrogen bonds between paired bases, encodes genetic information.

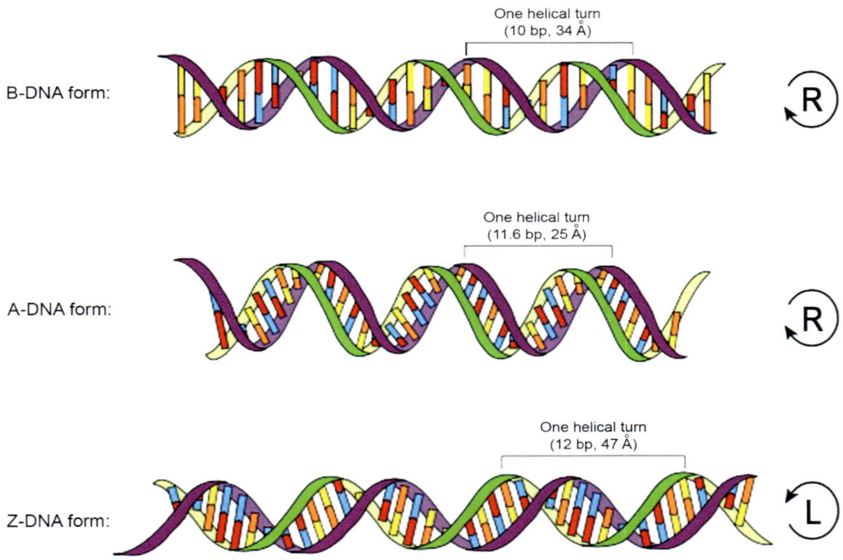

Figure 2.2. Three conformations of DNA. B-DNA is most common, A-DNA is likely an artifact of crystallization in vitro, and Z-DNA may form transiently in parts of the chromosome.

Source: E. V. Wong, DNA Structure, bio.libretexts.org.

Since prokaryotes have no other option, it is rather obvious that the DNA is found in the cytoplasm, whereas in eukaryotes, the DNA is found inside the nucleus. Proteins that aid in organizing and stabilizing the overall chromosome structure are packaged into both prokaryotic and eukaryotic DNA, despite the differences in their locations, degrees of protection from outside forces, and most importantly, sizes. Although some of the proteins found in prokaryotic and eukaryotic chromosomes share structural similarities, relatively little is known about the packing of prokaryotic chromosomes. Consequently, the majority of beginning cell biology courses continue to focus on eukaryotic chromosomal packing (Figure 2.3).

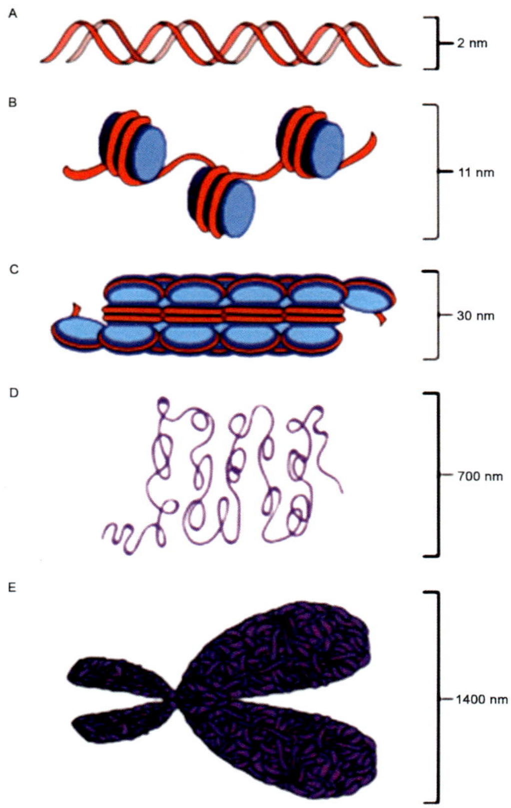

Figure 2.3. DNA packaging. (A) A naked strand of DNA is approximately 2 nm in diameter. (B) Histones, which are octameric proteins depicted here as roughly cylindrical protein, have positive charges distributed on the outer surface to interact with the negatively-charged DNA backbone. (C) Even the organization afforded by histone binding can leave an unmanageable tangle of DNA, especially with longer eukaryotic genomes, and therefore, the histone-bound DNA is packaged into the "30-nm strand." This is held together, in part, by histone interactions. (D) The 30-nm fibers are looped into 700-nm fibers, which are themselves formed into the typical eukaryotic chromosome (E).

Source: E. V. Wong, DNA Structure, bio.libretexts.org.

Prokaryotic or eukaryotic naked DNA is a very thin strand of material with a diameter of about 11 nm. However, if the DNA was kept in the nucleus in that manner, it would become uncontrollably tangled due to the size of eukaryotic genomes. Imagine a bucket that you have filled with 100 meters of yarn, and you haven't tried to coil or bunch it in any way. Now think about whether you could reach into that bucket and pull on a single strand, in which case you would probably pull up a single strand, or if you would probably pull up at least a small tangle of yarn. The cell does for the yarn basically what you would do for it: it is carefully wound into more manageable, smaller skeins. The nucleosome, the first unit of chromosomal organization in DNA, is formed by each chromosome looping around a histone complex (Figure 2.4).

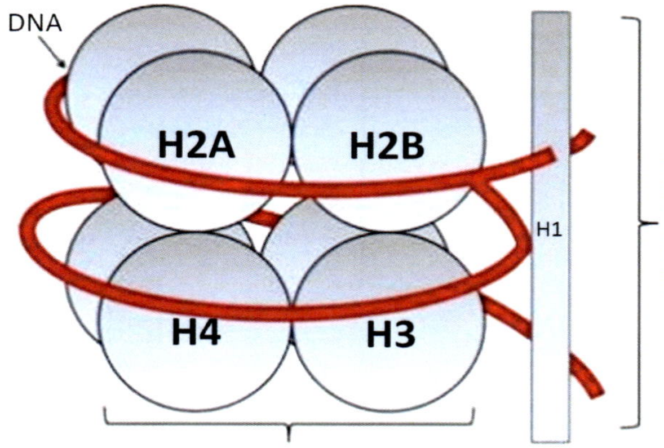

Figure 2.4. The nucleosome is composed of slightly over two turns of DNA around a histone core containing two copies each of H2A, H2B, H3, and H4 histones. The H1 histone is not part of the core unit and functions in coordinating interaction between nucleosomes.

Source: https://df0b18phdhzpx.cloudfront.net/ckeditor_assets/pictures/1272134/original_01.JPG.

A family of basic, positively-charged proteins is known as histones. Their primary role is to organize DNA. The nucleosome is created when DNA wraps around a core of eight histones, two of each of H2A, H2B, H3, and H4. This process occurs a little more than twice. Since DNA is extremely negatively-charged, as was previously mentioned, the quantity and orientation of the positive charges—mostly from lysines and arginines—are essential to their capacity to firmly bind DNA. It's not just a dating adage from the advice columns that "opposites attract."

Remember

Genes, specific sequences within DNA, play a central role in encoding instructions for the synthesis of proteins – the essential building blocks and functional entities of cells. This relationship between DNA and proteins is at the core of genetic expression and cellular function.

2.1.4. Functions of DNA

The functions of DNA are central to the growth, development, functioning, and reproduction of cells and organisms. Here are the key functions of DNA:

1. **Genetic Storage:** Function – DNA serves as the repository of genetic information. It contains the instructions for building and maintaining an organism. The genetic code is encoded in the sequence of nucleotide bases (adenine, thymine, cytosine, and guanine) along the DNA strands.
2. **Transmission of Genetic Information:** Function – DNA is passed from one generation to the next during reproduction.

This ensures the continuity of genetic traits and characteristics from parents to offspring.

3. **Protein Synthesis:** Function – DNA provides the instructions for the synthesis of proteins, which are essential for the structure, function, and regulation of cells. The process involves transcription of DNA into mRNA (messenger RNA) and translation of mRNA into proteins.
4. **Cellular Replication:** Function – DNA is duplicated during cell division (replication) to ensure that each daughter cell receives a complete set of genetic information. This process is crucial for growth, development, and tissue repair.
5. **Gene Expression Regulation:** Function – DNA plays a role in the regulation of gene expression. Various mechanisms, such as promoter regions and regulatory proteins, control when and to what extent specific genes are transcribed and translated into proteins.
6. **Epigenetic Information:** Function – DNA carries epigenetic information, including modifications such as DNA methylation and histone acetylation, which can influence gene expression without altering the underlying DNA sequence. Epigenetic modifications are crucial for cellular differentiation and the adaptation of cells to environmental factors.
7. **Chromosome Structure and Organization:** Function – DNA is organized into structures called chromosomes. Chromosomes provide a stable and organized way to package and segregate genetic material during cell division.
8. **Response to Environmental Stimuli:** Function – DNA can be influenced by environmental factors, and certain genes may be turned on or off in response to external stimuli. This adaptability allows organisms to respond to changes in their surroundings.
9. **Repair and Maintenance:** Function – DNA repair mechanisms exist to correct errors that may occur during DNA replication or damage caused by environmental factors (e.g., radiation, chemicals). Maintenance of DNA integrity is crucial for the prevention of mutations and the proper functioning of cells.
10. **Storage of Evolutionary Information:** Function – DNA serves as a record of evolutionary history. Changes in the DNA sequence over time contribute to the diversity of species and the adaptation of organisms to their environments.

2.1.5. Chemical Modification

Once a DNA molecule is put together, it can undergo chemical modification. This can happen inadvertently through oxidation, ionizing radiation, or the action of chemical carcinogens, or it can happen intentionally by specialized enzymes known as DNA methyltransferases. Enzymes known as nucleases are also capable of cleaving and breaking down DNA as described below.

2.1.5.1. Methylation

There are three known forms of spontaneous methylation in DNA. N4-methylcytosine is created when cytosine is modified on the exocyclic amino group or on the ring to create 5-methylcytosine. N^6-methyladenine can be created by modifying adenine. While

5-methylcytosine is widely distributed, N^4-methylcytosine and N^6-methyladenine are exclusively present in bacteria and archaea. This methylation is caused by specialized enzymes known as DNA methyltransferases, that identify particular DNA molecule sequences to modify only a portion of the bases. Sometimes, carcinogens induce other methylations of the bases or of the deoxyribose. These must be eliminated if they are not to become mutagenic because they typically cause mispairing of the bases during replication.

There are numerous biological uses for natural methylation. Methylation plays a crucial role in the immune system of bacteria and archaea by shielding DNA molecules from restriction endonuclease fragmentation. Methylation aids in the removal of incorrect base sequences added during DNA replication in some organisms. A cellular mechanism called the mismatch repair system differentiates between the correctly sequenced portion on the template strand and the newly replicated strand, where the errors occur by attaching a methyl group to the parental strand. By blocking DNA transcription, 5-methylcytosine regulates a variety of cellular processes in higher eukaryotes. Methylation is also thought to indicate imprinting, a process that results in the selective inactivation of certain genes inherited from one parent. Important genes that regulate embryonic development may also be repressed or activated by proper methylation. However, 5-methylcytosine may be mutagenic because the methylation process produces thymine, which changes C:G pairs into T:A pairs. Mammals exhibit selective methylation within the dinucleotide sequence CG, which is a rare sequence that has likely been lost through mutation. CG dinucleotide mutations are present in several malignancies in important genes.

Keyword

Methylation – It is a form of alkylation, with a methyl group, rather than a larger carbon chain, replacing a hydrogen atom.

2.1.5.2. Nucleases

The phosphodiester backbone of DNA is hydrolytically cleaved by enzymes called nucleases. Endonucleases cleave in the middle of chains, whereas exonucleases degrade from the end of the chain in a selective manner. There are known nucleases that can function on single- or double-stranded DNA.

Restriction endonucleases are a special class that recognizes and cleaves specific sequences in DNA. Type II restriction endonucleases always cleave at or near their recognition sites. They produce small, well-defined DNA fragments that help to characterize genes and genomes and that produce recombinant DNAs. Fragments of DNA produced by restriction endonucleases can be moved from one organism to another. In this way, it has been possible to express proteins such as human insulin in bacteria.

2.1.5.3. Mutation

Genetic material mutations can result from chemical modification of DNA. Anions that deaminate cytosine to form uracil, like bisulfite, can alter the genetic message by inducing C-to-T transitions. Purine residues are lost when exposed to acid, but cells have certain enzymes that can heal these wounds. While oxidative damage from free radicals or strong oxidizing agents can cause a variety of lesions that are mutagenic if not repaired, exposure to UV light can cause adjacent pyrimidines to dimerize. Direct reactions between uracil, adenine, and guanine and halogens like chlorine and bromine result in substituted bases, which are frequently mutagenic. In a similar way, primary amine groups interact with nitrous acid.

2.1.5.4. Supercoiling

The topologies of circular DNA molecules, like those in bacterial chromosomes and plasmids, can vary greatly. One is active supercoiling, in which a single DNA strand is cleaved, wound around the complementary strand one or more times, and then the molecule is sealed back up. The DNA can undergo one supercoiled turn for every full rotation, a process that can continue until the DNA is completely wound and collapses into a tight ball. It is also possible to reverse. Supercoiled DNA is wound and relaxed by specialized enzymes known as gyrases and topoisomerases. The DNA in eukaryotic linear chromosomes is typically tightly bound by proteins at different locations, which permits the supercoiling of the intervening stretches. The extreme compaction of DNA required to fit it inside the cell wall is partly caused by this characteristic. Although the DNA in a single human cell can reach a length of two to three meters, it is packed extremely densely to fit inside the 10 µm-diameter nucleus of a human cell.

2.1.5.5. Sequence Determination

Frederick Sanger and Walter Gilbert invented techniques for figuring out DNA base sequences in the 1970s; they were awarded the 1980 Nobel Prize for their contributions. The Sanger method is predicated on the in vitro enzymatic synthesis of DNA, whereas the Gilbert-Maxam method depends on the distinct chemical reactivities of the bases. Both techniques calculate the separation between a fixed DNA point, and each instance of base—A, C, G, or T. Gel electrophoresis is used to divide DNA fragments from a sequence of reactions into four "lanes" based on their length. Each lane represents a distinct base, and the sequence is read straight from the gel. Fluorescent dyes have made the Sanger method automated, allowing a single machine to generate tens of thousands of DNA base sequences in a single run.

2.1.6. DNA Transcription

DNA transcription is a fundamental process in cellular biology where the genetic information encoded in DNA is transcribed into messenger RNA (mRNA). This process is crucial for gene expression, allowing the genetic instructions in DNA to be translated into functional proteins. DNA transcription occurs in the nucleus of eukaryotic cells and in the cytoplasm of prokaryotic cells. The process is mediated by various enzymes and involves multiple steps to ensure accurate and efficient transfer of genetic information.

2.1.6.1. Cellular Components Involved in DNA Transcription

Several key cellular components are involved in DNA transcription:

- **DNA Template**: The DNA molecule that contains the gene

to be transcribed serves as the template. The specific sequence of nucleotides in the DNA determines the sequence of the transcribed RNA.

- **RNA Polymerase**: This enzyme is essential for synthesizing RNA from the DNA template. In eukaryotes, RNA polymerase II is primarily responsible for transcribing mRNA. Prokaryotes have a single type of RNA polymerase that transcribes all types of RNA.
- **Transcription Factors**: These are proteins that help initiate and regulate the transcription process. They bind to specific DNA sequences known as promoters and enhance or inhibit the binding of RNA polymerase to the DNA template.
- **Promoter Regions**: These are specific DNA sequences located near the start site of a gene. Promoter regions are critical for the initiation of transcription, as they provide a binding site for RNA polymerase and transcription factors.
- **Enhancers and Silencers**: These regulatory DNA sequences can increase (enhancers) or decrease (silencers) the rate of transcription. They are usually located some distance from the gene they regulate and can function in either orientation.
- **Ribonucleotides**: These are the building blocks of RNA. During transcription, ribonucleotides are added to the growing RNA strand in a sequence complementary to the DNA template.

2.1.6.2. Steps Involved in DNA Transcription

DNA transcription can be divided into three main stages: initiation, elongation, and termination.

- **Initiation**:
  - The process begins when transcription factors and RNA polymerase bind to the promoter region of the gene. In eukaryotes, the core promoter often includes a TATA box, a specific DNA sequence recognized by transcription factors.
  - RNA polymerase unwinds the DNA double helix to expose the template strand. This forms a transcription bubble, allowing RNA polymerase to access the DNA template.
  - The RNA polymerase then starts synthesizing RNA by adding ribonucleotides complementary to the DNA template strand.
- **Elongation**:
  - During elongation, RNA polymerase moves along the DNA template, synthesizing the RNA strand in the 5' to 3' direction.
  - As RNA polymerase progresses, it unwinds the DNA ahead of it and rewinds the DNA behind it. The growing RNA strand detaches from the DNA template as it is synthesized.
  - The RNA strand grows longer as ribonucleotides are continuously added. The sequence of the RNA

transcript is complementary to the DNA template strand, except that uracil (U) replaces thymine (T).

- **Termination**:
  - Termination occurs when RNA polymerase reaches a specific DNA sequence called a terminator. In prokaryotes, terminators are specific sequences that signal the end of transcription. In eukaryotes, termination involves additional factors and processes.
  - The RNA polymerase releases the newly synthesized RNA strand and detaches from the DNA template.
  - In eukaryotes, the primary RNA transcript (pre-mRNA) undergoes further processing, including capping, polyadenylation, and splicing, to become mature mRNA ready for translation.

Therefore, DNA transcription is a complex and highly regulated process that involves multiple cellular components and steps to ensure accurate gene expression. Through the coordinated action of these components, genetic information in DNA is transcribed into RNA, ultimately leading to protein synthesis and the execution of cellular functions.

Did you get it?

1. What are the key properties of DNA?
2. Name the different types of DNA.

2.2. RNA STRUCTURE AND FUNCTION

→Learning Objectives

- Understand the RNA structure;
- Explain the types of RNA;
- Define functions of RNA.

RNA is a complex, high-molecular-weight substance that is involved in the synthesis of proteins in cells and, in certain viruses, takes the place of DNA (deoxyribonucleic acid) as the carrier of genetic codes. Ribose nucleotides are nitrogenous bases joined to ribose sugars and joined by phosphodiester bonds to form strands of different lengths, are the building blocks of RNA. Adenine, guanine, cysteine, and uracil are the nitrogenous bases found in RNA; uracil takes the place of thymine in DNA.

Five carbons and one oxygen make up the cyclical structure of RNA's glucose sugar. RNA is susceptible to hydrolysis due to a chemically reactive hydroxyl (–OH) group linked to the second carbon group in the ribose sugar molecule. Since DNA lacks a reactive –OH group in the same location on the sugar moiety (deoxyribose), it is hypothesized that one of the reasons DNA evolved to be the preferred carrier of genetic information in the majority of organisms is because of its chemical liability.

2.2.1. RNA Structure

Phosphodiester bonds bind the mononucleotides that make up RNA, which is normally single stranded. In the RNA chain, a ribonucleotide consists of a phosphate group, one of the four nitrogenous bases (A, U, G, and C), and ribose, or pentose sugar. While RNA is more suited for its shorter-term functions due to its relative instability, DNA is more stable due to the subtle structural differences between the sugars and is, therefore, better suited for storing genetic information. In place of the thymine used in DNA, RNA-specific pyrimidine uracil forms a complementary base pair with adenine. Notwithstanding single stranded nature of the RNA, the majority of RNA molecules exhibit extensive intramolecular base pairing between complementary sequences, which forms a predictable three-dimensional structure vital to the molecules' functionality (Figures 2.5 and 2.6).

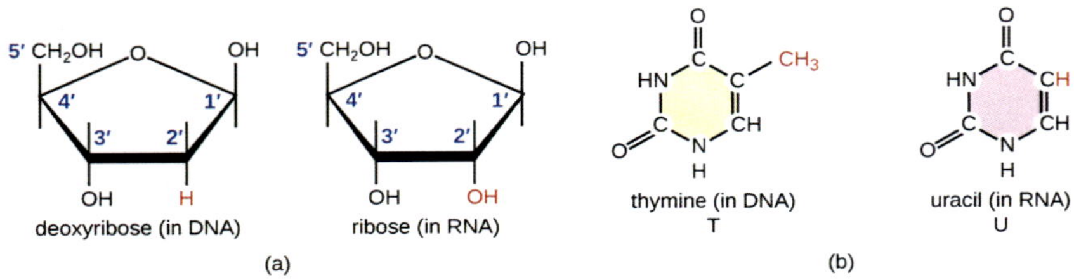

Figure 2.5. (a) Ribonucleotides contain the pentose sugar ribose instead of the deoxyribose found in deoxyribonucleotides. (b) RNA contains the pyrimidine uracil in place of thymine found in DNA.

Source: openstax.org/books/microbiology/pages/10-3-structure-and-function-of-rna.

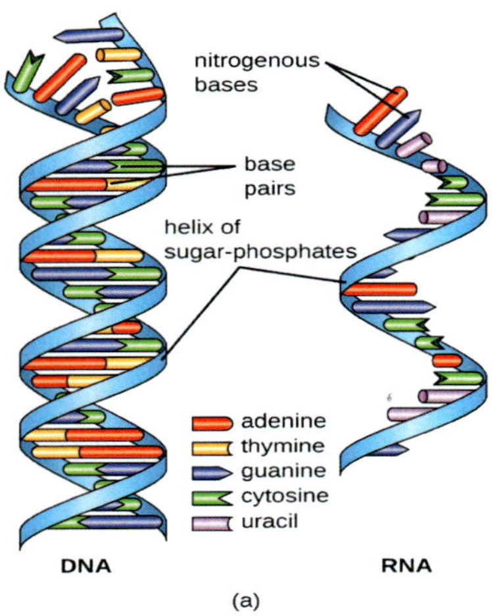

Figure 2.6 (a). DNA is typically double-stranded, whereas RNA is typically single-stranded.

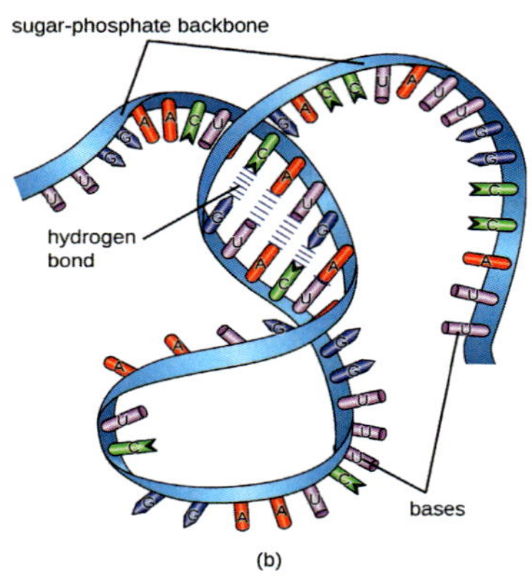

Figure 2.6 (b). Although it is single-stranded, RNA can fold upon itself, with the folds stabilized by short areas of complementary base pairing within the molecule, forming a three-dimensional structure.

Source: openstax.org/books/microbiology/pages/10-3-structure-and-function-of-rna.

2.2.2. Types of RNA

Of many types of RNA, the three well known and most commonly discussed and found in almost all organisms. These three types of RNA are (Figure 2.7):

1. mRNA (Messenger RNA):
   - It is a single-stranded RNA molecule that is complementary to one of the strands of DNA.
   - mRNA is the version of the genetic materials that leave the nucleus and move to the cytoplasm, where respective proteins are synthesized.
   - This RNA has the utmost importance during protein synthesis. When the ribosome moves along this mRNA, it reads the base sequences and uses the genetic code to translate them into specific proteins.
   - These codes are in the form of triplet sequences of nitrogenous bases and are often referred to as codons.
   - As we now know that mRNA is responsible for transferring genetic information into ribosomes, where by reading the base sequences on mRNA, the translation of proteins is made possible; thus, the name resembles its functions, i.e., messenger RNA.
   - We can even say mRNA uses genetic code present on a portion of DNA and makes proteins. If mRNA did not exist, then the information on DNA could never be used by our bodies.

Figure 2.7. mRNA (messenger RNA).

Source: Created with BioRender.com.

2. rRNA (Ribosomal RNA) (Figure 2.8):
   - It is a single-stranded RNA molecule found in cells that forms the part of the protein-synthesizing organelle, the ribosome.
   - It is synthesized inside the nucleus, particularly in the nucleolus where rRNA coding genes are present. The synthesized rRNA can be of varying sizes, commonly distinguished as small and large.
   - These newly synthesized rRNAs combine with ribosomal proteins and form smaller subunits and larger subunits of ribosomes respectively.
   - These rRNAs are vital in recognizing conserved regions of incoming mRNAs and tRNA thus facilitating their binding and carrying out protein synthesis.
   - Additionally, rRNA also has enzymatic activity (peptidyl transferase) and catalyzes the

formation of the peptide bond between two aligned proteins/amino acids during protein synthesis.

Figure 2.8. rRNA (ribosomal RNA).

Source: Created with BioRender.com.

Remember

Ribosomal RNA (rRNA) is an integral component of the ribosome, providing a structural framework for protein synthesis. The orchestration of these various RNA types in cellular processes underlines their significance in gene expression and the synthesis of functional proteins.

3. tRNA (Transfer RNA) (Figure 2.9):
   - It is a type of RNA molecule that helps to decode information present in mRNA sequences into specific proteins.
   - It is encoded by DNA in the cell nucleus and transcribed with the help of RNA polymerase III.
   - The structure of tRNA folds upon itself and creates an intra-complementary base pairing, which gives rise to hydrogen-bonded stems and associated loops that contain nucleotides with modified bases.
   - The two-dimensional structure resembles a cloverleaf having three loops and an open end, and are usually 75–90 ribonucleotides in length.
   - Each of these loops consisting of arms has a distinct name and function. The three-loop consisting arms are namely: DHU or D arm, which has a recognition site for specific enzyme amino-acyl tRNA synthetase; T arm, which consists of a ribosome recognition site, and the anticodon arm recognizes and binds to mRNA present in the ribosome.
   - The open end with no loop is the site for attachment of amino acid via 3’ OH bonding with the COOH- group of the amino acid.
   - In general, tRNA reads the code on the mRNA sequence in the ribosome and translates specific

amino acid; it does so along the length of the mRNA and gives out a polypeptide chain of amino acids (proteins) in association with other important enzymes like aminoacyl tRNA synthetase and peptidyl transferase.

Figure 2.9. tRNA (transfer RNA).

Source: Created with BioRender.com.

2.2.2.1. Some Other Types of RNA

1. Ribozymes:
   - These types of RNAs refer to those RNAs that are capable of showing enzymatic activities.
   - They were first discovered in introns of precursor ribosomal RNA of *Tetrahymena thermophilus*, where these noncoding sequences were capable of excising themselves without any protein or external source.
   - These RNAs play vital roles in major reactions like RNA splicing, viral replication, and tRNA biosynthesis.
   - rRNAs also show enzymatic activities and thus can be termed as Ribozymes.
2. Antisense RNAs:
   - Antisense RNAs are those RNAs that contain sequences that are complementary to protein-coding sequences of mRNA.
   - These are single-stranded like mRNA but cannot code for proteins.
   - However, they can interfere and inactivate their complementary mRNA sequences thus inhibiting protein synthesis.
   - This ability of antisense RNAs has led researchers to create artificial antisense RNAs that can inhibit the protein synthesis of potential disease-causing organisms or of infected cells, which then end up killing unwanted cells.

2.2.3. Functions of RNA

There are different types of RNA, each with specific functions. Here are some of the key functions of RNA:

- mRNA carries genetic information from the DNA in the cell nucleus to the ribosomes in the cytoplasm, where protein synthesis occurs. It serves as a template for protein production during the process of translation.
- rRNA is a structural component of ribosomes, which are cellular organelles responsible for protein synthesis. It helps in the binding of mRNA and tRNA (transfer RNA) during translation.

- tRNA plays a crucial role in protein synthesis by bringing amino acids to the ribosomes based on the information encoded in mRNA. Each tRNA molecule is specific to a particular amino acid.
- snRNA is involved in the splicing of pre-mRNA during the process of RNA maturation. It forms complexes with proteins to create small nuclear ribonucleoproteins (snRNPs) that help remove introns from the pre-mRNA.
- snoRNA is found in the nucleolus and is involved in the processing and modification of rRNA. It guides the chemical modifications, such as methylation and pseudouridylation, in rRNA molecules.
- miRNAs are small RNA molecules that regulate gene expression by binding to target mRNA molecules. This binding can result in mRNA degradation or inhibition of translation, thereby controlling the production of specific proteins.
- While the exact functions of many lncRNAs are still being discovered, some have been implicated in the regulation of gene expression, chromatin organization, and other cellular processes. They may act as scaffolds for protein complexes or regulators of transcription.
- siRNAs are involved in the regulation of gene expression by inducing the degradation of specific mRNA molecules. They are often used in RNA interference (RNAi) experiments for gene silencing.

Did you get it?

1. Explain the structure of RNA.
2. Identify the different types of RNA.

2.3. THE GENETIC CODE AND THE CONCEPT OF A GENE

→Learning Objectives

- Understand the genetic code;
- Evidence for the triplet code;
- Understand the cracking the code;
- Define the current concept of the gene;
- Explore the non-coding RNAs.

DNA is the genetic material found in cells. Through DNA replication, the information contained in a cell's genetic material can be copied and passed on to offspring. A significant amount of DNA data is translated into mRNA, which is subsequently converted into proteins. However, some RNAs also serve vital purposes even though they are never translated into proteins. In everyday speech, expressions like "it's in my genes" or "in my DNA" denote something fundamental to a person's identity.

Figure 2.10. The flow of genetic information.

Source: Steve, M., & Julia, L. Understanding Biochemistry: Structure and Function of Nucleic Acids: doi: 10.1042/EBC20180038.

During the early 1900s, the fundamental unit of heredity was named a gene. Genes were once believed as discrete loci arranged in a line on chromosomes. This theory was supported by breeding studies using the fruitfly Drosophila, which demonstrated that genes that are close to one another on a chromosome have a higher probability of being inherited together. The "one gene, one polypeptide" theory originated from the observation that gene mutations could result in changed phenotypes. Upon realizing that genes were composed of DNA, the term "central dogma of molecular biology" was established. This explains a two-step process whereby the DNA's genes are translated into RNA and then into the amino acid sequence that forms proteins. From DNA to RNA to protein is the information flow (Figure 2.10).

However, this has certain exceptions; first some viruses have RNA genomes, and in some cases, these are reverse-transcribed into DNA before the genes can be expressed. The retrovirus HIV is an example of this. The other exception is that not all functional RNAs are translated into proteins.

2.3.1. The Genetic Code

The set of instructions that living cells use to convert the information in the genetic material into proteins is known as the genetic code. Because nucleic acids are relatively simple molecules, many scientists were skeptical that they carried genetic information when DNA and RNA were first discovered. The question was how DNA, which only contains four different types of bases, could code for 20 different amino acids. Table 2.1. The Genetic Code

Keyword

Codon – It is a sequence of three DNA or RNA nucleotides that corresponds with a specific amino acid or stop signal during protein synthesis.

UUU	Phe	UCU	Ser	UAU	Tyr	UGU	Cys
UUC		UCC		UAC		UGC	
UUA	Leu	UCA		UAA	STOP	UGA	STOP
UUG		UCG		UAG	STOP	UGG	Trp
CUU	Leu	CCU	Pro	CAU	His	CGU	Arg
CUC		CCC		CAC		CGC	
CUA		CCA		CAA	Gln	CGA	
CUG		CCG		CAG		CGG	
AUU	Ile	ACU	Thr	AAU	Asn	AGU	Ser
AUC		ACC		AAC		AGC	
AUA		ACA		AAA	Lys	AGA	Arg
AUG	Met	ACG		AAG		AGG	
GUU	Val	GCU	Ala	GAU	Asp	GGU	Gly
GUC		GCC		GAC		GGC	
GUA		GCA		GAA	Glu	GGA	
GUG		GCG		GAG		GGG	

Source: Steve, M., & Julia, L. Understanding Biochemistry: Structure and Function of Nucleic Acids: doi: 10.1042/EBC20180038.

DNA could only encode four amino acids if there was a 1:1 correlation between bases and amino acids. There would still be an insufficient number of combinations (16) if bases were paired. However, there are 64 possibilities if you consider a triplet code, which is more than sufficient. This is the code that is known to us, in which each codon—a string of three nucleotides—designates a specific amino acid. Even so, the triplet code did not make sense because there are now far more codons than required.

The genetic code, as it is currently understood, is shown in Table 2.1. Because RNA is what cells translate into amino acids, it is written as RNA with a U rather than a T. Because multiple codons frequently code for the same amino acid, the code is referred to as redundant or degenerate. The codon's third nucleotide, sometimes known as the degenerate position, is typically the one that differs.

2.3.2. Evidence for the Triplet Code

Long before we knew the sequence of DNA, scientists conducted the experiments that led to deciphering the genetic code. At that point, it was feasible to ascertain the relative amounts of every amino acid in a protein, but not the sequence in which they occurred. Sydney Brenner and Francis Crick used bacteriophage virus mutants in an experiment to provide important answers to some important questions. The normal or wild-type phage will infect *E. coli* and grow. Crick and Brenner investigated mutants that would not grow on some strains of *E. coli*.

Frameshifts are caused by mutants, which are insertions or deletions. A single adenine base inserted into the DNA sequence modifies not only the amino acid at the insertion site but also all other amino acids translated from that sequence (see Figures 2.11(A) and (B)). This causes a one-base shift in the reading frame and produces a non-functional protein. Nevertheless, a wild-type or almost wild-type phenotype is frequently obtained by inserting three nucleotides. This is because a complete triplet codon has been inserted; as a result, one or two amino acids that were not in the original sequence will now be present, but the reading frame will remain unchanged (Figure 2.11(C)), and the remainder of the sequence will read normally.

```
(A) atgtacaatattatatgcctgtccaaaaac
    MetTyrAsnIleIleCysLeuSerLysAsn

(B) atgtacagatattatatgcctgtccaaaaac
    MetTyrArgTyrTyrMetProValGlnLys

(C) atgtacagacatattatatgcctgtccaaaaac
    MetTyrArgHisIleIleCysLeuSerLysAsn
```

Figure 2.11. DNA sequence showing the amino acid translation underneath.

Source: https://www.ncbi.nlm.nih.gov/pmc/articles/PMC6822018/bin/ebc-63-ebc20180038-g9.jpg.

Crick and Brenner searched for mutations known as suppressors, which could save the mutant and enable normal growth. They demonstrated that their suppressor mutants frequently added or subtracted one or more bases, rather than just reversing the original mutation. They discovered that a mutant phenotype results from the insertion or deletion of one, two, or four nucleotides. Nevertheless, there is little to no impact if three nucleotides are added or removed. This provided compelling evidence in favor of a triplet code. This further supports the idea of redundant coding, in which an amino acid can have multiple codes. There would be 20 codons that code for amino acids and 44 "nonsense" codons if the code were non-redundant. In this instance, adding three nucleotides would probably result in the introduction of a nonsense codon rather than the restoration of the wild-type. It was correctly proposed by Crick and Brenner that bases are read in groups of three and that the genetic code is read from a fixed starting point.

2.3.3. Cracking the Code

Approximately concurrently, two American scientists, Marshall Nirenberg and Heinrich

Matthaei, created a cell-free apparatus capable of producing proteins in a test tube upon receiving an RNA molecule. They demonstrated how the system produced a polypeptide made completely of phenyalanine residues when given an artificial RNA chain made only of uracil (polyuracil). With this newfound tool, they could now decipher the genetic code. Polyproline was synthesized by RNA made of cytosine (C) residues, while polylysine was made by RNA made of adenosine (A) residues. Nucleotide combinations were used in experiments to show that, for instance, RNA made from A and C can only contain six amino acids: asparagine, glutamine, histidine, lysine, proline, and threonine. Eight triplet codons can be formed from A and C; two of these are known to encode cysteine and proline. Other combinations of A and C must be used to encode the remaining four amino acids. Naturally, this offers more proof of the genetic code's redundancy.

These efforts, which employed RNA molecules made up of arbitrary pairings of two or three bases, were insufficient to completely decipher the genetic code. Additional significant information was added through the use of chemically synthesized RNA molecules with known repeating sequences. For instance, two alternating codons, CAC, and ACA, can be read from a synthetic RNA consisting of alternating A and G residues (AGAGAGAGAG). It produces a protein with threonine and histidine residues alternating.

2.3.4. Start and Stop Codons

Amino acids are encoded by 61 of the 64 potential codons. The three codons that are left, UAA, UAG, and UGA, are sometimes referred to as nonsense codons because they do not code for an amino acid. These are known as stop codons, and the ribosome halts protein synthesis in response to them. The most prevalent start codon is AUG, which also codes for the amino acid methionine.

2.3.5. Expanding the Genetic Code

As one of the goals of synthetic biology, it would be possible to create novel materials, some of which could have beneficial therapeutic properties, if we could engineer cells to use a wider range of building blocks with different physical and chemical properties than the small set of amino acids that nature uses to make proteins. Reprogramming the genetic code and designing the translation machinery to use these new combinations are necessary for this to be accomplished successfully. Some progress has been made, such as the ability to code for amino acids that are not typically found in proteins using both the AAG codon and the UAG stop codon for arginine.

2.3.6. Current Concept of the Gene

A gene is a sequence of bases on a DNA molecule that codes for a sequence of amino acids in a polypeptide chain or for an RNA molecule with a specific function, as was discovered after the genetic code was deciphered. Finding patterns typical of genes was made possible by the availability of DNA sequences of individual genes (see 'Recombinant DNA Technology and DNA Sequencing' in this issue of Essays in Biochemistry). An open reading frame is formed by a start codon, a string of codons encoding the amino acid sequence, and a stop codon in a protein-coding gene.

Unprecedented amounts of biological data have been made available through whole genome sequencing. The field of bioinformatics has developed as a result of the necessity to analyze sequence data in order to provide answers to biological questions. Homology is a fundamental idea in bioinformatics. A structure or a gene is considered homologous if it is shared by two organisms, and the same is true for a gene. For instance, pentadactyl limbs, which have five digits, are present in birds, amphibians, reptiles, humans, and other mammals. The homology of the limbs indicates that these animal groups shared a common evolutionary ancestor. Genes work in the same way. Haemoglobin is found in the red blood cells of all vertebrates; two α and two β globin molecules combine to form adult human hemoglobin. Since all vertebrate globin gene sequences are similar to one another, you can determine how long ago two animals shared an ancestor by comparing the DNA sequences of their globin genes. If a novel DNA sequence contains a segment that resembles a kn, this concept can also be utilized to locate genes in it.

A gene is more than just the sequence that codes for a protein; it also contains sequences that control how the gene is expressed, such as promoter sequences, which indicate the beginning of transcription and the locations where transcription-related proteins attach to DNA. Nearly every gene in bacteria is a single, continuous DNA sequence. The situation is more complex in eukaryotes because introns typically break the coding region. Precursor or pre-mRNA is the term used to describe the main transcript, which has both exons and introns. Exons are spliced together to form the mature mRNA, while the introns are eliminated during pre-mRNA processing before they exit the nucleus (Figure 2.12). The unique 5′ to 5′ linkage adds a methylated guanosine nucleotide to the 5′ end of eukaryotic mRNAs, known as the 5′ cap, which is crucial for starting translation.

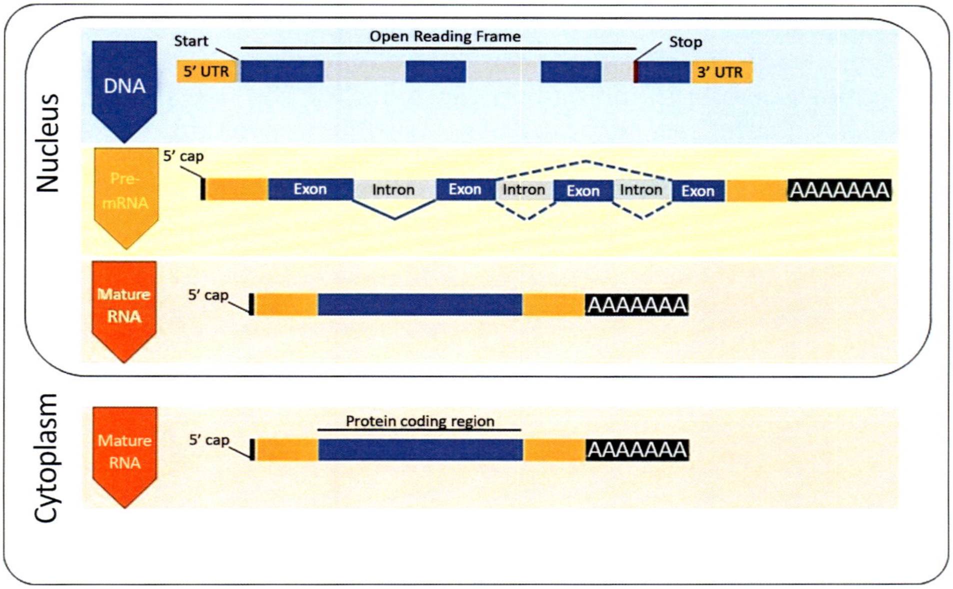

Figure 2.12. The structure of a protein-coding eukaryotic gene.

Source: https://www.ncbi.nlm.nih.gov/pmc/articles/PMC6822018/bin/ebc-63-ebc20180038-g9.jpg.

The poly-A tail, which is a chain of 100–250 adenine residues added to the mRNA to increase stability, is located at the 3′ end. Although there are many more distinct proteins than protein-coding genes, an analysis of the human genome sequence indicates that there are roughly 20,000–25,000 genes coding for proteins. This is because many genes can encode different protein variants. Genes can have multiple alternative promoters and alternative poly A sites in addition to allowing different exon combinations to be included in the mature mRNA through alternative splicing. It is estimated that alternative splicing occurs in 95% of human genes.

2.3.7. Non-Coding RNAs

Proteins are coded for in just 1.2% of the human genome. But when you look at the genomes of humans, mice, and dogs, you can see that since the species split, a lot more of the genome is subject to what is known as negative selection. Negative selection refers to the selection against unfavorable mutations. This implies that the fitness of the organism carrying the DNA is influenced by factors other than just the protein-coding regions. A portion of these are crucial DNA sequences that regulate the expression of certain genes (see the following section). On the other hand, a lot of processed RNA transcripts that do not encode proteins are being found through systematic screens. The most well-known are ribosomal and transfer RNAs, which are essential to protein synthesis, as we shall see in a later section. On the other hand, we are starting to realize that other non-coding RNAs perform crucial biological functions.

Two types of non-coding RNAs, small inhibitory RNAs (siRNAs) and microRNAs (miRNAs), participate in reducing gene expression after the mRNA has been transcribed from the DNA. They work by targeting a protein complex called RISC to specific mRNAs, which it then degrades. Expression of the gene is specifically knocked out or reduced and the phenotypic effect of this can then be observed. Another group of non-coding RNAs plays an important role in increasing the stability and correct folding of ribosomal RNAs. This process occurs in a compartment within the nucleus called the nucleolus; the RNAs are called small nucleolar RNAs (snoRNAs). These are mostly generated from intron RNA after it has been spliced out of the precursor mRNA and they function in association with proteins.

Did you get it?

1. What is the triplet code in genetics?
2. How was the genetic code “cracked” in molecular biology?

2.4. TRANSLATION OF RNA INTO PROTEINS

→Learning Objectives

- Describe the transfer RNA;
- Understand the structure of the ribosome;
- Discuss protein synthesis.

The ribosome, a complex structure made of proteins and RNA, is essential to the synthesis of proteins. The framework that the ribosome offers guarantees that the mRNA and tRNA are positioned correctly, facilitating the interpretation of the genetic code. Numerous other proteins are crucial for the process of protein synthesis; some of these are found within the ribosome, while others are likewise properly positioned by the ribosome's structure. As we shall see, ribosomal RNA, the small subunit, is an RNA molecule with catalytic properties akin to those of an enzyme, known as a ribozyme. A peptide bond can be created between two amino acids by ribosomal RNA.

2.4.1. Transfer RNA

The tRNA is the additional nucleic acid required for protein synthesis. The single-stranded tRNA molecule folds into a distinctive structure through base pairing (Figure 2.13). These function as adaptor molecules, carrying the amino acid that is specified by a particular mRNA codon and possessing an anticodon for that codon. On the mRNA, the anticodon and codon have complementary sequences.

A specific amino acid and its corresponding tRNA are recognized by the enzymes known as aminoacyl tRNA synthetases, which attach amino acids to tRNAs. The two-step reaction, which also needs ATP, proceeds as follows:

$$aa + ATP \rightarrow aa - AMP + PP$$

$$aa - AMP + tRNA \rightarrow aa - tRNA + AMP$$

The enzyme attaches the amino acid to the 3′ hydroxyl of the tRNA in the second step after hydrolyzing ATP to release pyrophosphate (PP) in the first. Aminoacyl tRNA synthetase enzymes are extremely selective; they will only bind to the appropriate tRNA after recognizing a given set of amino acids. This guarantees the proper coupling of tRNA molecules and amino acids, which is equally crucial for maintaining the integrity of protein synthesis as the ribosome's matching of the anticodon to the codon. Furthermore, because it generates the appropriate substrate for the ribosome and supplies a significant portion of the energy needed for the formation of peptide bonds during protein synthesis, this step is also known as activating the aminoacyl tRNA.

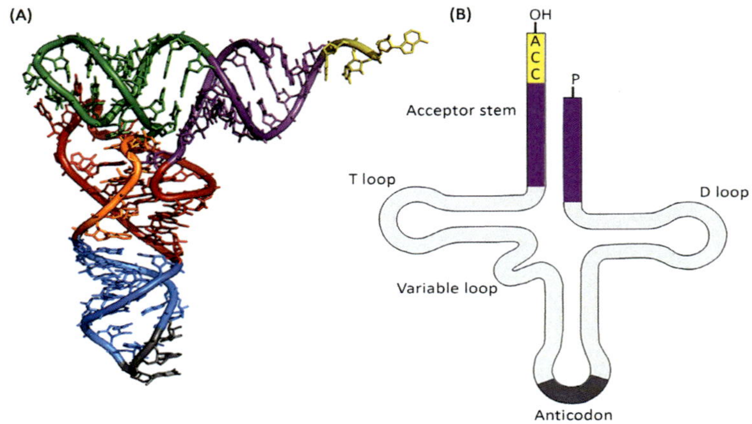

Figure 2.13. Transfer RNA: (A) Tertiary structure of the phenylalanine tRNA from yeast showing the anticodon (gray), the acceptor stem (violet) with the nucleotides CAA at the 3 OH end (yellow). (B) Cloverleaf representation of the secondary structure of tRNA.

Source: Steve, M., & Julia, L. Understanding Biochemistry: Structure and Function of Nucleic Acids: doi: 10.1042/EBC20180038.

Keyword

Protein Synthesis (Translation) – It is the production of a polymer of a chain of amino acids which produces a functioning protein. It involves reading the information from mRNA (messenger RNA) to put together a chain of amino acids.

2.4.2. Structure of the Ribosome

Every living organism possesses ribosomes. Although ribosomes in bacteria are somewhat smaller than those in eukaryotic cells (Table 2.2), their general makeup and mode of operation are nearly identical. Three scientists—Ada Yonath, Thomas Steitz, and Venkatraman Ramakrishnan—were granted the 2009 Nobel Prize in Chemistry for their use of X-ray crystallography to solve the three-dimensional structure of the bacterial ribosome. The two subunits that make up the ribosome are the large subunit, which adds amino acids to the expanding polypeptide chain, and the small subunit, which reads the messenger RNA. Three significant tRNA binding sites are located at the interface between the two subunits of the ribosome and are only formed when the two subunits come together. These locations, which are depicted in Figure 2.14, are known as the exit

(E) site, where spent tRNAs exit the ribosome; the peptidyl (P) site, where peptide bonds between amino acids are formed, and the acceptor or aminoacyl (A) site.

Table 2.2. Properties of Ribosomes

	Bacterial Ribosome	**Eukaryotic Ribosome**
Size	70S*, 2.3–2.6 MDa	80S, 3.3–4.5 MDa
Small subunit	30S	40S
	Approximately 20 proteins	32 proteins
	16S rRNA, ~1600 nucleotides	18S rRNA
Large subunit	70S	60S
	Approximately 33 proteins	79–80 proteins
	23S rRNA, ~2900 nucleotides	25S rRNA
	5S rRNA, ~120 nucleotides	5.8S & 5S

Source: Steve, M., & Julia, L. Understanding Biochemistry: Structure and Function of Nucleic Acids: doi: 10.1042/EBC20180038.

The ribosome, mRNA, and tRNA are not the only small proteins needed for protein synthesis. Initiation factors, elongation factors, and termination factors are among them. These proteins are not a part of the ribosome's structure. The inherited disease, Vanishing White Matter (VWM) disease, serves as an example of the significance of these factors. Mutations in one of the initiation factors cause this neurodegenerative disease, which is a serious condition that causes lesions in the brain's white matter.

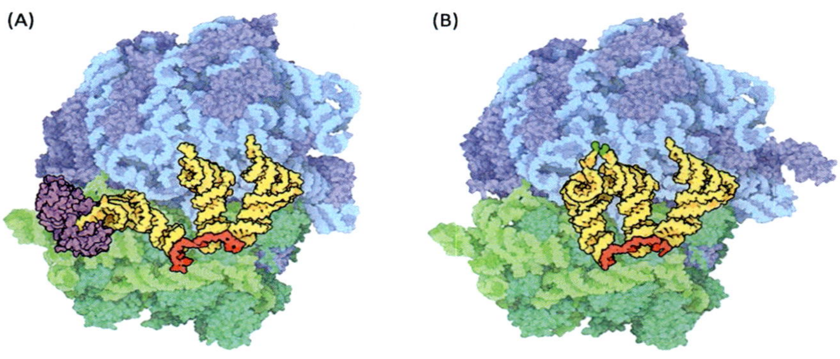

Figure 2.14. The structure of the ribosome of Thermus thermophilus showing the small subunit (green), large subunit (blue), mRNA (red), and three tRNAs in the acceptor, peptidyl and exit sites (yellow). In (A) the new tRNA is delivered to the ribosome by elongation factor EF-Tu (purple). In (B) the amino acid on the incoming tRNA is brought close to the amino acid on the tRNA in the peptidyl site to facilitate peptide bond formation (bright green).

Source: Steve, M., & Julia, L. Understanding Biochemistry: Structure and Function of Nucleic Acids: doi: 10.1042/EBC20180038.

2.4.3. Protein Synthesis

The ribosome assembles the mRNA and the charged tRNA during protein synthesis; codon and anticodon are matched, and the amino acids are combined in the right order. This process consists of three stages: elongation, which reads the triplet code and adds amino acids to the expanding peptide chain; termination, which marks the end of protein synthesis; initiation, which is when the ribosome assembles on the mRNA.

2.4.3.1. Initiation

The 5′ cap of the mRNA in the nucleus is bound by a complex of proteins known as the cap-binding complex. After that, the mRNA is exported to the cytoplasm, where it attracts the small (40S) ribosomal subunit, tRNA that has been charged with a methionine, and initiation factors.

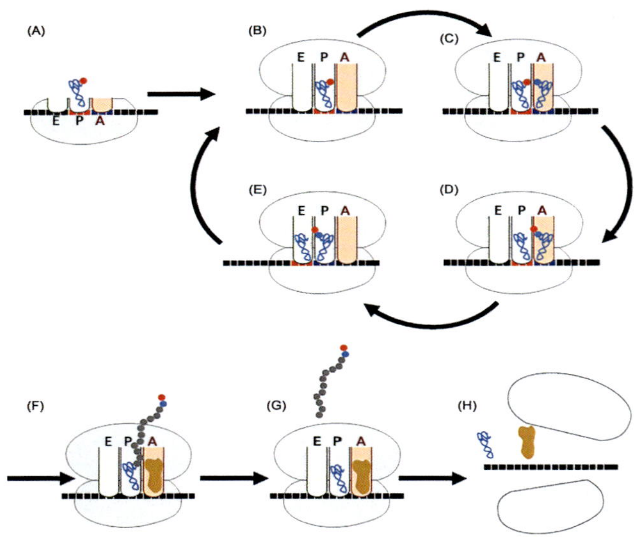

Figure 2.15. Protein synthesis.

Source: https://www.ncbi.nlm.nih.gov/pmc/articles/PMC6822018/pdf/ebc-63-ebc20180038.pdf.

When the small subunit comes across the first AUG start codon, it continues to scan along the 5′ untranslated region of the mRNA in the presence of initiation factors (Figure 2.15(A)). The big subunit then docks to produce the translation complex after the initiator tRNA's anticodon codon identifies this. The subsequent tRNA can now be accepted by the 80S ribosome, which has the tRNA charged with methionine at the P site (Figure 2.15(B)).

2.4.3.2. Elongation

After initiation is over, the mRNA is exposed to the next codon and is in the correct reading frame with the A site empty. An aminoacyl tRNA—a tRNA charged with an amino acid—enters the A site of the ribosome in a complex with an elongation factor during the elongation phase. It is properly positioned in the acceptor site, and GTP is hydrolyzed on the elongation factor if the anticodon it carries is complementary to the exposed codon (Figure 2.15(C)). The C terminus of an amino acid in the P site and the N terminus of an amino acid in the A site subsequently form a peptide bond (Figure 2.16). The peptidyl transfer center of the large subunit of the ribosome catalyzes this reaction.

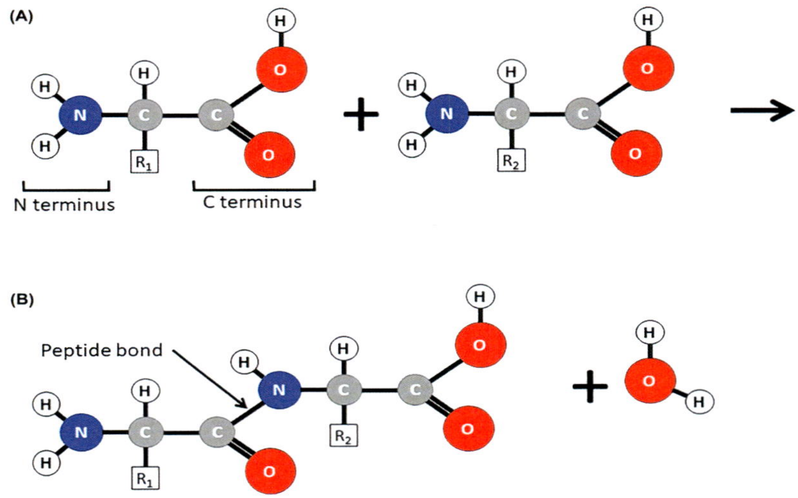

Figure 2.16. Amino acids and peptide bonds.

Source: https://www.ncbi.nlm.nih.gov/pmc/articles/PMC6822018/pdf/ebc-63-ebc20180038.pdf.

As a result, an empty or spent tRNA is left in the P site, and the growing peptide chain is transferred to the incoming aminoacyl tRNA in the A site (Figure 2.15(D)). Finally, the peptidyl tRNA with the growing peptide chain attached moves to the P site. This step is called translocation, and the energy is provided by the hydrolysis of GTP by the elongation factor EF-G. The spent tRNA moves to the exit site, from where it can leave the ribosome. The mRNA moves so that the next codon is exposed in the A site (Figure 2.15(E)), ready to accept a new aminoacyl-tRNA charged with another amino acid. During the elongation phase, the ribosome cycles through this process, adding amino acids to the growing peptide chain until a stop codon is exposed in the A site. The new protein emerges from the ribosome through an exit tunnel in the large subunit.

2.4.3.3. Termination

An anticodon on a tRNA does not recognize the stop codon, hence it cannot be decoded. Rather, proteins known as release factors or termination factors detect it. All three stop codons that enter the A site are recognized by a single release factor (RF1) in eukaryotes (Figure 2.15(F)). The peptide is released from the ribosome (Figure 2.15(G)) when the ester bond tying the peptide chain to the tRNA in the P site breaks (Figure 2.15(H)). The two ribosome subunits separate and are recycled. With a significant core of structurally conserved proteins and rRNAs present in both prokaryotic and eukaryotic ribosomes, the structure and function of ribosomes are highly conserved. However, there are some variations in the extra proteins involved in translation as well as in the rRNAs. Although there are significant differences in the initiation of protein synthesis, the elongation phase is highly conserved. A particular sequence found in bacterial mRNAs is known as the Shine–Dalgarno sequence or ribosome binding site. Shine-Dalgarno sequence binds to a complementary sequence of the 16S rRNA in the small subunit to guarantee that the mRNA is positioned correctly in the ribosome. In bacteria, N-formylmethionine, a modified amino acid, charges the initiator tRNA. Antibiotics, which are selective in that they influence protein synthesis in bacteria but not in mammalian cells, can take advantage of differences in the structure of bacterial and eukaryotic ribosomes. Protein synthesis is stopped by macrolide antibiotics such as erythromycin, which block the exit tunnel in the large subunit of bacterial ribosomes. Therefore, eukaryotic ribosomes are unaffected because their exit tunnel is marginally narrower. The antibiotic streptomycin binds to the 16S ribosome of bacteria, making it a valuable tool in the fight against tuberculosis. This leads to an incorrect reading of the mRNA by warping the structure of the decoding site.

Did you get it?

1. What is the role of transfer RNA (tRNA)?
2. Describe the structure of the ribosome.

2.5 POLYMERASE CHAIN REACTION (PCR)

→Learning Objectives

- Describe principles of PCR
- Explain the applications of PCR
- Define PCR Techniques and Variations
- Discuss advancements and Innovations in PCR
- Explain the future perspectives

Polymerase Chain Reaction (PCR) is a powerful molecular biology technique used to amplify specific DNA sequences, enabling researchers to study genes, diagnose diseases, and explore genetic relationships. Developed in the mid-1980s by Kary Mullis, PCR revolutionized genetic research and diagnostics by providing a method to amplify DNA exponentially from minute samples. This article explores the principles, applications, techniques, and advancements in PCR.

2.5.1 Principles of PCR

PCR relies on the ability of DNA polymerase enzymes to synthesize new DNA strands in a chain reaction. The process typically involves three main steps: denaturation, annealing, and extension.

- **Denaturation**: The DNA template containing the target sequence is heated to around 95°C, causing the double-stranded DNA to separate into two single strands.
- **Annealing**: The reaction temperature is lowered to allow short DNA primers (complementary to sequences flanking the target region) to bind (anneal) to their specific locations on the single-stranded DNA template.
- **Extension**: The temperature is raised to an optimal range (usually 72°C) for DNA polymerase to extend the primers by adding nucleotides complementary to the template strand, synthesizing new DNA strands.

These steps are repeated through cycles of heating and cooling in a thermal cycler machine, resulting in the exponential amplification of the target DNA segment. Each cycle doubles the amount of DNA, allowing billions of copies to be produced within a few hours.

2.5.2 Applications of PCR

PCR has a wide range of applications in research, medicine, forensic science, and agriculture:

- **Gene Cloning and Sequencing**: PCR facilitates the cloning and sequencing of specific DNA segments, enabling researchers to isolate and study genes of interest.
- **Diagnostic Testing**: PCR-based tests are used for diagnosing infectious diseases (e.g., COVID-19, HIV), genetic disorders (e.g., cystic fibrosis), and identifying pathogens in clinical samples.
- **Forensic Analysis**: PCR is crucial in forensic DNA profiling, where minute amounts of DNA from crime scenes can be amplified for analysis and comparison.
- **Environmental Studies**: PCR is employed to detect and identify microbial communities in environmental samples, monitor biodiversity, and study microbial ecology.
- **Genetic Engineering**: PCR plays a pivotal role in recombinant DNA technology, allowing specific DNA sequences to be amplified and manipulated for gene editing (e.g., CRISPR-Cas9).

2.5.3 PCR Techniques and Variations

- **Reverse Transcription PCR (RT-PCR)**:
  - RT-PCR is used to amplify RNA sequences by first converting RNA into complementary DNA (cDNA) using reverse transcriptase enzyme.
  - Applications include gene expression analysis, viral load quantification (e.g., HIV viral load testing), and detecting RNA viruses.
- **Quantitative PCR (qPCR)**:
  - Also known as real-time PCR, qPCR monitors the amplification of DNA in real-time using fluorescent dyes or probes.
  - It quantifies the initial amount of target DNA or RNA present in a sample, providing insights into gene expression levels, viral load, and pathogen detection.
- **Nested PCR**:
  - Nested PCR involves two rounds of PCR amplification, where the product of the first PCR serves as the template for a second PCR using nested primers.
  - It enhances specificity and sensitivity in detecting low-abundance targets and minimizing non-specific amplification.
- **Multiplex PCR**:
  - Multiplex PCR amplifies multiple target sequences in a single reaction using multiple primer sets specific to each target.
  - It is used for simultaneous detection of multiple pathogens (e.g., respiratory panel testing) and genotyping applications.
- **Digital PCR (dPCR)**:
  - dPCR partitions samples into thousands of individual

reactions, enabling absolute quantification of target DNA molecules.

- It provides precise quantification of rare sequences, such as detecting minimal residual disease in cancer patients or quantifying fetal DNA in maternal blood.

2.5.4 Advancements and Innovations in PCR

- **High-Throughput PCR Technologies**:
  - Automated systems and microfluidic devices enable high-throughput PCR, allowing rapid screening of large sample sets and integration with next-generation sequencing (NGS) technologies.
- **Isothermal Amplification Techniques**:
  - Isothermal amplification methods (e.g., loop-mediated isothermal amplification, LAMP) amplify DNA at a constant temperature without the need for thermal cycling equipment.
  - They offer advantages in point-of-care diagnostics and field-based applications where equipment and power are limited.
- **Digital Microfluidic PCR**:
  - This technology integrates droplet-based microfluidics with PCR, allowing single-cell analysis and precise control over reaction volumes.
  - It enhances sensitivity and reduces sample consumption, making it suitable for rare mutation detection and single-cell genomics.
- **CRISPR-based PCR Assays**:
  - CRISPR-Cas systems combined with PCR enable highly specific and sensitive detection of nucleic acids.
  - CRISPR-based assays like SHERLOCK (Specific High-sensitivity Enzymatic Reporter unLOCKing) and DETECTR (DNA Endonuclease-Targeted CRISPR Trans Reporter) offer rapid and portable diagnostics for infectious diseases and genetic disorders.

2.5.5 Future Perspectives

PCR continues to evolve with advancements in technology and applications. Future developments may focus on enhancing speed, sensitivity, and multiplexing capabilities, as well as integrating PCR with emerging technologies like artificial intelligence (AI) for data analysis and machine learning-based diagnostics. Applications in personalized medicine, environmental monitoring, and synthetic biology are also expected to expand, further cementing PCR's pivotal role in modern molecular biology and biotechnology.

ROLE MODEL

RICHARD JOHN ROBERTS

Nobel Prize-Winning English Biochemist and Molecular Biologist

Born Sept. 6, 1943, Derby, Eng., molecular biologist, the winner, with Phillip A. Sharp, of the 1993 Nobel Prize for Physiology or Medicine for his independent discovery of "split genes." Roberts received a PhD in organic chemistry from the University of Sheffield, Eng., in 1968. After postdoctoral research at Harvard University, he took a post at Cold Spring Harbor Laboratory in New York in 1972. In 1992 he joined New England Biolabs, a biotechnology firm.

In 1977, Roberts and a team including Thomas Broker, Louise Chow, and Richard Gelinas established that the genes of the adenovirus—one of the viruses that cause the common cold—are discontinuous: the segments of DNA that code for proteins are interrupted by lengthy stretches of DNA that do not contain genetic information. The coding segments are called exons; the noncoding ones are called introns. A research team working under Sharp at the Massachusetts Institute of Technology produced the same finding that same year. Previously, based on studies of bacterial DNA, biologists believed that genes consisted of unbroken stretches of DNA, all of which encoded protein structure. It has since been established that the discontinuous gene structure discovered by Roberts and Sharp is the most common structure found in higher organisms (eukaryotes). In addition to having important implications for the study of genetic diseases, this structure is believed to drive evolution by allowing information from different parts of the gene to be brought together in new combinations.

Richard John Roberts is an English biochemist and molecular biologist who was felicitated with the Nobel Prize in Physiology or Medicine for his discovery of the mechanism of gene-splicing. He shared the prize with Philip Sharp, who independently carried out similar work in the field. Roberts' fantasy with chemistry started as early as in high school. He was so interested in the subject that he chose to make it his career. Following his graduate studies and doctoral thesis, Roberts made his way to Harvard and, eventually, the Cold Harbor Laboratory. It was at the laboratory that Roberts first found success in restriction enzymes. By 1972, he discovered or characterized almost three-quarter of the world's first restriction enzymes. Later in 1977, Roberts demonstrated how RNA can be divided up into introns and exons, after which the exons can be joined together. The discovery was crucial as, until then, the scientific world believed that genes comprised unbroken sketches of DNA. The discovery had important implications for the study of genetic diseases.

- **Childhood and Early Life:**
  - Richard John Roberts was born on September 6, 1943, in Derby, England, to Edna Allsop and John Roberts. His father was an auto mechanic by profession, while his mother was a homemaker.
  - When young Robert was just four, the family shifted to Bath. There, he attended the Christ Church infant school and later St Stephen's Boys' School. It was there that Roberts developed a passion for logic and mathematics. He was encouraged by his headmaster, Mr Broakes, who would spend hours producing problems and later helping Roberts solve them.
  - Completing his elementary studies, he attended City of Bath Boys' School. Early on, Robert aimed to becoming a detective but when he was handed a chemistry play set, he changed his mind. His chemistry play set was transformed into a large in-house chemistry laboratory that comprised real apparatus and chemicals. He soon found his real calling in chemistry.
  - Following his preliminary education, he enrolled at the University of Sheffield. He graduated from the same with a Bachelor of Science degree in Chemistry in 1965. Four years later, he attained his PhD in the subject, his thesis was on phytochemical studies of neoflavonoids and isoflavonoids.
- **Career:**
  - While doing his PhD, Richard J. Roberts came across a book by John Kendrew that turned his focus to molecular biology. The book detailed the early history of crystallography and molecular biology, focussing on the MRC Laboratory in Cambridge.
  - For postdoctoral studies, Roberts looked for a laboratory that would provide him the pathway into molecular biology. At Harvard, Jack Strominger offered him an opportunity. He engaged himself in the research work from 1969 to 1972.
  - In September 1972, he moved to Cold Spring Harbor Laboratory, where he worked under James Dewey Watson, co-discoverer of the structure of DNA. The same year, he attended a seminar by Dan Nathans who described an enzyme Endonuclease R that could cleave DNA into specific pieces. Roberts realized that Nathans' restriction enzyme gave an immediate way to isolate small DNA molecules.
  - At Cold Harbor Laboratory, he made preparations of Endonuclease R and the few other restriction enzymes known then. By 1973, he had found superlative success in restriction enzymes. Almost three-quarters of the world's first restriction enzymes were discovered or characterized under

his guidance. He had compiled an enzyme collection that proved to be a valuable resource for scientists around the world.

- In 1974, Roberts was joined by Richard Gelinas to characterize the initiation and termination signals for an Adenovirus-2mRNA. The main purpose was to sequence the 5'-end of an mRNA, map its location on a restriction fragment, and then sequence the upstream region. Just before beginning the project, mRNA caps were discovered. The duo developed an assessment of the capped oligonucleotides.
- While working on the initiation and termination signals of Adenovirus-2mRNA, Roberts, and Gelinas found that all late mRNAs seemed to begin with the same capped oligonucleotide, which was not encoded on the DNA next to the main body of the mRNA. Though they had biochemical evidence for the same, they did not have real proof.
- In March 1977, Roberts came up with the experiment that displayed the proposed split structure for Adenovirus-2mRNAs was accurate. Soon, the duo was joined by Louise Chow and Tom Broker. The four scientists aimed at visualizing the split structure by hybridizing an intact mRNA to its two different coding regions. Eventually, they came up with appropriate DNA fragments that were visible as split genes in the electron microscope.
- Before the discovery made by Roberts, scientists believed that genes consisted of unbroken stretches of DNA, all of which encoded protein structure. However, with his discovery, it was found that the discontinuous gene structure is the most common structure found in higher organisms (eukaryotes). In addition to having important implications for the study of genetic diseases, this structure is believed to drive evolution by allowing information from different parts of the gene to be brought together in new combinations.
- In 1992, Roberts moved to New England Biolabs. Don Cromb, founder and president of the establishment, had started the establishment mainly to produce restriction enzymes for commercial sale. Roberts started off as a chief consultant at the establishment and was soon promoted as a Joint Research Director.

• **Major Works:** Richard J. Roberts' most distinguished work came in the 1970s, when he successfully discovered split genes. The discovery was crucial and important in the scientific world as, until then, scientists believed that the genes consisted of unbroken stretches of DNA, of which encoded protein structure. It was Roberts who established the discontinuous gene structure. Through his experiments, he proved that the segments of DNA that code for proteins are interrupted by lengthy stretches of DNA that do not contain genetic information. He demonstrated how RNA can be divided up into introns and exons, after which the exons can be joined together. This can occur in different ways, giving a gene the potential to form a number of different proteins.

- **Awards and Achievements:**
  - In 1993, Richard J. Roberts and Phillip A. Sharp were jointly awarded the Nobel Prize in Physiology or Medicine "for their discoveries of split genes." Sharp worked independently on the subject.
  - He was conferred with an Honorary Doctor of Science Degree by the University of Bath in 1994.
  - In 1995, he was elected a Fellow of the Royal Society.
- **Personal Life and Legacy:**
  - Roberts married Jean and has four children, namely Alison, Andrew, Christopher, and Amanda.
  - In 2005, the chemistry department at his alma mater, University of Sheffield, was later named after him.

SUMMARY

- Molecular biology is a field of biology that explores the intricate mechanisms underlying the structure, function, and interactions of biological molecules within cells. Molecular biology delves into the molecular processes that drive life, examining the fundamental building blocks of living organisms at the smallest scales.
- DNA, also known as deoxyribonucleic acid, is an organic chemical consisting of genetic instructions and information for protein synthesis. DNA can be seen in most cells of all living creatures. It is a vital part of reproduction, where genetic heredity occurs. Therefore, the main objective is to transfer the inheritance from the older generation to future generations.
- DNA is normally found as a double-stranded molecule in the cell, whereas RNA is mostly single-stranded. It is important to understand, though, that under the appropriate conditions, DNA could be made single-stranded, and RNA can be double-stranded.
- Nucleases are enzymes that hydrolytically cleave the phosphodiester backbone of DNA. Endonucleases cleave in the middle of chains, while exonucleases operate selectively by degrading from the end of the chain. Nucleases that act on both single- and double-stranded DNA are known.
- Chemical modification of DNA can lead to mutations in the genetic material. Anions such as bisulfite can deaminate cytosine to form uracil, changing the genetic message by causing C-to-T transitions.
- The genetic information in DNA is used as a basis to create messenger RNA (mRNA) by transcription. Single-stranded mRNA then acts as a template during translation. Ribosomes facilitate translation in the cytoplasm, by inducing the binding of complementary transfer RNA (tRNA) anticodon sequences to the mRNA.
- mRNA is used to convey information from DNA to the ribosome. It is a single-strand molecule, complementary to the DNA template, and is generated through transcription. Strands of mRNA are made up of codons, each of which signifies a particular amino acid to be added to the polypeptide in a certain order.
- RNA is a complex compound of high molecular weight that functions in cellular protein synthesis and replaces DNA (deoxyribonucleic acid) as a carrier of genetic codes in some viruses. RNA consists of ribose nucleotides (nitrogenous bases appended to a ribose sugar) attached by phosphodiester bonds, forming strands of varying lengths. The nitrogenous bases in RNA are adenine, guanine, cytosine, and uracil, which replace thymine in DNA.
- RNA is typically single-stranded and is made of ribonucleotides that are linked by phosphodiester bonds. A ribonucleotide in the RNA chain contains ribose (the pentose sugar), one of the four nitrogenous bases (A, U, G, and C), and a phosphate group.
- The genetic material in a cell is made of DNA and can be copied and passed

on to progeny through DNA replication, allowing for the inheritance of the information that it carries. A large proportion of the information on the DNA is first transcribed into mRNA and then translated into proteins.

- The genetic code is the set of rules used by living cells to translate the information encoded within genetic material into proteins. When DNA and RNA were first discovered, the relative simplicity of nucleic acids led many scientists to doubt that it carries genetic information.

REVIEW QUESTIONS

1. Elaborate on the structure and functions of DNA, highlighting its role in genetic information.
2. Discuss the structure and functions of RNA, emphasizing its diverse roles in cellular processes.
3. Explore the genetic code and provide insights into the concept of a gene.
4. Evaluate the impact of mutations in DNA on gene expression and protein synthesis.
5. Illustrate how the process of translation is regulated and its importance in cellular function.

MULTIPLE CHOICE QUESTIONS

1. What is the primary function of DNA?

a. Energy storage

b. Genetic information storage

c. Protein synthesis

d. Cellular respiration

2. Which nitrogenous base is not present in RNA?

a. Adenine

b. Thymine

c. Guanine

d. Uracil

3. What is the structure that carries the genetic code in DNA?

a. Ribosome

b. Nucleotide

c. Codon

d. Double helix

4. Which enzyme is responsible for synthesizing RNA during transcription?

a. DNA polymerase

   b. RNA polymerase
   c. Ligase
   d. Helicase

5. **What is the role of tRNA in protein synthesis?**
   a. Transcription
   b. Translation
   c. Replication
   d. Repair

6. **Which of the following represents a complementary base pair in DNA?**
   a. A-T
   b. G-C
   c. A-C
   d. G-T

7. **What is the function of mRNA in protein synthesis?**
   a. Carries amino acids to the ribosome
   b. Carries the genetic code from DNA to the ribosome
   c. Forms the structure of ribosomes
   d. Initiates translation

8. **Which of the following is a pyrimidine base in DNA?**
   a. Adenine
   b. Guanine
   c. Cytosine
   d. Thymine

9. **Where does translation occur in a eukaryotic cell?**
   a. Nucleus
   b. Mitochondria
   c. Endoplasmic reticulum
   d. Ribosome

10. **What is the purpose of the anti-codon in tRNA?**
   a. Binds to mRNA during translation
   b. Carries amino acids to the ribosome
   c. Recognizes the start codon
   d. Ensures correct base pairing with mRNA

Answer to Multiple Choice Questions

1. (b); 2. (b); 3. (d); 4. (b); 5. (b); 6. (b); 7. (b); 8. (d); 9. (d); 10. (d)

REFERENCES

1. Alberts, B., Johnson, A., Lewis, J., Morgan, D., Raff, M., Roberts, K., & Walter, P., (2014). *Molecular Biology of the Cell; Garland Science* (6th edn., pp. 1–10).
2. Bellelli, R., & Boulton, S. J., (2021). Spotlight on the replisome: Aetiology of DNA replication-associated genetic diseases. *Trends Genet., 37*(4), 317–336.
3. Bentley, D., (2002). The mRNA assembly line: Transcription and processing machines in the same factory. *Current Opinion in Cell Biology, 14*(3), 336–342.
4. Bertone, P., Stolc, V., Royce, T. E., Rozowsky, J. S., Urban, A. E., Zhu, X., et al., (2004). Global identification of human transcribed sequences with genome tiling arrays. *Science, 306*(5705), 2242–2246.
5. Carninci, P., Kasukawa, T., Katayama, S., Gough, J., Frith, M. C., Maeda, N., et al., (2005). The transcriptional landscape of the mammalian genome. *Science, 309*(5740), 1559–1563.
6. Cheng, J., Kapranov, P., Drenkow, J., Dike, S., Brubaker, S., Patel, S., et al., (2005). Transcriptional maps of 10 human chromosomes at 5-nucleotide resolution. *Science, 308*(5725), 1149–1154.
7. Govindarajan, R., Duraiyan, J., Kaliyappan, K., & Palanisamy, M., (2012). Microarray and its applications. *Journal of Pharmacy & Bioallied Sciences, 4*(Suppl 2), S310–312.
8. Kokate, C., Jalalpure, S. S., & Hurakadle, P. J., (2016). Textbook of pharmaceutical biotechnology. *Expression Cloning* (p. 125). Elsevier.
9. Lessard, J. C., (2013). Molecular cloning. Laboratory methods in enzymology: DNA. *Methods in Enzymology, 529*, 85–98.
10. Minchin, S. D., & Busby, S. J. W., (2013). Transcription factors. In: Maloy, S., & Hughes, K., (eds), *Brenner's Encyclopedia of Genetics.* Elsevier, U.S.A.
11. Roeder, R. G., (2003). The eukaryotic transcriptional machinery: Complexities and mechanisms unforeseen. *Nat. Med., 9*, 1239–1244. 10.1038/nm938.
12. Simpson, B., Tupper, C., & Al Aboud, N. M., (2023). *StatPearls* [Internet]. StatPearls Publishing; Treasure Island (FL). Genetics, DNA Packaging.
13. Tian, J., (2013). *Molecular Imaging: Fundamentals and Applications* (pp. 550, 552). Springer-Verlag Berlin & Heidelberg GmbH & Co.K. ISBN 9783642343032.

CHAPTER

3

Food Biotechnology

LEARNING OBJECTIVES

After studying this chapter, you will be able to:

- Gain a basic understanding of food biotechnology;
- Identify microorganisms associated with food;
- Explain the fermentation biotechnology;
- Understand the genetically modified foods;
- Define enzymes in food industry.
- Production of Additives, Colorants, and Flavorings by Biotechnology
- Biotechnological Methods for the Detection of Pathogens and Contaminants in Foods for the
- The Most Recent Trend: Lab-grown Meats

INTRODUCTORY EXAMPLE

Genetically Modified Organism (GMO):

The current connotation of food biotechnology—also known as GMO or Genetically Engineered (GE), among others—is a food product developed through the genetic modification of a plant, animal, or microorganism in a laboratory by scientists. However, conventional crossbreeding techniques for improving crop yield through trial and error have been used for thousands of years—from ancient civilizations to the 19th-century teachings of Gregor Mendel, which led to the 20th-century green revolution in which Norman Borlaug crossbred wheat varieties to avert starvation.

More recently, in the 1980s and 1990s, scientists developed new plant varieties, such as an herbicide-resistant tobacco plant that reduced weed growth without harm to the plant. The first FDA-approved use of food biotechnology was the transgenic Flavr Savr tomato, developed in the US in 1994 to delay ripening until after harvest. Over the next 20 years, food biotechnology has been used to develop many products. In 2012, more than 80 percent of US corn and cotton were developed through food biotechnology, with the US leading other countries.

UNIT INTRODUCTION

Food biotechnology is a branch that encompasses the application of biological techniques and processes to improve and enhance various aspects of food production, processing, and consumption. This interdisciplinary field combines principles from biology, genetics, microbiology, and other related disciplines to address challenges and opportunities within the food industry. The primary goals of food biotechnology include increasing crop yields, developing new and improved food products, enhancing nutritional content, and optimizing food processing methods.

One key aspect of food biotechnology is genetic engineering, which involves manipulating the genetic makeup of plants and animals to achieve desired traits. GMOs are developed to exhibit characteristics such as resistance to pests, tolerance to harsh environmental conditions, and improved nutritional profiles. This technology has the potential to address global food security issues by creating crops that are more resilient and productive.

In addition to genetic engineering, food biotechnology encompasses other techniques, such as fermentation, which has been used for centuries to produce a variety of food products, including bread, cheese, and yogurt. Modern biotechnological methods have expanded the possibilities within fermentation, allowing for the development of new and innovative food products with improved taste, texture, and nutritional value.

The field also plays a crucial role in the development of functional foods – products designed not only to provide basic nutrition but also to offer health benefits beyond traditional nutrients. Through the incorporation of bioactive compounds, probiotics, and other health-promoting ingredients, food biotechnology contributes to the creation of foods that may positively impact human health and well-being.

Despite its potential benefits, food biotechnology raises ethical, environmental, and safety concerns. Issues related to labeling, consumer acceptance, and potential environmental impacts are subjects of ongoing debate and research. As the field continues to evolve, the responsible and sustainable application of food biotechnology will be essential to meeting the growing demands of a global population while ensuring the safety and quality of our food supply.

3.1. BASIC OF FOOD BIOTECHNOLOGY

→Learning Objectives

- Explain the history of food biotechnology;
- Discuss the benefits of biotechnology in food;
- Define biotechnology contributes to the social, economic, and environmental sustainability of agriculture.

The process of altering the genes in our food sources through technology is known as food biotechnology. Microorganisms, plants, and animals are our food sources. By using food biotechnology, we can produce new species of plants and animals—that is, plants and animals that are intended for human consumption. These novel species possess desirable traits for marketing, production, and nutrition.

3.1.1. History of Food Biotechnology

The history of food biotechnology stretches back thousands of years, starting with the domestication and selective breeding of plants and animals for agriculture. However, modern food biotechnology, which involves the application of scientific techniques to improve the production, quality, and nutritional value of food, has evolved significantly over the past few decades. Here's a brief overview:

1. **Early Agriculture:** The origins of food biotechnology can be traced back to the dawn of agriculture, when humans began to selectively breed plants and animals to enhance desirable traits such as yield, taste, and resistance to pests and diseases. This process involved simple techniques such as seed selection and cross-breeding.
2. **Discovery of Microorganisms:** The discovery of microorganisms and their role in fermentation processes laid the foundation for modern food biotechnology. Ancient civilizations used microbial fermentation to produce foods like bread, cheese, yogurt, and beer. This knowledge formed the basis for later developments in food processing and preservation techniques.
3. **Genetic Engineering:** The advent of genetic engineering in the 1970s revolutionized food biotechnology. Scientists developed techniques to manipulate the genetic material of organisms, allowing them to introduce or modify specific

traits. The first genetically engineered food product, the Flavr Savr tomato, was approved for commercialization in the United States in 1994. It was engineered to have a longer shelf life.

4. **Transgenic Crops:** The commercialization of transgenic crops, also known as genetically modified (GM) crops, began in the 1990s. These crops are engineered to possess traits such as herbicide resistance, pest resistance, and improved nutritional content. Some of the most widely cultivated GM crops include soybeans, corn, cotton, and canola.
5. **Biotechnological Applications in Food Production:** Food biotechnology encompasses various applications beyond genetic engineering, including the use of enzymes, probiotics, and other biologically active substances to enhance food quality, safety, and functionality. For example, enzymes are used in processes like cheese making and bread baking, while probiotics are added to dairy products for their health benefits.
6. **Safety and Regulation:** The safety of GM foods has been a subject of debate and scrutiny. Regulatory agencies in different countries, such as the Food and Drug Administration (FDA) in the United States and the European Food Safety Authority (EFSA) in Europe, assess the safety of GM foods before they can be commercialized. Labeling regulations vary globally, with some countries requiring mandatory labeling of GM foods.
7. **Emerging Technologies:** Recent advancements in biotechnology, such as gene editing techniques like CRISPR-Cas9, offer new opportunities for enhancing crop traits and developing novel food products. These technologies have the potential to address challenges related to food security, sustainability, and nutritional value.

Keyword

Nutritional Value – This of a food describes the amount of carbohydrates, fats, proteins, and energy that can be used during digestion.

3.1.2. Benefits of Biotechnology in Food

Biotechnology has brought numerous benefits to the food industry, impacting various aspects of food production, processing, and consumption. Some of the key benefits include:

1. **Increased Crop Yields:** Biotechnology has enabled the development of GM crops with traits such as pest resistance, disease resistance, and tolerance to environmental stressors like drought and salinity. These traits help farmers increase crop yields, reduce losses, and improve agricultural productivity, ultimately contributing to food security.

2. **Improved Nutritional Content:** Biotechnology allows for the enhancement of the nutritional content of food crops. For example, biofortification programs aim to increase the levels of essential micronutrients such as vitamin A, iron, and zinc in staple crops like rice, wheat, and maize. This helps address nutritional deficiencies and improve public health, especially in developing countries.
3. **Enhanced Food Quality and Safety:** Biotechnological methods are employed to improve the quality and safety of food products. For instance, enzymes derived from microorganisms are used in food processing to enhance flavor, texture, and shelf life. Biotechnology also facilitates the rapid detection of foodborne pathogens and contaminants, leading to improved food safety standards and regulations.
4. **Reduced Environmental Impact:** Biotechnology offers sustainable solutions for agriculture by reducing the environmental impact of food production. GM crops engineered for pest resistance require fewer chemical pesticides, leading to decreased pesticide use and reduced environmental pollution. Additionally, crops engineered for drought tolerance or nitrogen fixation can help conserve water and soil nutrients, promoting sustainable farming practices.
5. **Enhanced Crop Resilience:** Climate change poses significant challenges to global agriculture, including shifts in temperature, precipitation patterns, and the proliferation of pests and diseases. Biotechnology provides tools to develop crops with enhanced resilience to changing environmental conditions, helping farmers adapt to climate variability and mitigate the impacts of climate change on food production.
6. **Improved Food Processing and Preservation:** Biotechnological innovations have revolutionized food processing and preservation techniques. Enzymes, microbial cultures, and fermentation processes are used to produce a wide range of food products with improved taste, texture, and nutritional value. Biopreservation methods, such as the use of beneficial bacteria and antimicrobial peptides, help extend the shelf life of perishable foods and reduce food waste.
7. **Development of Functional Foods and Nutraceuticals:** Biotechnology enables the development of functional foods and nutraceuticals with health-promoting properties. These products may contain bioactive compounds, probiotics, prebiotics, vitamins, and antioxidants that offer various health benefits, including improved digestion, immune function, and disease prevention.

Overall, biotechnology plays a crucial role in addressing global food-related challenges, including food insecurity, malnutrition, environmental degradation, and foodborne illnesses. By harnessing the power of biotechnology, the food industry can innovate and adapt to meet the growing demands of a rapidly expanding global population while ensuring sustainability, safety, and nutritional adequacy.

3.1.3. Biotechnology Contributes to the Social, Economic, and Environmental Sustainability of Agriculture

By enhancing the safe and efficient application of pesticides, lowering the quantity of insecticide applied to crops, cutting greenhouse gas emissions, maintaining, and enhancing soil quality, and minimizing crop losses during and after harvest, biotechnology helps make agriculture more environmentally sustainable. Along with other precision agricultural technologies, biotechnology (e.g., The amount of food that can be harvested per acre of land or per animal is increased by conservation tillage, integrated pest management (IPM), and automated farming equipment systems using computerized global positioning system (GPS) technology. This decreases the need to use more and more land to feed a growing population. Good agricultural practices and biotechnology enable farmers to till (or mechanically work the soil) less frequently or not at all, improving soil quality and lowering pollution. Agriculture's "carbon footprint" is lowered employing biotechnology, as more carbon is kept in the soil and less is released into the atmosphere. To guarantee a positive social impact, agricultural biotechnology initiatives are pursued in developing countries under the direction of and in collaboration with the local communities.

In order to feed a growing global population, biotechnology enables farmers to harvest more food with the amount of farmland that is available. Crops protected by biotechnology may be more resilient to adverse weather, drought, and poor soil. These developments are vital in developing countries where crop losses can have disastrous effects on the economy and public health.

Did you get it?

1. Summarize the history of food biotechnology.
2. Identify the benefits of biotechnology in the food industry.

3.2. MICROORGANISMS ASSOCIATED WITH FOOD

→Learning Objectives

- Explain the bacteria, yeast, and molds;
- Identify primary sources of microorganisms commonly associated with food;
- Recognize microorganisms important in the food industry;
- Understand the moisture and air-loving molds;
- Appreciate the importance of bacteria and yeasts in the food industry.

Food provides nutrients to humans, animals, and microorganisms, acting as a medium for interactions between different living species. This is an inevitable byproduct of living together. The primary sources of food for humans and animals are plants and animals. Microorganisms can grow and function as a medium in food that is safe for human consumption. Therefore, food for humans is always linked to a wide range of microbes. Microorganisms use food from humans or animals as a source of nutrients for their growth and activity because their main purpose is self-preservation. Although microbial activity in food can be advantageous in some circumstances, it usually causes the food to deteriorate and become unfit for human consumption. Fermented foods can be produced with the aid of microorganisms as processing aids. Certain species of microorganisms can be used in fermentation to produce a wide range of food chemicals and additives.

Food that has pathogenic microorganisms growing in it and using its nutrients to produce toxins is bad for the consumer's health. Certain pathogens that cause food infections and diseases can also be carried by food or used as a medium. Food deterioration is brought on by the unfavorable enzymatic changes that occur as a result of the metabolic activity of different microorganisms, which also use the nutrients in food.

3.2.1. Bacteria

Single-celled, microscopic organisms called bacteria can survive in a variety of conditions. They are capable of surviving in the ocean, on land, and in the human stomach. The interaction between humans and bacteria is nuanced. They can assist with our digestion or curdle milk into yogurt, among other things. Occasionally, they can be harmful, leading to illnesses like pneumonia and Methicillin-resistant *Staphylococcus Aureus* (MRSA).

Eukaryotes and prokaryotes are the two broad categories into which all living things fall, according to the relative complexity of their cells. As prokaryotes, bacteria exist. One cell with a basic internal structure makes up the entire organism. Bacterial

DNA floats freely in a twisted mass resembling a thread called a nucleoid, in contrast to eukaryotic DNA, which is neatly packed into a cellular compartment called the nucleus.

Separate, circular DNA fragments known as plasmids are also present in bacterial cells. The specialized cellular structures known as membrane-bound organelles, which are intended to carry out a variety of cellular tasks, including energy production and protein transport, are absent from bacteria. Nonetheless, ribosomes are present in both eukaryotic and bacterial cells. Proteins are assembled from individual amino acids into these spherical units by means of a messenger RNA strand containing information.

Keyword

DNA Fragments – These are hybridized from a whole-genome library to complementary sequences that have been synthesized and combined into a mixture of probes designed with high specificity for the matching regions in the genome.

3.2.2. Yeasts

Yeasts are unicellular organisms that fall under the taxonomic category known as fungi, which also includes molds. Yeasts are found in numerous species. *S. cerevisiae* is the most widely used yeast in the baking and brewing industries. Additionally, yeasts are essential to the production of kefir, wine, and a few other goods. The majority of round yeasts used in the food industry divide themselves through budding. Their budding is a feature that helps identify them under a microscope. During budding, the cells take on the shape of an eight.

To grow, yeasts require sugar. They use sugar to make carbon dioxide and alcohol. Why yeast is so important to the food industry is because of this reaction. Yeasts also yield components of a pleasant aroma. The final product's flavor is greatly influenced by these aroma compounds. Yeast is necessary for the production of both the alcohol and the carbon dioxide in beer. Both carbon dioxide and alcohol are produced in the bread industry; the alcohol evaporates during baking.

Yeasts are present in nature everywhere, particularly on fruits and plants. Fruits that fall off the tree rot due to the action of mold, which converts the sugars in the fruit to carbon dioxide and alcohol. In the industry, yeast is grown in large tanks with sugary water and oxygen. The liquid is pumped out once the required amount of yeast has been reached, and the yeast is subsequently dried. In the process of making yeast, nothing more is added.

3.2.3. Molds

Microscopic fungi called molds feed on plant or animal debris. Although the exact number of fungal species is unknown, estimates range from tens of thousands to possibly 3,00,000 or more. The majority are filamentous, or thread-like, organisms, and fungi in general are known for their spore-producing abilities. These spores can travel by insects, water, or the air. Molds, in contrast to bacteria, consist of multiple cells and are occasionally visible to the unaided eye.

Under a microscope, they look like skinny mushrooms. In many molds, the body consists of:

- Root threads that invade the food it lives on;
- A stalk rising above the food;
- Spores that form at the ends of the stalks.

The color of the mold is caused by spores. The mold is dispersed by the airborne spores, which resemble dandelion seeds blowing across a meadow. The branches and roots of molds resemble extremely thin threads. When food is growing mold, the roots may be deeply ingrained in the food and difficult to see. Invisible bacteria may grow alongside mold on foods that have developed mold. Almost every ingredient in food is attacked by bacteria, yeasts, and molds. Certain enzymes ferment sugars and hydrolyze starches and cellulose, while others hydrolyze fats and cause rancidity, all depending on the type of enzyme and the availability of the enzyme.

3.2.4 Microorganisms Important in Food Industry

Every food may be contaminated from external sources on the way from the field to the processing plant, as well as during storage, transit, and distribution, in addition to the natural microflora that is determined by the type of plant or animal and environmental conditions. There are thousands of different kinds of microorganisms in the air, soil, and water, which have an impact on food and the digestive systems of both humans and animals. Thankfully, most microorganisms serve valuable purposes in the environment and in specific areas of the food industry, like the manufacturing of dairy, wine, beer, and other beverages. However, microorganisms are typically to blame for unintended food spoilage, and pathogen contamination of food results in issues with food safety.

Practically speaking, the microorganisms that grow on and/or in food can be categorized into three groups: bacteria, yeast, and molds. Molds are typically associated with food spoilage; however, their application in the food industry is restricted (e.g., cheese ripened by mold). Because of their capacity to ferment carbohydrates into ethanol and carbon dioxide, yeasts are the most extensively utilized microorganisms in the food industry. Certain yeasts, like baker's yeasts, are farmed commercially, and other types can be utilized as sources of protein, mostly in animal feed.

Food microbiologists can categorize bacteria into groups based on the fermentation product, e.g., bacteria that produce lactic, acetic, and propionic acids. Differentiating between proteolytic, saccharolytic, and lipolytic bacteria can be done by considering the food component that was attacked and utilized as sustenance by microorganisms. Primarily, their physiological and morphological characteristics (e.g., bacteria that form gases, both anaerobic and aerobic, etc.). The dairy industry uses lactic acid bacteria extensively, and the vinegar production uses acetic acid bacteria. Numerous bacteria are recognized as spoiling microorganisms, and some of them are pathogens (e.g., *Staphylococci*, *Salmonellae*, etc.).

3.2.4.1. Importance of Molds in Foods

1. Main Groups:

   i. Deuteromycota (Fungi Imperfecti): This is a class of septate fungi in which a sexual stage (perfect state) of the life cycle has not been observed. The Fungi imperfecti are essentially a provisional taxonomic grouping; if a sexual stage is observed in a fungus assigned to this group, the organism will then be assigned a new name that reflects its perfect state.

      a. Aspergillus: Very important genus that includes species used in industrial fermentations and the production of enzymes and organic acids. It also includes several spoilage microorganisms and two species, Aspergillus flavus and A. parasiticus, which produce carcinogenic aflatoxins.

      b. Botrytis: Cause gray mold rot in fruits.

      c. Penicillium: Like Aspergillus, this genus includes industrially important species as well as others that are involved in food spoilage and the production of mycotoxins.

   ii. Ascomycota: Septate mycelium, sexual reproduction characterized by the formation of a sac-like structure called an ascus. The ascus eventually ruptures, and the enclosed spores are liberated. This is a very large class of fungi that includes the genera Byssochlamys and the perfect states of some fungi formerly classified under Aspergillus.

      a. Byssochlamys: Form heat-resistant spores that can survive heat treatment and result in the spoilage of high-acid canned foods, especially fruits.

   iii. Zygomycota: A diverse group of lower filamentous fungi that have aseptate mycelium and are capable of rapid growth. This group has sexual and asexual modes of reproduction. Members include the genus Rhizopus, which holds several species that cause bread molds, as well as Rhizopus oligosporus, a species used in oriental soybean fermentations (tempeh).

      a. Yeasts: Classified into the same group as molds, most do not live in soil but are instead found in environments with high sugar content such as the nectar of flowers or the surface of fruits.

2. Ascomycota:

   i. Debaromyces: One of the most prevalent genera in dairy products. Debaryomyces hansenii is an important food spoilage species, it can grow in 24% NaCl and at a_w as low as 0.65.

ii. Saccharomyces: Includes the important industrial species, S. cerevisiae, used in bread and alcoholic fermentations. S. bailli is another important species because it is involved in the spoilage of several types of foods and is resistant to the antifungal preservatives benzoate and sorbate.

3. Deuteromycota:

i. Candida: Includes the pathogen Candida albicans, this genus also includes the most common species of yeasts in fresh ground beef and poultry, and a few species are involved in industrial fermentations.

3.2.5. Moisture and Air Loving Molds

Molds, which are perhaps the most well-known microorganisms, are found all over nature and can grow in a range of environments that include both moisture and air. They belong to the fungal family and are also plants. Almost everyone has witnessed the growth of mold on worn-out shoes and damp clothes. Thus, it may come as a surprise to many that mold is a microorganism. On the other hand, the mold that is visible to the unaided eye is actually a colony of millions of mold cells coexisting. Molds have different looks. Some are slimy, some are glossy and wet, and some are fluffy and filamentous.

3.2.5.1. Growth and Importance in Food Industry

In contrast to bacteria, molds consist of multiple cells. Through the consumption of food materials for energy and the synthesis of new cell material, vegetative cells maintain the organism. Spores are tiny "seed" cells produced by reproductive cells. Mold spores are the source of new mold organisms, as opposed to bacterial spores. Spores of bacteria typically only form in unfavorable environmental conditions.

Molds generate a stem that is made up of multiple cells. These cells come together to form a "fruiting body.." The spores are produced by the fruiting body and, when the conditions are right, they separate, get carried away by air currents, and land to establish new colonies of mold. There are lots of mold spores in the air. Consequently, if sufficient moisture is present, any food left out in the open will soon become contaminated with mold. Certain molds are also psychrophiles, which means they can rot food that has been refrigerated.

The food industry relies heavily on molds. Adding flavor and color to cheeses and preparing soy sauce are just two of their numerous contributions. They are also involved in the synthesis of numerous enzymes as well as chemicals like lactic and citric acid. Food-related issues can also arise from mold. Some varieties can yield toxins known as mycotoxins. Molds are most commonly used in the pharmaceutical industry, where they aid in the production of antibiotics like penicillin.

3.2.6. Importance of Bacteria in the Food Industry

Food-relevant bacteria are frequently categorized according to how they behave in food rather than according to a systematic classification. The most common group of microorganisms is called bacteria. They are frequently only associated with germs and the harm they cause. The number of pathogenic (disease-causing) bacteria is quite small. Many are beneficial and the majority are safe. Bacteria and other microorganisms can also be categorized based on how much oxygen they require. Aerobes are those that can only grow in the presence of oxygen. Others, known as anaerobes, can only grow in the absence of oxygen. Some are facultative anaerobes, meaning they can grow both with and without oxygen. Eating habits of bacteria: In order to develop and proliferate, bacteria, and other microorganisms require food. Their dietary requirements differ, but almost everything we eat can be consumed by certain bacteria.

To be used by bacteria, a food substance must pass into the cell where it can be processed into energy and new cell material. Because most foods are too complex to move into a bacterial cell, they must be broken down into simpler substances. Enzymes do this by increasing the rate of biochemical reactions. Produced within the bacterial cell, enzymes move through the cell wall to break down the food on the outside into a form bacteria can use.

3.2.6.1. Growth and Importance in Food Industry

Binary fission is the process by which bacteria divide into two cells from one parent. Some can reproduce extremely quickly when the right circumstances are met. Certain bacteria can proliferate in as little as 20 minutes if the conditions are right—that is, if there is enough food, moisture, and temperature. One cell multiplies to 2 in 20 minutes, to 4 in 40 minutes, and so on. The original cell will have multiplied into close to 17 million new bacteria in just eight hours. Naturally, the environment isn't conducive to this kind of reproduction rate for very long. We might be buried under bacterial cells if they did.

3.2.6.2. Lactic Acid Bacteria and Its Uses in Food

The term "lactic acid bacteria" describes a broad category of helpful bacteria that share characteristics and yield lactic acid as a byproduct of fermentation. They are present in our digestive systems and are widely distributed in the natural world. They are used to pickle vegetables, baking, making wine, and curing fish, meats, and sausages, but they are most recognized for their use in the preparation of fermented dairy products.

People used lactic acid bacteria thousands of years ago, without knowing the scientific basis, to produce foods that were better preserved and had distinct flavors and textures from the original food. Similar to this, a broad range of fermented milk products are used today, including semi-solid or firm products like yogurt and cheese and liquid drinks like kefir. These microbial allies are quite valuable. A microbial process is used in the manufacturing process to convert lactose, the milk sugar, into lactic acid. The structure of the milk protein changes (curdling) as the acid builds up, changing the product's texture in the process. The specific characteristics of various products are also influenced by other factors like temperature and milk composition.

3.2.7. Importance of Yeasts in Foods

The yeasts that are most commonly used are Saccharomyces species. The primary species *S. cerevisiae* is used to produce a wide range of foods; specific strains are utilized to leaven bread and produce glycerol, invertase, wine, ale, and alcohol. Top yeasts grow quickly at 20°C and are active fermenters. The term "top yeasts" comes from the way the cells cluster and the quick evolution of carbon dioxide, pushing the cells to the surface. The term "bottom yeasts" refers to yeasts that do not clump and sink to the bottom. They ferment at lower temperatures, between 10 and 15°C, and grow more slowly.

S. cerevisiae var *ellipsoideus* is a high alcohol-yielding variety used in the production of industrial alcohol, wine, and distiled liquors. *S. uvarum* is a bottom yeast used in beet manufacture. *S. fragilis* and *S. lactis* have the ability to ferment lactose and hence are important in milk products.

Zygosaccharomyces species are osmophilic and are involved in the spoilage of honey, syrups, and molasses. They find use in the fermentation of soy sauce and some wines.

Candida species spoil foods high in acid salt content. *C. utilis* is grown for food and *C. krusei* is used with dairy starter cultures to maintain the activity and longevity of lactic acid bacteria. *C. lipolytica* spoils butter and oleomargarine.

Brettanomyces species are involved in the fermentation of Belgian and English beers and in French wines, while Kloeckera species are found commonly on fruits and flowers and in soil. Trichosporon species (*Trichosporon pullulans*) grow at low temperatures and are found in breweries and on chilled beef. Rhodotorula species cause discoloration of foods by forming colored spots on meat and sauerkraut.

3.2.7.1. Sugar-Loving Yeasts

Yeasts are tiny, unicellular plants. They belong to the same family as mushrooms: the fungi (plural: fungus). The absence of chlorophyll sets fungi apart from other types of plants. Bacteria can be found growing on a wide variety of foods. However, most yeasts are limited to eating starches and sugars. They generate alcohol and carbon dioxide gas from these. Because of this, they have been used by humans for centuries to produce a variety of foods and drinks. They are in charge of the fermentation of wines, brandy, whiskey, and bread dough as well as beer. They are also the first in the vinegar-making process.

3.2.7.2. Growth and Importance in Food Industry

Certain yeasts have the ability to grow at relatively low temperatures because they are psychrophilic. In actuality, temperatures close to 40°F are frequently used for the fermentation of wines and beers. Some types can cause spoiling in meat and other refrigerated storage areas because they are psychrophiles.

In contrast to bacteria, which reproduce through binary fission, yeasts use a process known as budding. On the parent cell, a tiny knob or bud appears, develops, and eventually splits off to form a new yeast dell. Despite being the most popular mode of reproduction, yeasts can also proliferate through spore formation. They can cause the spoiling of some foods where bacteria would not grow because they can grow in environments with high salt or sugar content. Jellies, honey, maple syrup, and sweetened condensed milk are a few examples. Pickles and sauerkraut are examples of foods made by the bacterial fermentation process that can go bad due to yeasts interfering with the regular fermentation process.

Certain yeasts are pathogenic. However, yeast infections are much less common than bacterial infections. Today, the impact of yeasts on food and beverage production extends beyond the original and popular notions of bread, beer, and wine fermentations by *S. cerevisiae*. The commercial significance of yeasts in food and beverage production are shown in points below:

- Production of fermented foods and beverages;
- Production of ingredients and additives for food processing;
- Spoilage of foods and beverages;
- Biocontrol of spoilage microorganisms;
- Probiotic and biotherapeutic agents;
- Source of food allergens;
- Source of opportunistic, pathogenic yeasts.

Positively, they aid in the fermentation of a wide variety of other commodities, where different yeast species may cooperate with filamentous fungi and bacteria. Nowadays, yeasts are used to make a wide range of useful food ingredients and processing aids. Strong antifungal activity in certain yeasts makes them useful as novel agents in the biocontrol of food spoilage. Another unique characteristic of some yeasts that is drawing more attention is their probiotic activity. Regretfully, yeast activity also has a negative aspect. The food and beverage industries are well aware of their capacity to cause significant economic losses due to the spoiling of various commodities. Additionally, the potential health risks associated with yeasts in food and beverages are a developing concern.

Did you get it?

1. Name three types of microorganisms found in food.
2. What are the primary sources of microorganisms commonly associated with food?

3.3. FERMENTATION BIOTECHNOLOGY

→Learning Objectives

- Discuss the history of fermented foods;
- Explain the types of fermentation.

Food can typically be preserved through the process of fermentation, which has been employed for centuries. Nowadays, though, the primary goal of food fermentation is to create a wide range of fermentation products with distinct tastes, flavors, aromas, and textures rather than to preserve, as there are other known methods of preservation. Utilizing different microbial strains, the parameters of the fermentation process (microorganisms, substrates, temperature, duration of fermentation, etc.) and advances in chemical engineering allow us to produce hundreds of varieties of dairy products (cheeses, fermented milk products), vegetables (pickles, sauerkraut, olives), meat products (fermented sausages), breads, alcoholic beverages (wine, beer, cider), vinegar, and other food acids, as well as oils. It's astonishing that only two types of fermentation—lactic acid and ethanolic fermentation—are used most of the time in a wide range of products, tastes, and textures.

Both modify the environment to prevent harmful or spoiling microorganisms from proliferating and changing the food. Food products have historically been the primary products of fermentation, but in recent years, there has been a noticeable increase in interest in the production of bulk chemicals (such as ethanol and other solvents), specialty chemicals (such as pharmaceuticals and industrial enzymes), biofuels, and food additives (such as flavor modifiers). Agriculture also makes use of fermentation processes.

There are two types of fermentation processes: induced and spontaneous (e.g., combining flour and baking yeast to make bread dough). Because fermentation products have chemical energy, they are not completely oxidized and need oxygen to fully mineralize. Since only substrate-level phosphorylation produces ATP, oxidative phosphorylation is more energetically efficient than fermentation. Aerobic respiration produces 36 ATP molecules, while fermentation produces 2 ATP molecules from 1 glucose molecule. Although it doesn't produce energy, the last stage of the fermentation process—converting pyruvate into final products—produces NAD+, which is needed for glycolysis.

Fermentation processes utilize microorganisms to convert solid or liquid substrates into various products. The substrates used vary widely, any material that supports microbial growth being a potential substrate. Similarly, fermentation-derived products show tremendous variety. Commonly consumed fermented products include bread, cheese, sausage, pickled vegetables, cocoa, beer, wine, citric acid, glutamic acid, and soy sauce.

3.3.1. History of Fermented Foods

The use of fermentation in food processing dates back thousands of years (Table 3.1). In antiquity, the history of fermented foods is lost. It appears that the Indian Subcontinent, specifically the communities that existed before the great Indus Valley civilization, is where the art of fermentation first emerged. In the fertile Crescent of Iraq, between the Tigris and Euphrates rivers, the craft of making cheese dates back to 8000 years ago, when domestication of plants and animals was just beginning. Later, it was believed that the Egyptians and Sumerians invented the alcoholic fermentations used in winemaking and brewing between 4000–2000 BCE. Back in 4000–3500 BCE, the Egyptians also invented dough fermentations, which are used to make leavened breads. But fermentation's scientific justification began when van Leeuwenhoek and Hooks identified microorganisms in 1665. Louis Pasteur revoked the "spontaneous generation theory" around 1859 AD by elegantly designed experimentation. The role of a sole "bacterium" lactis (*Lactococuus lactis*), in fermented milk, was shown around 1877 by Sir John Lister. Fermentation, from the Latin word Fevere' was defined by Louis Pasteur as "la vie sans l'air" (life without air). Coincidentally, this was the time of the industrial revolution in Europe, which resulted in large scale migration of populations from villages to larger cities. There was, therefore, a dramatic shift from food production for local communities to large scale food production, necessary to meet the requirements of expanding and more distant markets. This, in turn, led to the development of large scale fermentation processes for commercial production of fermented foods and alcoholic beverages, with the most widely used microorganisms including yeast for the production of beer, wine, and spirits, and LAB, for a variety of dairy, vegetable, and meat fermentations. The undefined strain mixture that was previously utilized to produce these goods has been replaced almost entirely by defined strain starters in the large-scale production of fermented foods and beverages that occurs today. A smaller number of strains are now heavily utilized and depended upon by the food and beverage industries, despite the fact that culture performance, product quality, and consistency have all significantly improved as a result of the shift to defined strains. However, there are some disadvantages to this heavy reliance on particular starters, and it may cause production issues that ultimately result in subpar strain performance. The performance of cheese starters in lactococcal fermentations may be impacted by bacteriophage proliferation.

The LAB produces nisin, which Rogers and Whittier identified in 1928 CE and showed to have antagonistic properties against other food-borne bacterial pathogens. The International Dairy Federation (IDF) published a comprehensive list of microbes in 2002 that are

Keyword

Ethanol Fermentation – It also called alcoholic fermentation, is a biological process that converts sugars such as glucose, fructose, and sucrose into cellular energy, producing ethanol and carbon dioxide as by-products.

safe to use as microbial food cultures in the dairy industry. A de facto reference for food cultures in practical use is now the "2002 IDF inventory.." The members of the IDF Task force created an updated inventory of microorganisms (fungi, filamentous fungi, bacteria, and yeasts) utilized in food fermentations covering a variety of food matrices in 2002.

Table 3.1. *Milestones in the History of Fermented Foods*

Milestone	Development/Location
6000–4000 BCE	Dahi—Coagulated sour milk eaten as a food item in India
7000 BC	Cheese production in Iraq, following the domestication of animals
6000 BC	Wine making in the Near East
5000 BC	Nutritional and health value of fermented milk and beverages described
4000 BC	Egyptians discovered how to use yeasts to make leavened bread and wine
2000 BCE-1200 CE	Different types of fermented milks from different regions
1750 BCE	Sumerians fermented barley to beer
1500 BCE	Preparation of meat sausages by ancient Babylonians
500 BCE	Mouldy soyabean curd as antibiotic in China
300 BCE	Preservation of vegetables by fermentation by the Chinese
500–1000 CE	Development of cereal-legume based fermented foods
1276 CE	First whisky distillery established in Ireland
1500 CE	Fermentation of sauerkraut and yoghurt
1851 CE	Louis Pasteur developed pasteurization
1877 CE	*Bacterum lactis* (*Lactococcus lactis*) was shown in fermented milk by John Lister
1881 CE	Published literature on koji and sake brewing
1907 CE	Publication of book *Prolongation of Life* by Eli Metchnikoff describing therapeutic benefits of fermented milks

1900–1930 CE	Application of microbiology to fermentation, use of defined cultures
1928 CE	Discovery of nisin—antagonism of some lactococci to other LAB shown by Rogers and Whittier
1970 CE–present	Development of products containing probiotic cultures or friendly intestinal bacteria
1953 CE	Nisin marketed in UK and since approved for use in over 50 countries
1990 CE–Present	Deciphering of genetic code of various LAB isolated from fermented foods
2002 CE	First authoritative list of microorganisms to be used in dairy culture was released by IDF and EFFCA
2012 CE	The list of Microbial Food Cultures regarded as GRAS to be used in all types of food fermentations has been released by IDF and EFFCA

Source: Ross et al. (2002); Prajapati & Nair (2003), updated.

3.3.2. Types of Fermentation

Most commercially useful fermentations may be classified as either solid-state or submerged cultures. In solid-state fermentations, the microorganisms grow on a moist solid with little or no 'free' water, although capillary water may be present. Examples of this type of fermentation are seen in mushroom cultivation, bread-making, and the processing of cocoa, and in the manufacture of some traditional foods, e.g., miso (soy paste), sake, soy sauce, tempeh (soybean cake) and gari (cassava), which are now produced in large industrial operations. Submerged fermentation may use a dissolved substrate, e.g., sugar solution or a solid substrate, suspended in a large amount of water to form a slurry. Submerged fermentations are used for pickling vegetables, producing yogurt, brewing beer and producing wine and soy sauce.

Both submerged and solid-state fermentations can be further classified as anaerobic processes, which must be carried out in the absence of oxygen, and aerobic processes, which need oxygen. Examples of aerobic fermentations are solid-state koji fermentations (which are used to make soy sauce) and *Aspergillus niger* submerged-culture citric acid production. Acid-forming bacteria, specifically Lactobacillus, Pediococcus, and Micrococcus species, are used in solid-state anaerobic fermentations to produce fermented meat products like salami, pepperoni, and bologna sausage (polony). Making yogurt involves a submerged-culture anaerobic fermentation.

It might only take one species of microorganism for a fermentation to produce the required chemical change. In this instance, sterilizing the substrate before inoculating it with the intended microorganism might eliminate undesirable species. The majority of food fermentations, however, are not sterile. Several microbial species must typically participate in simultaneous or sequential fermentations used in food processing in order to produce a product with the appropriate texture, flavor, aroma, and appearance. The culture environment in non-sterile fermentations can be specifically adjusted to support the desired microorganisms. For instance, there could be a lot of salt present, a low pH, or additives like sugar or salt reducing the water activity.

Factors influencing fermentations: A fermentation is influenced by numerous factors, including temperature, pH, nature, and composition of the medium, dissolved O_2, dissolved CO_2, operational system (e.g., batch, fed-batch, continuous), feeding with precursors, mixing (cycling through varying environments), and shear rates in the fermenter. Variations in these factors may affect: the rate of fermentation, the product spectrum and yield, the organoleptic properties of the product (appearance, taste, smell, and texture), the generation of toxins, nutritional quality, and other physicochemical properties.

The yield, rate, and product profile are impacted by the fermentation medium's formulation. The required concentrations of nitrogen, carbon, trace elements, and micronutrients (e.g., vitamins). There might be a need to regulate the carbon-to-nitrogen ratio and use certain kinds of carbon and nitrogen sources.

Comprehending fermentation biotechnology is crucial to creating a medium with the right composition. To get the desired outcome during a fermentation, it might be necessary to adjust the concentrations of some nutrients in a particular way. Certain trace elements may need to be avoided; for instance, *A. niger* produces less citric acid when small amounts of iron are present. The choice of medium is influenced by additional variables as well, including cost, availability, and batch-to-batch variability.

3.3.2.1. Submerged Fermentations

3.3.2.1.1. Fermentation Systems

Industrial fermentations can be conducted as continuous cultures, fed-batch operations, or batch-wise processes (Figure 3.1). Continuous fermentations are comparatively uncommon, with batch and fed-batch processes being quite prevalent. While most beer breweries use batch processes, continuous brewing is used in the commercial sector. In batch processing, a microorganism known as the "starter culture" is added to a batch of culture medium in a fermenter. The product is harvested once the fermentation has continued for a predetermined amount of time, also known as the "fermentation time" or "batch time." While most batch fermentations take four to five days, some traditional food fermentations can take up to several months. The sterile culture medium is continuously or sporadically added to the inoculation fermentation batch in fed-batch fermentations.

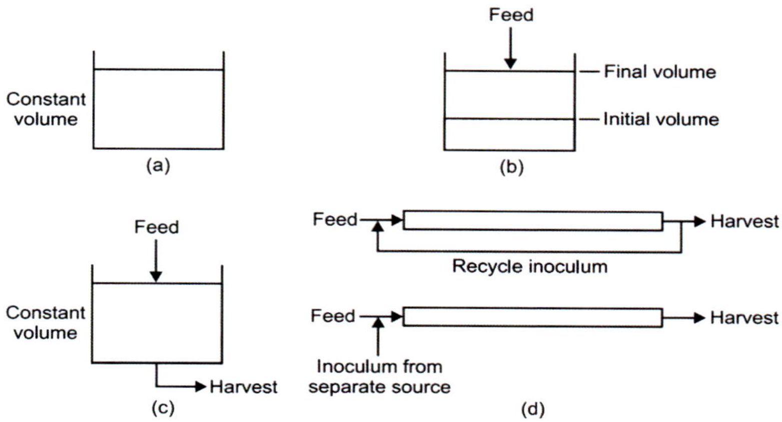

Figure 3.1. *Fermentation methodologies: (a) Batch fermentation; (b) fed-batch culture; (c) continuous-flow well-mixed fermentation; and (d) continuous plug flow fermentation, with and without recycling of inoculum.*

Source: S. C. Bhatia. Food Biotechnology. Woodhead Publishing India Pvt. Ltd. ISBN: 978-93-85059-18-6.

With each addition of medium, the volume of the fermenting broth grows, and the fermenter is harvested after the batch time. Continuous fermentations maintain an unchanged fermentation volume by continuously feeding a sterile medium into a fermenter and continuously withdrawing the fermented product. Continuous fermentations are usually initiated as batch cultures, and feeding is initiated once the concentration of the microbial population is reached. A tiny portion of the harvested culture may be recycled in some continuous fermentations to continuously inoculate the sterile feed medium that enters the fermenter (Figure 3.1(D)). The kind of mixing used in the fermenter determines whether ongoing inoculation is required. Fermentation devices that "plug flow" (Figure 3.1(D)), which include lengthy tubes that prevent back mixing, need to be continuously vaccinated. A plug flow device's flowing fluid components behave like little batch fermenters. Therefore, continuous operations in plug-flow fermenters can be easily converted from true batch fermentation processes, particularly when aeration and pH control are not needed. Microbial contamination is especially prone to affect continuous cultures; however, in certain instances, the fermentation conditions might be chosen (e.g., low pH, a lot of alcohol, or salt) to make the intended microorganisms more favorable than possible contaminants.

In a 'well-mixed' continuous fermenter (Figure 3.1(C)), the feed rate of the medium should be such that the dilution rate, i.e., the ratio of the volumetric feed rate to the constant culture volume, remains less than the maximum specific growth rate of the micro-organism in the particular medium and at the particular fermentation conditions. If the dilution rate exceeds the maximum specific growth rate, the microorganism will be washed out of the fermenter. Industrial fermentations are mostly batch operations. Typically, a pure starter culture (or seed), maintained under carefully controlled conditions, is used to inoculate sterile Petri dishes or liquid medium in the shake flasks. After sufficient growth, the pre-culture is used to inoculate the 'seed' fermenter. Because industrial fermentations tend to be large (typically 150–250 m^3), the inoculum is built up through several successively larger stages to 5–10% of the working volume of the production fermenter. A culture in rapid exponential growth is normally used for inoculation. Slower-growing microorganisms require larger inocula, to reduce the total duration of the fermentation. An excessively long fermentation time (or batch time) reduces productivity (amount of product produced per unit time per unit volume of fermenter) and increases costs. Sometimes inoculation spores, produced as seeds, are blown directly into large fermentation vessels with the ingoing air.

3.3.2.1.2. Microbial Growth

Microbial growth in a newly inoculated batch fermenter typically follows the pattern shown in Figure 3.2. Initially, in the lag phase, the cell concentration does not increase very much. The length of the lag phase depends on the growth history of the inoculum, the composition of the medium, and the amount of culture used for inoculation. An excessively long lag phase ties up the fermenter unproductively–hence, the duration of the lag phase should be minimized. Short lag phases occur when the composition of the medium and the environmental conditions in the seed culture and the production vessel are identical (hence, less time is needed for adaptation), the dilution shock is small (i.e., a large amount of inoculum is used), and the cells in the inoculum are in the late exponential phase of growth. The lag phase is essentially an adaptation period in a new environment. The lag phase is followed by exponential growth, during which the cell mass increases exponentially. Eventually, as the nutrients are exhausted and inhibitory products of metabolism buildup, the culture enters a stationary phase.

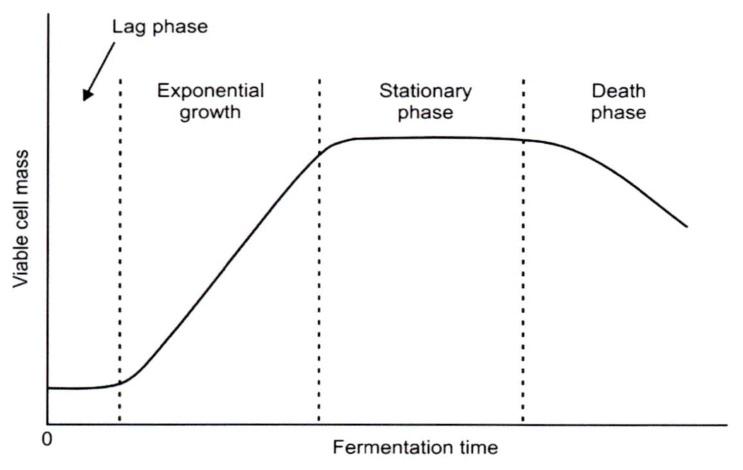

Figure 3.2. *Typical growth profile of microorganisms in a submerged culture.*

Source: Sciencedirect.com/topics/medicine-and-dentistry/microbial-growth.

3.3.2.2. Aeration and Oxygen Demand

The most popular method for aerating submerged cultures is bubbling with sterile air. The maximum aeration rate in small fermenters is usually limited to one volume of air per unit volume of culture broth. The maximum superficial aeration velocity in large bubble columns and stirred vessels is typically < 0.1 m s^{-1}. The volume flow rate of air divided by the fermenter's cross-sectional area is the superficial aeration velocity. Airlift fermenters allow for significantly higher aeration rates. These involve forcing aeration gas through perforated plates, perforated pipes, or single-hole spargers situated close to the fermenter bottom. O_2 is only sporadically soluble in aqueous culture broths, so even a brief cessation of aeration quickly depletes the available O_2, harming the culture irreversibly. Therefore, constant aeration is required. Microporous membrane filters are used to filter out any suspended particles, microbes, and spores from the gas before it is used for aeration.

The O_2 requirements of a fermentation depend on the microbial species, the concentration of cells, and the type of substrate. O_2 supply must at least equal O_2 demand, or the fermentation will be O_2-limited. O_2 demand is especially difficult to meet in viscous fermentation broths and in broths containing a large concentration of O_2-consuming cells.

Remember

Choosing the right microorganism for the desired fermentation process is crucial. It must possess the necessary metabolic pathways to convert the chosen substrate into the desired product.

As a general guide, the capability of a fermenter in terms of O_2 supply depends on the aeration rate, the agitation intensity, and the properties of the culture broth. In large fermenters, supplying O_2 becomes difficult when demand exceeds 4–5 kg m^{-3} h^{-1}.

At concentrations of dissolved O_2 below a critical level, the amount of O_2 limits microbial growth. The critical dissolved O_2 level depends on the micro-organism, the culture temperature and the substrate being oxidized. The higher the critical dissolved O_2 value, the greater the likelihood that O_2 transfer will become limiting. Under typical culture conditions, fungi such as *Penicillium chrysogenum* and *A. oryzae* have a critical dissolved O_2 value of about 3.2×10^{-4} kg m^{-3}. For bakers' yeast and *Escherichia coli*, the critical dissolved O_2 values are 6.4×10^{-5} kg m^{-3} and 12.8×10^{-5} kg m^{-3}, respectively.

Foam is produced when fermentation broths are aerated. Generally, to make room for the foam and permit gas disengagement, 20–30% of the fermenter's volume needs to be left empty. Furthermore, chemical antifoaming agents and mechanical "foam breakers" are frequently employed. Vegetable oils, silicone oils, and compounds based on low-molecular-weight polypropylene or polyethylene glycol are common antifoams. Because they disperse more readily in the fermenter, emulsified antifoams work better. A foam sensor sends signals, and antifoams are added in response. Hydrophobic silicone antifoams are especially problematic when used excessively since they can obstruct certain downstream separations, like membrane filtrations.

3.3.2.3. Heat Generation and Removal

Every fermentation produces heat. Microbial activity typically provides 3–15 kW m^{-4} in submerged cultures. Furthermore, the broth's mechanical agitation generates a maximum of 15 kW m^{-3}. A fermenter needs to be cooled to avoid a temperature spike that could harm the culture. The cooling water's temperature is usually only a few degrees lower than the fermentation broth's, therefore, heat removal is typically challenging. As a result, the ability of industrial fermentations to transfer heat is often a limitation. The surface area available for heat exchange, the temperature differential between the cooling water and the broth, the characteristics of the coolant and the broth, and the turbulence in these fluids all affect the capacity to remove heat. The geometry of the fermenter determines the surface area that can be provided for heat exchange. Heat generation due to metabolism depends on the rate of O_2 consumption, and heat removal in large vessels becomes difficult as the rate of O_2 consumption approaches 5 kg m^{-3}.

A fermenter needs to be able to remove metabolic heat and transfer heat during sterilization and cooling. Both batch and continuous methods can be used to sterilize liquid medium, or slurry, for batch fermentation. By heating the medium inside the fermenter, batch processes typically sterilize the medium, or some of its components, and the fermenter in one step. Heating can occur through the fermenter wall or by injecting steam directly into the medium. When sterilizing, heating to high temperatures (usually 121°C) frequently causes unfavorable reactions between the medium's constituent parts. These reactions either destroy nutrients or produce compounds that impede growth, lowering the yield. By sterilizing only some of the medium's ingredients in the fermenter and subsequently adding additional, independently sterilized ingredients, this heat damage can be avoided or at least minimized. Sterilization of sugars and nitrogen-containing components is frequently done in isolation.

Particularly vulnerable dissolved nutrients can be sterilized by passing them through hydrophilic polymer filters, which hold particles with a diameter of at least 0.45 μm. Further refinement (e.g., retaining particles with a diameter of 0–2 μm) is also offered. A large fermentation batch takes time to heat and cool, which ties up a fermenter in an inefficient manner. Additionally, thermal degradation or nutrient loss increases with the length of time a medium is kept at a high temperature. Therefore, even for batch fermentations, it is preferable to continuously sterilize the culture medium while it is being transported to a pre-sterilized fermenter.

3.3.2.4. Solid-State Fermentations

The term "solid-state (substrate) fermentation" (SSF) refers to a fermentation process that takes place without or almost without free water. Natural raw materials are typically used in SSF processes as a source of energy and carbon. SSF can also use an inert material as a solid matrix, but doing so will require adding a carbon source and necessary nutrients to a nutrient solution. On the other hand, a solid substrate (matrix) needs adequate moisture.

Depending upon the nature of the substrate, the amount of water absorbed could be one or several times more than its dry weight, which leads to relatively high water activity (a_w) on the solid/gas interface in order to allow a higher rate of biochemical process. Low diffusion of nutrients and metabolites takes place in lower water activity conditions whereas compaction of substrate occurs at higher water activity. Hence, maintenance of adequate moisture levels in the solid matrix along with suitable water activity are essential elements for SSF processes. Solid substrates should generally have a large surface area per unit volume (say in the range of 10^3–10^6 m^2/cm^3 for the ready growth on the solid/gas interface). Smaller substrate particles provide larger surface area for microbial attack but pose difficulty in aeration/respiration due to limitation in inter-particle space availability. Larger particles provide better aeration/respiration opportunities but provide lesser surface area. In bioprocess optimization, sometimes it may be necessary to use a compromised size of particles (usually a mixed range) for the reason of cost-effectiveness. For example, wheat bran, which is the most commonly used substrate in SSF, is obtained in two forms, fine and coarse. The former contains particles of smaller size (mostly smaller than 500–600 µ) and the latter, mostly larger than these.

Most of SSF processes use a mix of these two forms at different ratios for optimal production. Solid substrates typically offer beneficial living conditions for the microbial flora, which includes fungi, yeast, and bacteria. Because of their hyphal growth, which can both grow on and through the surface of substrate particles, filamentous fungi are the most studied among them for SSF. Many agricultural products, including barley, cassava, etc., as well as agricultural waste products like rice bran, wheat bran, sugarcane bagasse, cassava bagasse, different types of oil cakes (e.g., cakes made with coconut oil, palm kernels, soy, ground nuts, etc., fruit pulps (e.g., apple pomace), sawdust, corn cobs, seeds (e.g., tamarind, jack fruit), coffee pulp and husk, leftover brewing grains, tea waste, etc., are the substrates that are most frequently and widely used in SSF processes. As the microorganisms grow on these substrates, they produce hydrolytic exo-enzymes that are expeled from the cells. These enzymes enable the cells to access and create simple products, such as nutrients and carbon sources. This further encourages microbial activity and biosynthesis. In addition to these, a number of other significant factors need to be considered when developing SSF processes. These include biological and physicochemical elements like the substrate's type, the type of microorganism used, the pH of the medium, the temperature and duration of incubation, the age, size, the kind of inoculum, etc.

3.3.2.4.1. Significance of SSF

Because of the many benefits it offers, SSF has been deemed superior to submerged fermentation (SMF) in a number of areas. It uses minimal water, which results in the release of little to no effluent and lowers concerns about pollution. It is also cost-effective because it uses simple growth and production media made of agro-industrial residues. SSF procedures are straightforward, require less expensive, low-volume equipment, and produce concentrated products with high titers, making them effective. Furthermore, because there is a higher rate of oxygen diffusion into moistened solid substrate, which supports the growth of aerial mycelium, the aeration process—the availability of atmospheric oxygen to the substrate—is made easier because oxygen limitation does not occur.

These are also appropriate for rural areas because they can be used at smaller scales with success.

3.3.2.4.2. General Aspects of SSF

The development of any bioprocess in SSF should take into account a number of significant factors in general. These consist of the appropriate microorganism and substrate selection, process parameter optimization, and product isolation and purification. Based on their water activity, only yeast and fungi were considered suitable microorganisms for SSF in the theoretical classification. It was believed that bacterial cultures might not be appropriate for SSF because of the high water activity requirement. Experience has demonstrated, nevertheless, that bacterial cultures can be effectively controlled and manipulated for SSF procedures. Product yields are generally higher in SSF compared to SMF, according to common claims. Nevertheless, there isn't currently a recognized scale or technique to accurately compare product yields in SMF and SSF. The precise cause of the higher product titers in SSF is presently unknown. The rationale behind this is that because SSF microbial cultures are more closely aligned with their natural environment, their activity levels are likely higher.

Choosing the right substrate is yet another crucial component of SSF. Non-soluble solid material serves as a source of nutrients and physical support in SSF. A naturally occurring solid substrate, such as agricultural crops, agro-industrial residues, or inert support, can be considered solid material. Nevertheless, by employing a nutritionally inert material soaked in a nutrient solution, it is possible to replicate the conditions of low water activity and high oxygen transference without having to combine the roles of support and substrate. Two main factors may need to be taken into account when choosing a substrate: first, there may be a particular substrate that needs appropriate value-adding and/or disposal. The second may be connected to the objective of creating a particular product from an appropriate substrate. It would be necessary to screen various substrates and select the most suitable one. Similarly, it would be important to screen suitable microorganisms and select the most suitable one. If inert materials such as polyurethane foam are used, product isolation could be relatively simpler and cheaper than using naturally occurring raw materials such as wheat bran because while extracting the product after fermentation, along with the product, several other water-soluble components from the substrate also leach out and may pose difficulties in the purification process. Inert materials have often been used for studying modeling or other fundamental aspects of SSF.

The choice and optimization of process parameters may present additional pertinent issues in this case. Particle size, initial moisture, pH, substrate pretreatment, relative humidity, incubation temperature, agitation, and aeration, age and size of the inoculum, nutrient supplementation (N, P, and trace elements), supplemental carbon source and inducers, product extraction and purification, etc., are some examples of these physicochemical and biochemical parameters. The kind, degree, and application of the experimentation will determine whether to use a single variable parameter optimization approach or one with multiple variables.

3.3.2.4.3. Biochemical Engineering Aspects of SSF

The fundamentals of SSF have been thoroughly covered in a number of excellent reports that have surfaced in recent years. Additionally, our understanding of the effects of heat and mass transfer in SSF processes—which have long been thought to be the primary challenges in managing SSF systems—has improved. But there's still a lot of work to be done in this area. Significant heat is produced during SSF, and this heat is directly correlated with the microorganism's metabolic activity. Due to the low thermal conductivities of the solid materials/matrices used in SSF, heat removal from the process may proceed very slowly. The product that forms and accumulates in the bed is sometimes denatured by high heat accumulation. Certain parts of the bed may have temperatures up to 20°C above the incubation temperature. Early on in the SSF process, the temperature and oxygen concentration in the substrate are constant. However, as the fermentation process advances, heat is produced due to oxygen transfer. The aeration of the fermentation system is intimately linked to the heat transfer that occurs within or outside of the SSF system. In SSF, the substrate's temperature is also very important since it eventually influences the microorganism's growth, spore formation and germination, and product formation. Elevated moisture levels lead to a reduction in the porosity of the substrate, thereby impeding the entry of oxygen. This could aid with bacterial overgrowth. Conversely, low moisture content can result in poor nutrient accessibility, which would hinder microbial growth.

Keyword

Microbial Growth – It is defined as an increase in cellular constituents, resulting an increase in a microorganism size, population number, or both.

3.3.2.5. Koji Fermentations

Fermentations using koji are very common. Koji, which has been used for thousands of years to prepare food in the East, is made of soybeans or grains that have mold growing on them. Fungal enzymes, which break down proteins, carbohydrates, and fats into nutrients that other microbes can use in later fermentations, can be found in koji. There are numerous varieties of koji that vary depending on the mold, substrate, preparation technique, and harvest stage. Koji fermentation is used in the making of sake, miso, and soy sauce. In Japan, citric acid is also produced using koji technology. The following provides a detailed explanation of a typical industrial solid-substrate fermentation process: the making of soy sauce or shoyu in Japanese.

The koji for soy sauce is made from soybeans and wheat. Soybeans, or defatted soybean flakes or grits are moistened and cooked (e.g., for 0.25 min or less, at about 170°C) in continuous pressure cookers. The cooked beans are mixed with roasted, cracked wheat, the ratio of wheat to beans varying with the variety of shoyu.

The mixed substrate is inoculated with a pure culture of *Aspergillus oryzae* (or *A. sojae*), the fungal spore density at inoculation being about 2.5 × 10^8 spores per kilogram of wet solids. After a 3-day fermentation, the substrate mass becomes green-yellow because of sporulation. The koji is then harvested for use in a second submerged fermentation step. Koji production is highly automated and continuous—processes producing up to 4150 kg h^{-1} of koji have been described. Similar large-scale operations are used to produce koji for miso and sake in Japan.

3.3.2.5.1. Solid-State Fermenters

The technological complexity of solid-state fermentation devices varies, ranging from extremely basic methods such as banana-leaf wrapping, bamboo baskets, and substrate heaps to highly automated machines primarily utilized in Japan. Some 'less sophisticated' fermentation systems, e.g., massive fermentation of cocoa bean, proves to be highly efficient when processing on a large scale. Additionally, some of the highly automated, continuous processes used in Japan to ferment soy sauce successfully are inappropriate for use in less developed parts of Asia. As a result, fermentation techniques need to be customized for the region.

The use of pressure vessels is not the norm for solid-state fermentation. The commonly used devices are:

- Tray fermenter;
- Static-bed fermenter;
- Tunnel fermenter;
- Rotary disc fermenter;
- Rotary drum fermenter;
- Agitated-tank fermenter;
- Continuous screw fermenter.

Large concrete or brick fermentation chambers, or koji rooms, may be lined with steel, typically Type 304 stainless steel, and are corrosion-resistant.

3.3.2.5.2. Tray Fermenter

A common fermenter type in small- and medium-sized Koji operations is this straightforward one (*see* Figure 3.3). In order to achieve better aeration, the trays—which can be made of wood, metal, or plastic—often have a wire-mesh or perforated base. The substrate ferments in thin layers that are 0–15 m deep. Although processing is not sterile, the trays may be covered with cheesecloth to lessen contamination. Trays can be placed singly or in stacks in areas with ventilation or in chambers with controlled humidity and temperature. Automation may be used for trays' handling, filling, emptying, and washing, but manual inoculation and occasional mixing are still done. Tray fermenters are labor-intensive and need a large production area, even with some automation. Therefore, there is little chance of increasing production.

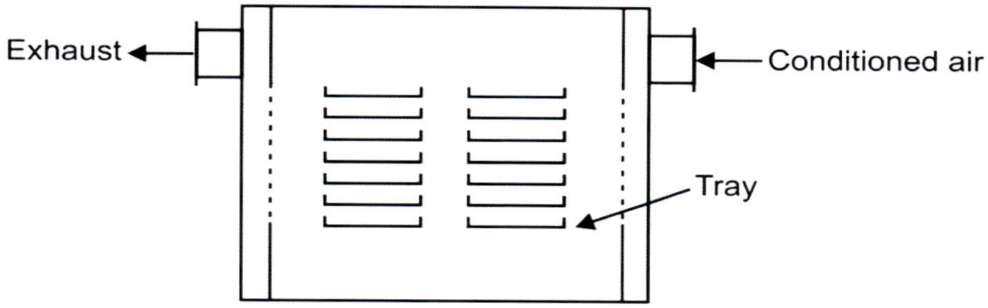

Figure 3.3. *Tray fermenter.*

Source: S. C. Bhatia. Food Biotechnology. Woodhead Publishing India Pvt. Ltd. ISBN: 978-93-85059-18-6.

3.3.2.5.3. Static-Bed Fermenter

This is a tray fermenter modification (Figure 3.4). It uses a single, deeper, bigger static bed of substrate that is housed in an insulated room. Forced aeration through the substrate bed provides O_2.

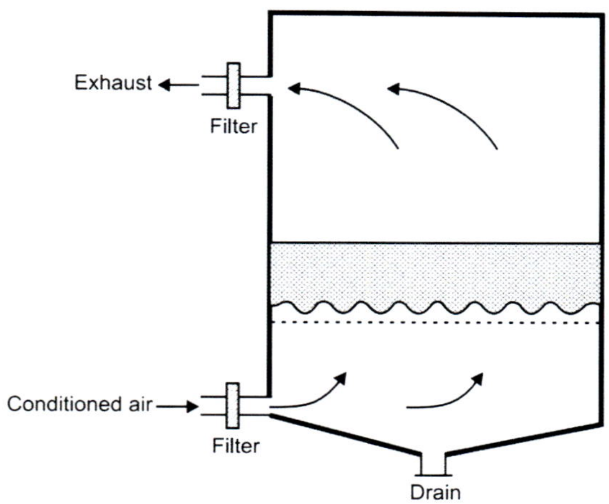

Figure 3.4. *Static-bed fermenter.*

Source: S. C. Bhatia. Food Biotechnology. Woodhead Publishing India Pvt. Ltd. ISBN: 978-93-85059-18-6.

3.3.2.5.4. Tunnel Fermenter

This device is a modification of the static-bed apparatus (Figure 3.5). The bed of solids is usually quite long, but it doesn't go any deeper than 0.5 m. With this equipment, fermentation can be highly automated through mechanisms for substrate harvesting, inoculation, mixing, and continuous feeding.

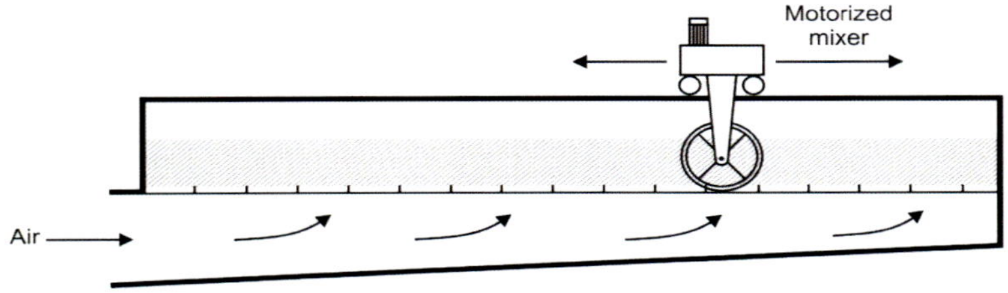

Figure 3.5. *Tunnel fermenter.*

Source: S. C. Bhatia. Food Biotechnology. Woodhead Publishing India Pvt. Ltd. ISBN: 978-93-85059-18-6.

3.3.2.5.5. Rotary Disc Fermenter

The upper and lower chambers of the rotary disc fermenter are each equipped with a circular perforated disc to support the substrate bed. The discs are rotated by a common central shaft. The upper chamber is filled with the inoculated substrate, which is then gradually transferred to the transfer screw. The partially fermented solids are transferred by the upper screw to the lower chamber, where additional fermentation takes place, via a mixer. The lower transfer screw is used to harvest the fermented substrate. Air that has been temperature-regulated and humidified is used to aerate both chambers. In Japan, koji is produced on a large scale using rotary disc fermenters.

3.3.2.5.6. Rotary Drum Fermenter

The cylindrical drum of the rotary drum fermenter is supported on rollers and rotated at 1–5 rpm around the long axis (Figure 3.6). Rotation may be intermittent, and the speed may vary with the fermentation stage.

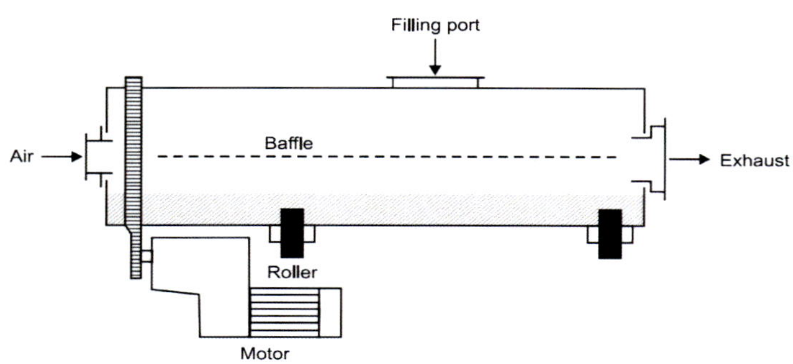

Figure 3.6. *Rotary drum fermenter.*

Source: S. C. Bhatia. Food Biotechnology. Woodhead Publishing India Pvt. Ltd. ISBN: 978-93-85059-18-6.

Straight or curved baffles inside the drum aid in the tumbling of the substrate, hence improving aeration and temperature control. Sometimes, the drum can be inclined, causing the substrate to move from the higher inlet end to the lower outlet during rotation. Aeration occurs through coaxial inlet and exhaust nozzles.

3.3.2.5.7. Agitated-Tank Fermenter

This kind of fermenter uses cylindrical or rectangular tanks with one or more helical-screw agitators installed to stir the fermented substrate (Figure 3.7). Occasionally, mobile trolleys that travel on horizontal rails above the tank send screws into the tank. A paddle fermenter is another stirred-tank configuration. Similar to a rotary drum device, this one has a stationary drum and uses motor-driven paddles supported by a concentric shaft to achieve periodic mixing.

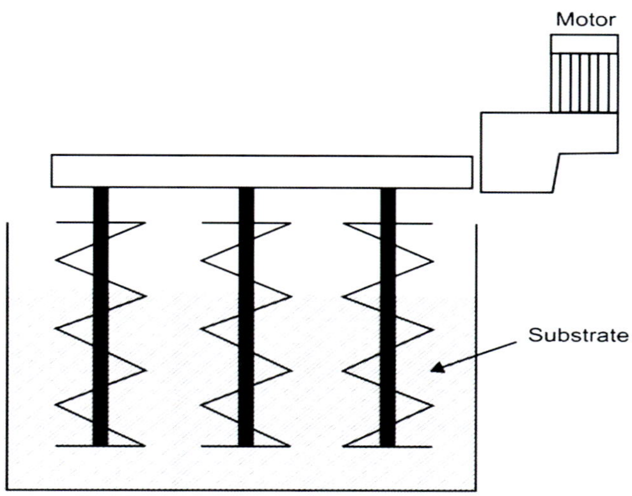

Figure 3.7. *Agitated-tank fermenter.*

Source: S. C. Bhatia. Food Biotechnology. Woodhead Publishing India Pvt. Ltd. ISBN: 978-93-85059-18-6.

3.3.2.5.8. Continuous Screw Fermenter

Sterilized, cooled, and inoculated substrate is fed into this type of fermenter through the non-aerated chamber's inlet (Figure 3.8). The screw moves the solids in the direction of the harvest port, and its length and rotational speed regulate how long the fermentation takes. Continuous anaerobic or microaerophilic fermentations can be conducted with this kind of fermenter.

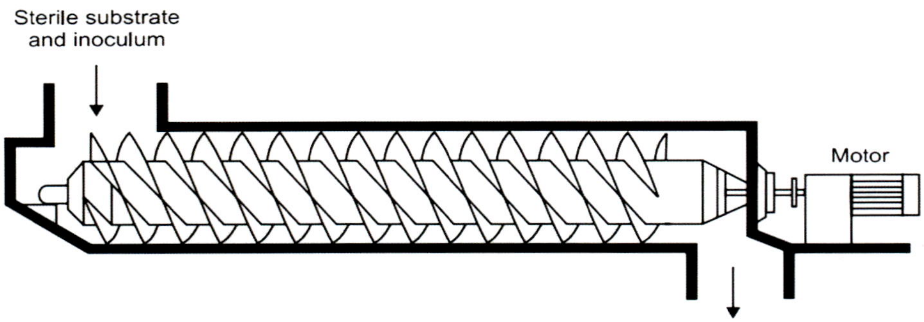

Figure 3.8. Continuous screw fermenter.

Source: https://www.researchgate.net/publication/325393615/figure/fig7/AS:631039638659074@1527462982302/Continuous-screw-fermenter.png.

Did you get it?

1. What is solid-state fermentations?
2. Explain the process of koji fermentations.

3.4. GENETICALLY MODIFIED FOODS

→Learning Objectives

- Understand the advantages of GM foods;
- Examine criticisms against GM foods;
- Evaluate human health risks associated with GM foods;
- Explore the regulation of GM foods and role of government;
- Define labeling of GM foods;
- Identify the most common GM goods.

By increasing yield and lowering reliance on chemical pesticides and herbicides, GM foods have the potential to help protect and preserve the environment and address many of the world's problems related to hunger and malnutrition. Governments will, however, face numerous obstacles in the future, particularly with regard to food labeling, international policy, safety testing, and regulations. A lot of people believe that genetic engineering is the way of the future and that we simply cannot overlook a technology with so many promising applications. However, because of our enthusiasm for this powerful technology, we must proceed cautiously to avoid unintentionally harming human health and the environment.

Using the most recent molecular biological techniques, crop plants intended for human or animal consumption are most commonly referred to as GM organisms, or GMOs, or GM foods. These plants have undergone laboratory modification to improve desired characteristics like greater herbicide resistance or better nutritional value. Traditionally, breeding has been used to enhance desired traits, but traditional plant breeding techniques can be very time-consuming and inaccurate. On the other hand, plants with the precise desired trait can be quickly and accurately created through genetic engineering. Plant geneticists, for instance, can separate a gene that confers drought tolerance and transfer it to a different kind of plant. Additionally, the newly genetically altered plant will become resistant to drought. Genes from non-plant organisms can also be utilized, in addition to genes that are transferred from one plant to another. The application of *Bacillus thuringiensis* genes in corn and other crops is the most well-known illustration of this. A naturally occurring bacterium known as *B. thuringiensis*, or Bt, creates crystal proteins that are fatal to insect larvae.

3.4.1. Advantages of Genetically Modified Foods

GM foods, also known as GMOs, have been a subject of considerable debate and research. While opinions on GM foods vary, proponents argue that there are several potential advantages associated with the development and use of GM crops. Some of the commonly cited advantages include:

1. **Increased Crop Yields:** Genetic modification can enhance the resistance of crops to pests, diseases, and harsh environmental conditions, leading to increased productivity and higher yields. This can contribute to food security by ensuring a more reliable and abundant food supply.
2. **Improved Nutritional Content:** Genetic engineering allows for the enhancement of the nutritional content of crops. For example, scientists can modify crops to contain higher levels of essential nutrients such as vitamins and minerals, potentially addressing malnutrition issues in certain populations.
3. **Reduced Need for Pesticides:** Some GM crops are engineered to be resistant to certain pests and diseases, reducing the need for chemical pesticides. This can lead to lower production costs, decreased environmental impact, and reduced exposure of farmers and consumers to harmful pesticides.
4. **Enhanced Crop Tolerance to Environmental Stress:** Genetic modifications can make crops more resilient to environmental stresses such as drought, extreme temperatures, or soil salinity. This is particularly relevant in the context of climate change, where unpredictable weather patterns may affect traditional crop production.
5. **Extended Shelf Life:** Genetic modifications can be employed to slow down the ripening process of fruits and vegetables, potentially reducing food waste by extending the shelf life of perishable goods.
6. **Development of Stress-Resistant Crops:** Genetic engineering allows the creation of crops that can thrive in conditions where traditional crops might struggle. This can be especially beneficial in regions with challenging growing conditions.
7. **Cost Savings for Farmers:** The increased resilience of GM crops to pests and diseases can result in cost savings for farmers by reducing the need for expensive pesticides and other inputs.
8. **Improved Crop Quality:** Genetic modification can be used to improve the overall quality of crops, including attributes such as taste, texture, and nutritional value.

Did you know?

The first GM crop, the Flavr Savr tomato, designed for improved flavor and – longer shelf life, was approved for commercial sale in 1994.

3.4.2. Criticisms Against Genetically Modified Foods

Concerns regarding GM foods have been voiced by environmental activists, religious organizations, public interest groups, professional associations, other scientists, and government officials. They have also criticized agribusiness for pursuing profit at the expense of potential hazards and the government for not exercising sufficient regulatory oversight. Everybody seems to have a strong view on GM foods. Three main areas of concern exist regarding GM foods: risks to human health, environmental hazards, and economic issues.

3.4.2.1. Environmental Hazards

- Unintended harm to other organisms: Recently, a research study was published. Transgenic pollen harms monarch larvae, showing that pollen from *B. thuringiensis* corn caused high mortality rates in monarch butterfly caterpillars. Monarch caterpillars consume milkweed plants, not corn, but the fear is that if pollen from *B. thuringiensis* corn is blown by the wind onto milkweed plants in neighboring fields, the caterpillars could eat the pollen and perish. The results seemed to support this viewpoint. Unfortunately, *B. thuringiensis* toxins kill many species of insect larvae indiscriminately; it is not possible to design a *B. thuringiensis* toxin that would only kill crop-damaging pests and remain harmless to all other insects.
- Reduced effectiveness of pesticides: Just as some populations of mosquitoes developed resistance to the now-banned pesticide DDT, many people are concerned that insects will become resistant to *B. thuringiensis* or other crops that have been GM to produce their own pesticides.
- Gene transfer to non-target species: Another concern is that crop plants engineered for herbicide tolerance and weeds will cross-breed, resulting in the transfer of the herbicide resistance genes from the crops into the weeds. These 'super weeds' would then be herbicide tolerant as well. Other introduced genes may cross over into nonmodified crops planted next to GM crops.

The three issues raised above have a number of potential fixes. Pollen is the means by which plants exchange genes. Creating GM plants that are male sterile or altering the GM plant such that the pollen does not carry the introduced gene are two ways to guarantee that non-target species will not receive introduced genes from GM plants. There would be no cross-pollination, and if innocuous insects like monarch caterpillars ate pollen from GM plants, they would live.

The establishment of buffer zones around GM crop fields is an additional potential remedy. For instance, a field of *B. thuringiensis* GM corn would be surrounded by non-GM corn that would not be harvested. The non-GM corn would serve as a haven for beneficial or harmless insects, while insect pests could be allowed to decimate the non-GM corn without developing a resistance to *B. thuringiensis* pesticides. Because the wind-blown pollen would not cross the buffer zone, gene transfer to weeds and other crops would not happen. The recommended buffer zone width is estimated to be between 6 and 30 meters or more. Should the buffer zones require an excessive amount of acreage, this planting method might not be practical.

3.4.3. Human Health Risks

1. **Allergenicity:** Many children in the US and Europe have developed life-threatening allergies to peanuts and other foods. There is a possibility that introducing a gene into a plant

may create a new allergen or cause an allergic reaction in susceptible individuals.

2. **Unknown Effects on Human Health:** There is a growing concern that introducing foreign genes into food plants may have an unexpected and negative impact on human health.

On the whole, with the exception of possible allergenicity, scientists believe that GM foods do not present a risk to human health.

3.4.4. Economic Concerns

Bringing a GM food to market is a lengthy and costly process, and of course agri-biotech companies wish to ensure a profitable return on their investment. Many new plant genetic engineering technologies and GM plants have been patented, and patent infringement is a big concern of agribusiness. Yet consumer advocates are worried that patenting these new plant varieties will raise the price of seeds so high that small farmers and third-world countries will not be able to afford seeds for GM crops, thus widening the gap between the wealthy and the poor.

Patent enforcement may also be difficult, as the contention of the farmers that they involuntarily grew Monsanto-engineered strains when their crops were cross-pollinated shows one way to combat possible patent infringement is to introduce a 'suicide gene' into GM plants. These plants would be viable for only one growing season and would produce sterile seeds that do not germinate. Farmers would need to buy a fresh supply of seeds each year. However, this would be financially disastrous for farmers in third-world countries who cannot afford to buy seed each year and traditionally set aside a portion of their harvest to plant in the next growing season. In an open letter to the public, Monsanto has pledged to abandon all research using this suicide gene technology.

3.4.5. Regulation of Genetically Modified Foods and the Role of Government

The regulation of GM foods varies among countries, and governments play a crucial role in establishing and enforcing these regulations. The regulatory frameworks are designed to ensure the safety of GM foods for human consumption, assess environmental impacts, and address ethical and societal concerns. Here are key aspects of the regulation of GM foods and the role of governments:

1. **Risk Assessment and Approval Process:** Governments typically require developers of GM crops to undergo a comprehensive risk assessment process before these products can be approved for commercialization. This process involves evaluating the potential risks to human health, the environment, and biodiversity. Regulatory authorities may require developers to submit data from scientific studies to support their claims of safety.
2. **Transparency and Public Consultation:** Many countries involve the public in the decision-making process regarding the approval of GM crops. This often includes public consultations, where stakeholders, including consumers, farmers, and environmental groups, can express their opinions and concerns. Transparency in the

regulatory process helps build public trust and ensures that a diverse range of perspectives is considered.

3. **Labeling Requirements:** Some countries have implemented mandatory labeling of GM foods to inform consumers about the presence of GM ingredients. This allows consumers to make informed choices and has become a subject of debate, with some arguing for clear and transparent labeling, while others emphasize the need to avoid stigmatization or misinformation.
4. **Monitoring and Post-Market Surveillance:** Regulatory authorities play a role in monitoring GM crops and foods after they have been approved for commercialization. This post-market surveillance helps identify any unexpected adverse effects or unintended consequences that may not have been evident during the pre-market risk assessment.
5. **International Coordination and Standards:** As the global food supply chain is interconnected, there are efforts to establish international standards for the regulation of GM foods. Organizations such as the Codex Alimentarius Commission, a joint initiative of the Food and Agriculture Organization (FAO) and the World Health Organization (WHO), work to develop international food standards, guidelines, and codes of practice, including those related to GM foods.
6. **Enforcement and Compliance:** Governments are responsible for enforcing compliance with GM food regulations. This includes taking action against unauthorized or non-compliant GM products and ensuring that developers adhere to the conditions stipulated during the approval process.
7. **Ethical and Socioeconomic Considerations:** Beyond safety assessments, regulatory frameworks may also address ethical and socioeconomic considerations related to GM foods. This can include issues such as intellectual property rights, potential impacts on small farmers, and the concentration of seed ownership within a few biotechnology companies.

3.4.6. Labeling of GM Foods

Labeling of GM foods is a significant aspect of regulatory frameworks in many countries. The purpose of GM food labeling is to provide consumers with information about whether a product contains GMOs. Labeling requirements, however, vary globally, and different countries have adopted different approaches. Here are some common aspects of GM food labeling:

1. Mandatory vs. Voluntary Labeling:
    - i. **Mandatory Labeling:** Some countries have implemented mandatory labeling laws, requiring food products that contain a certain percentage of GM ingredients to be labeled as such. This provides consumers with clear information about the presence of GMOs in the food they purchase.
    - ii. **Voluntary Labeling:** In other regions, labeling may be voluntary, allowing manufacturers to choose whether to disclose the presence of GM ingredients. This approach relies on

market forces, assuming that consumer demand for transparency will drive producers to label their products.

2. Thresholds for Labeling:
   i. **Thresholds for GM Content:** Mandatory labeling often involves setting a threshold for the percentage of GM content in a product that triggers the requirement for labeling. Below this threshold, a product may not be required to be labeled as containing GMOs. The specific threshold varies from country to country.
3. Scope of Labeling:
   i. **Whole Foods vs. Processed Foods:** Some countries focus on labeling whole foods, such as fresh produce or specific grains, while others extend labeling requirements to processed foods that contain derivatives of GM crops, such as oils or sugars derived from GM plants.
4. Labeling Terminology:
   i. **Clear Terminology:** Where labeling is required, regulations often specify the terminology that must be used on labels. This may include terms such as "genetically modified," "genetically engineered," or specific language indicating the presence of GMOs.
5. Exemptions:
   i. **Certain Products Excluded:** Some regulations may exclude certain categories of products from mandatory labeling. For example, animal products from animals fed with GM feed may be exempt in some regions.
6. Enforcement:
   i. **Penalties for Non-Compliance:** Governments typically establish penalties for non-compliance with labeling regulations. These penalties may include fines or other consequences for companies that fail to adhere to the labeling requirements.
7. International Standards:
   i. **Harmonization Efforts:** International organizations, such as the Codex Alimentarius Commission, work towards harmonizing food standards globally. While international standards may not mandate labeling, they can influence the development of national regulations.
8. Consumer Education:
   i. **Accompanying Educational Initiatives:** In some regions, labeling requirements are complemented by educational initiatives to inform consumers about the nature and safety of GM foods. This aims to enhance consumer understanding and reduce misconceptions.

3.4.7. Most Common Genetically Modified Goods

GM material sounds a little bit like science fiction territory, but in reality, much of what we eat on a daily basis is a GMO. Whether or not these modified foods are actually healthy is still up for debate—and many times, you don't even know that you are buying something GM. It is not required to label GMOs in the U.S. and Canada, but there are substantial restrictions, and even outright bans, on GMOs in many other countries. However, by 2018, Whole Foods Market will start labeling GMOs in the United States. The location of this grocery chains in Britain already provide GMO-labeled products, as required by the European Union. According to the EU, GMO refers to plants and animals 'in which the genetic material has been altered in a way that does not occur naturally by mating and/or natural recombination.'

Some of the most common GM foods are briefly discussed below:

1. **Corn:** Almost 85 percent of corn grown in the U.S. is GM. Even Whole Foods' brand of cornflakes was found to contain GM corn. Many producers modify corn and soy so that they are resistant to the herbicide glyphosate, which is used to kill weeds.
2. **Soy:** It is the most heavily GM food in the United States. The largest US producer of hybrid seeds for agriculture, Pioneer HiBred International, created a genetically engineered soybean, which was approved in 2010. It is modified to have a high level of oleic acid, naturally found in olive oil. Oleic acid is a monounsaturated omega-9 fatty acid that may lower LDL cholesterol (traditionally thought of as 'bad' cholesterol) when used to replace other fats.
3. **Alfalfa:** Cultivation of genetically engineered alfalfa was approved in 2011, and consists of a gene that makes it resistant to the herbicide roundup, allowing farmers to spray the chemical without damaging the alfalfa.
4. **Canola:** It is a genetically engineered form, approved in 1996, and as of 2006, around 90 percent of U.S. canola are GM crops.
5. **Sugar Beets:** A very controversial vegetable, sugar beets were approved in 2005, banned in 2010, then officially deregulated in 2012. GM sugar beets make up half of the U.S. sugar production and 95 percent of the country's sugar beet market.
6. **Milk:** To increase the quantity of milk produced, cows are often given rBGH (recombinant bovine growth hormone), which is also banned in the European Union, as well as in Japan, Canada, New Zealand, and Australia.
7. **Zucchini:** GM zucchini contains a toxic protein that helps make it more resistant to insects. These introduced proteins have recently been found in human blood, including that of pregnant women and fetuses. This indicates that some of the insecticide is making its way into our bodies rather than being broken down and excreted.
8. **Yellow Squash:** Yellow squash has also been modified with toxic proteins to make it insect-resistant. This plant is very similar to zucchini, and both have also been modified to resist viruses.
9. **Papaya:** GM papaya trees have been grown in Hawaii since 1999. These Papayas are sold in the United States and Canada for human consumption. These papayas have been modified to be naturally resistant to Papaya ringspot virus, and also to delay the maturity of

the fruit. Delaying maturity gives suppliers more time to ship the fruit to supermarkets.

3.4.8. Pros and Cons of Genetically Modified Foods

Many people today take for granted exactly where the foods they eat come from. In fact, GM foods have become a common place thing in America, even though few people understand what 'GM' means. While there are some benefits that GM foods may offer, there are also some risks and negative effects that these foods can cause as well.

Keyword

Genetic Makeup – It refers to the specific set of genes or genetic material present in an organism. It encompasses the entire genetic information encoded in the DNA (deoxyribonucleic acid) of an individual.

When the term 'genetically modified' is used to describe a food, it means that the genetic makeup of one of the ingredients in that food has been altered. This is achieved by a very special set of technologies that combine the genes from different organisms, with the resulting organism being called a GM food. In most cases, the specific genes that are combined have been hand-picked for the specific traits that they have. Those traits could include everything from the resistance to insects to specific nutritional value. These GM foods can be in anything from corn to canola oil, which are quite common ingredients in many foods found on the market today such as snacks, cereals, and sodas.

3.4.8.1. Pros of Genetically Modified Foods

Genetically Modified (GM) foods offer numerous advantages. These benefits span various areas, including agriculture, environment, nutrition, and economic aspects.

1. Enhanced Nutritional Content
   - **Biofortification**: GM foods can be engineered to contain higher levels of essential vitamins and minerals. For example, Golden Rice is enriched with beta-carotene, a precursor of vitamin A, to combat vitamin A deficiency in developing countries.
   - **Improved Protein Content**: Certain GM crops are designed to have a higher protein content, which can be beneficial in regions where protein deficiency is prevalent.
2. Increased Agricultural Productivity
   - **Pest and Disease Resistance**: GM crops can be engineered to be resistant to specific pests and diseases, reducing the need for chemical pesticides.

For instance, Bt crops produce a toxin that is harmful to certain insects but safe for human consumption.

- **Herbicide Tolerance**: Some GM crops are designed to tolerate specific herbicides, making weed control more efficient and reducing crop losses due to weed competition.

3. Environmental Benefits
   - **Reduced Pesticide Use**: By incorporating pest resistance, GM crops can significantly lower the amount of chemical pesticides needed, reducing environmental contamination and promoting biodiversity.
   - **Conservation of Resources**: GM crops often require less water and fertilizer, contributing to more sustainable agricultural practices and conservation of natural resources.
4. Economic Advantages
   - **Lower Production Costs**: With improved resistance to pests, diseases, and environmental conditions, GM crops can reduce the costs associated with agricultural inputs like pesticides, herbicides, and fertilizers.
   - **Increased Yields**: Higher crop yields contribute to greater food production and can help address food security challenges, particularly in regions with rapidly growing populations.
5. Improved Food Quality and Shelf Life
   - **Extended Shelf Life**: Genetic modifications can enhance the shelf life of certain foods, reducing waste and providing consumers with fresher products. For example, GM tomatoes can be engineered to delay ripening, extending their marketable period.
   - **Enhanced Taste and Appearance**: GM technology can improve the taste, texture, and appearance of food products, making them more appealing to consumers.
6. Potential Health Benefits
   - **Allergen Reduction**: Genetic engineering can be used to remove or reduce allergenic proteins in certain foods, making them safer for individuals with food allergies.
   - **Medicinal Properties**: Some GM foods are being developed to contain pharmaceutical compounds or vaccines, potentially providing an innovative way to deliver medical treatments.
7. Adaptation to Climate Change
   - **Drought and Heat Tolerance**: GM crops can be designed to withstand extreme weather conditions such as drought and high temperatures, making agriculture more resilient to climate change.
   - **Salinity Tolerance**: Genetic modifications can enable crops to grow in saline soils, expanding the arable land available for food production.

3.4.8.2. Cons of Genetically Modified Foods

1. **Environmental Concerns:** There are concerns about the potential environmental impact of GM crops, including the unintended spread of modified genes to wild populations, the development of resistant pests and weeds, and disruptions to ecosystems.
2. **Health and Safety Concerns:** Some individuals and groups express concerns about the safety of consuming GM foods. While many scientific studies have not found conclusive evidence of harm, the long-term effects of consuming GM foods are still under investigation.
3. **Biodiversity Loss:** The widespread adoption of GM crops, especially those engineered for pest resistance, can lead to a reduction in biodiversity as certain species of plants and insects are disproportionately affected.
4. **Potential Allergenicity:** There are concerns that introducing genes from one species into another could potentially lead to the development of new allergens in GM foods. Rigorous testing is necessary to mitigate this risk.
5. **Ethical and Social Issues:** The concentration of intellectual property rights and seed ownership in a few large biotechnology companies raises ethical concerns about access to technology, the rights of farmers, and the potential for corporate control over the food supply.
6. **Consumer Acceptance and Labeling Issues:** Public perception of GM foods varies, and there is ongoing debate about the need for clear labeling. Some consumers demand transparency, while others argue that labeling may stigmatize GM products without scientific justification.
7. **Cross-Contamination:** There is a risk of cross-contamination between GM and non-GM crops, especially in regions where both are grown. This can pose challenges for organic farming and the production of non-GM crops.

Did you get it?

1. Describe the regulation of GM foods and the role of government.
2. Explain the process of labeling GM foods.

3.5. ENZYMES IN FOOD INDUSTRY

→Learning Objectives

- Define the production of food enzymes from microorganisms;
- Explain the applications of microbial enzymes in the food industry.

All living cells produce enzymes to act as catalysts in particular chemical reactions. Enzymes have been utilized in the food industry to produce, improve, and diversify food. Products that use enzymes include cheese, yogurt, bread syrup, and others. Traditional ancient crafts like cheese-making, brewing, using papaya leaves to tenderize meat, and preparing condiments (e.g., fish sauce and soy sauce) depend on proteolysis, even though the processes were invented before we knew about enzymes. Early proteolysis-related food processes were typically an unintentional result of the food's microbial or endogenous enzyme activity.

3.5.1. Production of Food Enzymes From Microorganisms

Enzymes were first isolated from living cells in the 20th century, which paved the way for their widespread commercial production and use in the food sector. These days, the primary source of commercial enzymes comes from microorganisms. Despite the fact that microorganisms do not have the same enzymes as plants or animals, it is usually possible to identify a microbe that produces an enzyme that is related to the desired reaction and can catalyze it. Manufacturers of enzymes have used traditional breeding methods in conjunction with natural selection to enhance microorganisms for the production of enzymes.

The most exact techniques for enhancing microorganisms for the synthesis of enzymes are included in direct genetic modification (biotechnology). High-yielding production organisms are obtained using these techniques. Another benefit of biotechnology is the ability to transfer a genetic sequence from a plant, animal, or microorganism whose commercial enzyme production is insufficient to a microorganism with a safe history of producing enzymes for food use.

The enzyme that is produced by the GM production organism is not. The enzymes created by biotechnology are exactly like the ones that occur in nature. Additionally, before being utilized in the production of food, the enzymes that microorganisms produce are extracted and purified. Commercially speaking, GM microorganisms are valuable, but they would not exist in the natural world.

3.5.2. Applications of Microbial Enzymes in Food Industry

Since ancient times, food has been fermented using microorganisms, and many food items are still prepared using fermentation techniques. Due to their greater stability compared to plant and animal enzymes, microbial enzymes are used extensively in the food industry. Due to their high consistency, process modification and optimization can be accomplished with ease. They can be produced through fermentation techniques in a way that is both cost-effective and requires less time and space. Numerous industries use these enzymes for a wide range of purposes, including, e.g., the food, detergent, paper, and textile industries all use amylolytic enzymes. High fructose corn syrups, maltose syrups, crystalline glucose, glucose syrups, and other products are prepared using them. They are also added to detergents to help remove stains that are based on starch. In the paper industry, these enzymes are employed to lower the viscosity of starch so that paper can be coated appropriately. Amylases are used in the textile industry to size textile fibers for the warp.

Similarly, enzymes like proteases, lipases or xylanases have wide applications in food sectors. The following sections give detailed and updated information about various food enzymes of microbial origin. Table 3.2 gives an overview of the applications of microbial enzymes in the food industry.

Table 3.2. *Applications of Microbial Enzymes in the Food Industry*

Microbial Enzyme	Application
α-Amylase	• Baking, brewing, starch liquefaction • Bread quality improvement • Rice cakes • Clarification of fruit juice
Glucoamylase	• Beer production • Bread quality improvement • High glucose and high fructose syrups
Protease	• Brewing • Meat tenderization • Coagulation of milk • Bread quality improvement

Lactase (β-galactosidase)	• Lactose intolerance reduction in people • Prebiotic food ingredients
Lipase	• Cheese flavor development • Cheddar cheese production
Phospholipase	• Cheese flavor development • Production of lipolyzed milk fat
Esterase	• Enhancement of flavor and fragrance in fruit juice • De-esterification of dietary fiber • Production of short-chain flavor esters
Cellulase	• Animal feed • Clarification of fruit juice
Xylanase	• Clarification of fruit juice • Beer quality improvement
Pectinase	• Clarification of fruit juice
Glucose oxidase	• Food shelf life improvement • Food flavor improvement
Laccase	• Polyphenol removal from wine • Baking
Catalase	• Food preservation (with glucose oxidase) • Removal of hydrogen peroxide from milk prior to cheese production
Peroxidase	• Development of flavor, color, and nutritional quality of food
α-Acetolactate dehydrogenase	• Shortening maturation of beer
Asparaginase	• Reduction of formation of acrylamide during baking
Debittering enzymes – naringinase	• Removal of bitter taste in fruit juice • Wine aroma enhancement

3.5.2.1. α-Amylases

α-Amylases (EC 3.2.1.1) are starch-degrading enzymes capable of hydrolyzing α-1,4-glycosidic bonds of polysaccharides, which results in the production of short-chain dextrins. These enzymes are widely distributed in all living organisms. Majority of α-amylases are metalloenzymes and require calcium ions for their activity, stability as well as integrity. A-amylases are widely used in the food industry for liquefaction of starch, baking, brewing, and as a digestive aid. They are extensively employed in the baking industry as an antistaling agent and flavor enhancer to raise the caliber of bread. A-amylases are added to the dough during baking to help break down starch into smaller dextrins, which is then fermented by yeast. It enhances the bread's flavor, color, and toasting capabilities.

α-Amylases are also employed in the production of branched dextrins with a high molecular mass. When making rice cakes and other powdered foods, they are utilized as a glazing agent. Starch liquefaction, which turns starch into glucose and fructose syrups, is also used in the starch industry. Three processes are involved in the enzymatic conversion of starch: saccharification, liquefaction, and gelatinization. As starch granules dissolve, a viscous suspension is created during the gelatinization process. A liquefaction process that involves partial hydrolysis and lowers viscosity comes next. Saccharification results in the further production of glucose and maltose. Highly thermostable enzymes are needed for this, and *B. amyloliquefaciencs*, *B. stearothermophilus*, or *B. licheniformis* α-amylases are used for most of the starch saccharification process.

α-amylases break down starch into fermentable sugars, which *S. cerevisiae* then ferments further to produce alcohol. This process produces ethanol. α-amylases are also used in the clarification of fruit juices, where they work in tandem with cellulases and pectinases to increase yield and lower processing costs.

3.5.2.2. Glucoamylases

Glucoamylases (EC 3.2.1.3) are exo-acting enzymes that catalyze the hydrolysis of polysaccharide starch from the non-reducing end, releasing β-glucose. They are also called saccharifying enzymes and are widely distributed in all living organisms. These enzymes are produced mainly by *A. niger* and *A. awamori,* but the one produced by *Rhizopus oryzae* is widely used for industrial applications. The majority of glucoamylases are stable at low temperature. At higher temperatures, they lose activity due to conformational change. Glucoamylases find a wide range of applications in the food industry, such as for the production of high-glucose syrups and high-fructose syrups. They also find applications in the baking industry to improve flour quality, reduce dough staling, as well as to improve bread crust color and the quality of high-fiber baked products. Glucoamylases convert the starch present in the flour to maltose and fermentable sugars. Fermentation by yeast leads to dough rise. These enzymes are also used for the production of glucose, which upon fermentation with *S. cerevisiae* yields ethanol. Glucoamylases play an important role in the production of sake and soy sauce, as well as in the production of light beer. They metabolize dextrins and convert them to fermentable sugars with reduced calorific value and alcohol content in the beer.

3.5.2.3. Proteases

Enzymes called proteases are responsible for catalyzing the breakdown of peptide bonds found in proteins and polypeptides. The food and detergent industries are the ones that use them the most. They make up 60% of the commercially available industrial enzymes. Between 2014 and 2019, the market for protease enzymes was expected to expand at a compound annual growth rate (CAGR) of 5.3% worldwide. As they find uses in bioremediation and leather processing, their demand is predicted to rise significantly.

Proteases can be categorized based on their origin, catalytic activity, and the type of reactive group present in the catalytic site. Protease enzymes are primarily found in plants, animals, and microorganisms (both bacterial and fungal). Based on where they act on polypeptide chains, proteases are classified as either endopeptidases or exopeptidases. Endopeptidases randomly act in the inner regions of polypeptide chains, while exopeptidases act on the ends of polypeptide chains. Based on the catalytic residue found in the active site, the endopeptidases are further divided into six groups: serine, aspartic, cysteine, metallo, glutamic acid, and threonine protease. Plant proteases, including ficin, papain, and bromelain, are extensively employed in the food industry for a variety of purposes, including milk coagulation, meat tenderization, brewing, and digestive aids. Proteases are also used to change the functional characteristics of food proteins, such as coagulation and emulsification, as well as to enhance their flavor, nutritional value, solubility, and digestibility. In the baking industry, proteases are frequently utilized in the creation of bread, baked goods, crackers, and waffles. These enzymes are used to shorten the time needed for mixing, lessen the homogeneity and consistency of the dough, control the strength of the gluten in bread, and enhance its flavor and texture. Aspergillus usamii's acid protease has been effectively utilized to enhance the functional characteristics of wheat gluten. Protease could be added to the wort to release enough peptides and amino acid levels to enable a healthy fermentation. Because acidic fungal proteases balance the amino acid profile of beer and are effective even at low pH levels, they are used to improve beer fermentation. Proteases have a significant role in the dairy industry as well. Proteases found in nature play a major role in the flavor profiles of cheese. They are used to change the functional characteristics of milk products, lessen their allergenic qualities, and quicken the ripening process of cheese. Proteases are also utilized in the production of cheese to hydrolyze a particular peptide bond and produce paracasein and macropeptides.

3.5.2.4. Lactase (B-Galactosidase)

Lactose hydrolysis is a crucial biotechnological procedure used in the food industry. Lactose hydrolysis is catalyzed by the enzyme β-galactosidase. It is a member of the hydrolase family. Numerous biological systems, including plants, animals, and microorganisms, can provide β-galactosidase. This method is preferred because microorganisms like yeast, fungi, and bacteria can produce β-galactosidase with a higher yield at a relatively low cost. The ultimate use of the β-galactosidase enzyme determines the source to use, i.e., β-galactosidase, derived from yeasts with pH optima between 6.5–7.0, is commonly utilized for the hydrolysis of lactose in whey milk. In the case of acidic whey hydrolysis, fungal β-galactosidase with pH optima of 3.0–5.0 is suitable. Thus, the selection of β-galactosidase depends on the final application of the enzyme or industry. β-Galactosidase produced from yeast *Kluyveromyces lactis* requires ions such as Mn^{2+} or Na^{+}, whereas *Kluyveromyces fragilis* requires Mn^{2+}, Mg^{2+} or K^{+}.

In industrial applications, two major classes of β-galactosidase are of prime importance. They are cold-active and thermostable β-galactosidase. On a commercial scale, β-galactosidase is produced using microorganisms with GRAS status for their application in milk and dairy products. Lactase is used with milk and milk-based products to reduce lactose intolerance in people. The scoopability and creaminess of ice creams improved significantly after the hydrolysis of lactose with lactase. Additional advantage of hydrolyzing lactose into monomers is the reduction requirement of sweeteners as they could improve the sweetness of the products. Another major application of lactase is the lactose hydrolysis in whey. Whey is a byproduct of cheese production and its main components are lactose, proteins, and minerals. This causes critical environmental issues associated with the dairy industry, as lactose is associated with high biological oxygen demand (BOD) and chemical oxygen demand (COD). Another application of lactase is the formation of galactooligosaccharides (GOS) from lactose hydrolysis due to the transglycosylation activity of β-galactosidase. The GOS could be used as prebiotic food ingredients.

Remember

Enzymes must be selected based on their specific functions and applications in food processing. The choice of enzymes should align with the desired outcome, such as flavor enhancement, tenderization, or dough conditioning.

3.5.2.5. Lipases

Enzymes called lipases are responsible for catalyzing the breakdown of long-chain triglycerides. They are naturally found in the stomach and pancreas of humans and other animal species to aid in the digestion of fats and lipids. Bacteria, fungi, and yeast all produce microbial lipases. Enzymes found in microbes contribute approximately to 90% of the worldwide lipase market. This enzyme is used in the food, biofuel, detergent, and animal feed industries, among others. It is also employed in applications related to paper, leather, and textile processing. Lipases are widely used in the dairy, baking, fruit juice, beer, and wine sectors of the food and beverage industry. Lipase has a market share of less than 10% of the global industrial enzyme market despite its wide range of applications across multiple industries.

Commercial lipases are mainly used for flavor development in dairy products and processing of other foods containing fat. They can improve the characteristic flavor of cheese by acting on the milk fats to produce free fatty acids after hydrolysis. Different types of cheese can be made by using lipases from various sources, e.g., Romano cheese using kid/lamb pre-gastric lipase, Camembert cheese using lipase from *Penicillium camemberti* and cheddar cheese using *A. niger* or *A. oryzae*. Lipase catalysis could improve the texture and softness of cheese. Lipases are also used as flavor development agents in butter and margarine, also to prolong the shelf life of various baking products. In alcoholic beverages such as wine, the aroma can be modified using lipase. They are used to improve the quality of cocoa butter, which has a melting point of 37 °C due to the presence of palmitic and stearic acids and can easily melt at 37 °C. A patent has been filed by Unilever using immobilized *Rhizopus miehei* lipase, which can replace palmitic with stearic acid to give desired stearic-oleic-stearic triglyceride. Functionalized phenols were esterified for the synthesis of lipophilic antioxidants for the application in sunflower oil using immobilized lipase from *C. antarctica* (CALB), *C. cylindracea* Ay30, *Helvina lanuginosa, Pseudomonas* sp. and *Geotrichum candidum*. Lipases also find application as a biosensor in the food industry. Immobilized lipase was successfully used for the determination of organophosphorous pesticides with a surface acoustic wave impedance sensor by lipase hydrolysis. It may also be used in the determination of triglycerides and blood cholesterol if the lipase is immobilized onto pH/oxygen electrodes in combination with glucose oxidase. Microbial lipases such as lipase from *Candida rugosa* have many applications that cannot be met by chemical synthesis. This lipase finds application in the production of ice cream, single-cell protein, carbohydrate esters, and amino acid derivatives. In addition to this, lipase could also be used in the processing of different waste streams that are released from food industries.

3.5.2.6. Phospholipases

Phospholipases convert phospholipids into fatty acids and other lipophilic compounds on a selective basis. A, B, C, and D are the four main classes into which they can be separated according to their mode of action. Phospholipase A1 (PLA1), phospholipase A2 (PLA2), and phospholipase B act on the carboxylic ester bonds of phospholipids, thus displacing and replacing the acyl group chain through various chemical reactions like hydrolysis, esterification, and transesterification. Phospholipases C (PLC) and D (PLD) are phosphodiesterases because they recognize the phosphodiester linkage and modify the polar head group. Phospholipases are extensively utilized in the food industry, particularly in the dairy, oil, and bread manufacturing sectors. They are also used in the production of cheese, bread, and various vegetable oils by degumming.

3.5.2.7. Esterases

Esterases have the ability to help split esters into acid and alcohol in an aqueous solution. Furthermore, esterases differ from lipases in that they hydrolyze short-chain acylglycerols as opposed to long-chain ones. Esterases are widely used in the food and beverage industries, primarily for producing fragrances and flavors as well as altering the oil and fat content of different fruit juices. An essential class of enzymes within the esterase family, feruloyl esterases, hydrolyze the ester link between ferulic acid and various plant cell wall polysaccharides. Feruloyl esterases are essential for waste management because they hydrolyze lignocellulosic biomass. Cheng and colleagues found a protease-resistant feruloyl esterase that can release ferulic acid from wheat straw after screening a metagenomic library derived from the microbial population of a cow rumen for feruloyl esterase activity. 'The high pH, thermal stability, and protease resistance of this esterase make it highly useful in the commercial world.

In cheese manufacture, the fruity flavors are the result of different methyl or ethyl esters of short-chain fatty acids. Bacterial production of ethyl esters and thioesters has been reported. Alvarez-Macarie and Baratti reported the production of a novel thermostable esterase from the highly thermotolerant *B. licheniformis* heterologously expressed in *E. coli* for the production of short-chain flavor esters. Feruloyl esterase is a key enzyme in the biosynthesis of ferulic acid, which is the precursor for vanillin, an aroma compound used in foods and beverages. Several researchers have reported the microbial production of feruloyl esterase.

3.5.2.8. Lipoxygenases (LOX)

In lipids containing a cis-1,4-pentadiene, LOX play a role in the dioxygenation of polyunsaturated fatty acids. They are made up of a single polypeptide chain that is assembled into the catalytic β-barrel domain and the N-terminal domain. LOX enzymes are iron-containing non-haem digesting enzymes. Different precursors for the synthesis of various volatile and aroma-producing chemical substances in plants are produced by the LOX-catalyzed reaction. LOXs are utilized in the food industry to create aromas and in the baking process. The most researched is soy bean LOX. The specificity of bacterial LOXs toward fatty acids varies. The LOX from *Nostoc* sp. oxygenates linoleic acid at a particular location, but the LOX from *Anabena sp.* demonstrates a range of specificity.

The main applications of LOXs in the dough are based on their ability to bleach the flour pigment carotenoid, by co-oxidation of the pigment with fatty acids. LOXs are also employed to improve tolerance to mixing and different handling properties of dough. This effect is due to the oxidation of the thiol group in gluten, which may lead to the redistribution of different disulfide bonds, tyrosine cross-linking, and subsequent strengthening of the gluten. This also leads to the improvement in dough rheology. Recently, Patel et al. purified LOX from *Lasiodiplodia theobromae* by different chromatography techniques and fully characterized the enzyme. *L. theobromae* was reported to contain two types of LOXs with a molecular mass of 93 and 45 kDa.

3.5.2.9. Cellulases

Cellulases are enzymes that act on polymeric cellulose and hydrolyze β-1,4 linkages to liberate glucose units. The three major classes of cellulases are endo-(1,4)-β-d-glucanase (EC 3.2.1.4), exo-(1,4)-β-d-glucanase (EC 3.2.1.91) and β-glucosidases (EC 3.3.1.21). The catalytic modules of cellulases belong to the glycoside hydrolase (GH) family and have been classified into different groups based on differences in amino acid sequences and three-dimensional structural features. GH family enzymes mainly use acid–base catalysis mechanism for cleaving glycoside bonds in cellulose. The catalysis is achieved by two major residues (a proton donor and a nucleophile) of the enzyme in the active site region. The hydrolysis occurs *via* retention or inversion mechanism depending on the spatial position of these catalytic residues in the enzyme. Endoglucanases cleave β-1,4-bonds in the amorphous region of cellulose and expose the non-reducing and reducing ends of cellulosic polymer. Endoglucanases from various sources belong to different GH families, among which the major are 5–9, 12, 44, 45, 48, 51, and 74. Most of the fungal endoglucanases contain a catalytic module with a carbohydrate-binding module (CBM), but a catalytic module without CBM was also reported from fungal species. Multiple catalytic modules and CBMs are present in bacterial endoglucanases. A cleft/grove-shaped active site is present in the catalytic module of most of the endoglucanases. Exoglucanases or cellobiohydrolases (CBHs) act on available reducing or non-reducing ends of cellulose polymer and liberate cellobiose. Fungal and bacterial CBHs show diversity in catalytic modules and belong to GH families 5, 6, 7, 9, 48 and 74. A tunnel-shaped catalytic module is observed in most of the CBHs. Β-Glucosidases catalyze the final step in cellulose breakdown by cleaving the non-reducing terminal β-dglucosyl residues and removing β-d-glucose. The catalytic modules belonging to GH families 1, 3, and 9 are reported from various β-glucosidases. The cellulolytic machinery of microbes is mainly regulated through feedback inhibition of b-glucosidases by their reaction product glucose. The major difference from CBHs is the absence of CBM in their structure. A pocket-shaped active site region of β-glucosidases helps them to attach the glucose molecule to a non-reducing end and release glucose units from cellodextrins or cellobioses. A large diversity of microorganisms is reported to produce cellulases during their growth on cellulosic materials. The industrial making of cellulases is mainly from microbial sources, bacteria, and fungi, and these microorganisms can be diverse in their habitat. The aerobic bacteria show a similar mechanism of cellulose degradation to that of aerobic fungi. In anaerobic bacteria, cellulosomes are located on the cell surface and operate *via* a different system. Cellulases from fungi (*Aspergillus* and *Trichoderma*) and bacteria (*Bacillus* and *Paenibacillus*) are potentially used in the production of food. They are also widely used for various industries, such as textile, paper, detergent, and the food. In the juice industry, cellulases are applied in combination with other macerating enzymes to increase process performance and yield, improving the extraction methods, clarification, and stabilization of juices. They can also reduce the viscosity of nectar and puree from fruits such as apricots, mangoes, plums, papayas, pears, and peaches, and are used for the extraction of flavonoids from flowers and seeds. The preferences for cellulase-mediated extraction over conventional methods is due to higher yield, less heat damage, and short processing time. Cellulases are utilized for the extraction of phenolic compounds from grape pomace. β-Glucosidases, in combination with pectinase, alter the structure, flavor, and aroma of fruits and vegetables. They are also reported to reduce the bitterness of citrus fruit and improve aroma and taste. Cellulases are used with other enzymes for efficient olive oil extraction. In wine production, cellulases are used in combination with other enzymes to increase yield and quality. The main advantages of using these enzymes are improved maceration, better color development, must clarification, and finally, wine stability and quality.

3.5.2.10. Xylanases

Xylanases are produced by microorganisms to cleave xylans, a major constituent of hemicellulose. Three major enzymes, endoxylanases, exoxylanases, and β-xylosidases, act synergistically and are required for the breakdown of the xylan backbone in hemicellulose. Endoxylanases (EC 3.2.1.8) cleave the β-1,4 bonds of xylan backbone. Exoxylanases hydrolyze β-1,4 bonds of xylan from the non-reducing ends and release xylooligosaccharides. β-Xylosidases cleave the xylobiose and xylooligosaccharides to release xylose. The major functions of xylanases are performed by a catalytic module, and few classes possess an additional CBM for binding to substrates. The two major catalytic modules of hemicellulases are GHs and carbohydrate esterases. Endoxylanase hydrolyzes the xylan backbone and has catalytic cores belonging to GH families 8, 10, 11, 30, and 43, with the most common ones being GH 10 and 11. These differ in their substrate specificities, and the GH10 is more active on substituted xylan. Similar to cellulases, they may also contain CBMs. Exoxylanases randomly cleave the xylan backbone from the inside, releasing long-chain xylooligomers on which the β-xylosidases act. The catalytic module of these enzymes belongs to the GH families 3, 30, 39, 43, 52, and 54. These two enzymes are often collectively called xylanases. B-Xylosidase or xylan-1,4-β-xylosidase act on the xylooligosaccharides and xylobiose to release xyloses.

Xylanases are produced by microbes like actinomycetes, bacteria, and fungi. The major actinomycete and bacterial species producing xylanase are *Streptomyces* sp., *Bacillus* sp., and *Pseudomonas* sp. Those produced by bacteria and actinomycetes are effective in a broader range of pH (5.0–9.0), with the optimum temperature for xylanase activity between 35 and 60°C. Fungi are major sources of xylanase due to their high content and extracellular release of the enzyme. The major fungal species producing xylanase are *Aspergillus* sp., *Fusarium* sp., and *Penicillium* sp. Fungal xylanases have higher activity than bacteria or yeast.

Carbohydrate-hydrolyzing enzymes are usually used in the bread-making industry. Rheological properties of dough are improved through enzymatic hydrolysis of non-starch polysaccharides. Xylanases are widely used in the bread-making industry with other enzymes. The potential effectiveness of xylanolytic enzymes increases its use in bread making. They can increase the specific bread volume, and this improves the quality of bread. The hemicellulose in wheat flour is broken down by xylanase, which increases the binding of water in the dough. The dough becomes softer and crumb formation is delayed, allowing the dough to grow. Xylanase is used to improve texture, tastiness, and palatability in biscuits. They also play an important role in juice production by improving extraction, clarification, and stabilization. In combination with other enzymes, xylanases lead to a better yield of juice and increased recovery of aromas, essential oils, vitamins, mineral salts, pigments, etc. In beer-making industries, xylanases are used for hydrolyzing the cellular wall of barley. Hydrolysis leads to the release of arabinoxylans and lower oligosaccharides, which reduces the muddy appearance and viscosity of the beer.

3.5.2.11. Pectinases

Enzymes called pectinases catalyze the breakdown of glycosidic bonds found in pectic polymers. The natural substrate for this enzyme is pectic substances, which are present in citrus fruits such as tomato, pineapple, orange, apple, lemon pulp, orange peel, and others. Pectinases can be classified functionally into four groups: pectin lyase, pectate lyase, polygalacturonases, and pectin esterases. Polygalacturonases hydrolyze glycosidic α-(1-4) bonds.

Both naturally occurring and recombinant microbes can produce pectinases, with efforts being made to improve their yield and thermostability. Pectinases have the ability to act on hairy or smooth regions of pectin. Acidic and alkaline pectinases are also classified as endopectinases when the enzyme cleaves randomly and as exopectinases when the terminal ends are targeted, according to pH.

Pectinases are used in many different industrial processes, including paper bleaching, food processing, remediation, and more. Pectinase-added juices look clearer and are easier to filter than their counterparts with reduced enzyme levels. Pectinases not only lighten the turbidity and haze that naturally occur in fruit juices like apple and banana, but they also enhance the flavor and color of beverages. Juices become viscous and turbid after adding gelatin and pectin, and the most expensive step in the process is clearing the haze. For optimal results, the use of biogenic enzymes, like pectinases, in juices would perform nearly nine times better than mechanical maceration.

3.5.2.12. Glucose Oxidase

The enzyme glucose oxidase is a member of the oxidoreductases, a broad family of enzymes. Müller identified glucose oxidase as a flavoprotein in 1928. He said that the enzyme can change glucose into gluconic acid when there is dissolved oxygen present. Gluconolactone is produced by oxidizing β-d-glucose, while hydrogen peroxide is produced by reducing molecular oxygen. After that, glucanolactone hydrolyzes spontaneously to produce gluconic acid. The enzyme has two similar polypeptide chain subunits (80 kDa) and is homodimeric. One flavin adenine dinucleotide (FAD) molecule is non-covalently bound to the active site region of each subunit, and the subunits are covalently linked by disulfide bonds. Glucose oxidase production has been reported from various microorganisms, and it was first discovered in *Aspergillus niger* and *Penicillium glaucum. A. niger* species is widely used for the production of glucose oxidase, and its strains can produce higher amount of glucose oxidase. *Penicillium adametzii* is a widely used fungus for the production of extracellular glucose oxidase. The different bacterial species are also reported to produce glucose oxidase. Although many species of bacteria and fungi are reported to produce this enzyme, fungi are considered for the industrial production of glucose oxidase.

Glucose oxidase finds extensive application in a variety of industries, including biofuel cells and the food and pharmaceutical sectors. Because of its oxidizing properties, which strengthen dough, its use in the baking industry is growing. In the food industry, it improves the taste, aroma, and stability of food items by taking the oxygen and glucose out of egg whites and diabetic beverages. Glucose oxidase keeps food from spoiling and enhances its flavor, color, texture, and shelf life. Glucose oxidase is used in food packaging to increase storage life by removing oxygen.

3.5.2.13. Laccase

The largest subgroup of multicopper enzymes is represented by the cluster of oxidases known as laccases. Known by their common name, blue oxidases, they are employed in many industrial sectors and are studied for their ability to oxidize phenolic compounds. These enzymes are powerful biocatalysts that can be used in textile finishing, chemical synthesis, bioremediation, biobleaching of paper pulp, biosensing, and wine stabilization. Their broad range of oxidizable substrates and varying substrate specificities are further influenced by the kind of microbial sources that produce the enzyme. Numerous substances, including ascorbate, aromatic amines, and phenolics, are catalyzed to undergo oxidation by laccases. In the presence of four copper atoms found in laccases, these enzymes combine reducing substrate with four oxidized electrons with four reduced electrons to cleave dioxygen bonds.

Laccases are secreted extracellularly by several fungi as a product of their secondary metabolism during fermentation, but their production is limited to a few fungal species. Well-known producers of laccases belong to Deuteromycetes, Ascomycetes, and Basidiomycetes. *Funalia trogii* is a white rot fungus capable of producing laccase through absorbent fermentation. The maximum laccase production by *F. trogii* reached 11,900 U/L, which was 4.97 times higher than that of normal fermentation. *B. licheniformis* produces recombinant laccases for industrial applications. Recently, heterologous expressions have been used for laccase production. *B. vallismortis* fmb-

103 genes were cloned and heterologously expressed in *E. coli* BL21 (DE3) cells.

Laccase is used to stabilize wine instead of using physical or chemical adsorbents or to change the color and appearance of products for the food and beverage industries. Wine should not have its polyphenols removed in order to prevent unfavorable changes to its organoleptic properties, such as stability in acidic environments and reversible inhibition from sulfite. Additionally, the industry that makes cork stoppers uses this enzyme.

One of the issues facing the brewing sector is haze formation. Several researchers have used laccases to replace the conventional method for polyphenol oxidation in order to prevent it. This enzyme is also utilized in the last stage of beer production to remove oxygen, extending beer's shelf life. The brewing industry uses a commercial laccase called Flavourstar, produced by Novozymes, to eliminate off-flavor formation. Because it can form a cross-link with biopolymers, it is used in baking. When laccase is used in baking, bread batter is more machinable because it increases stability, strength, and reduces stickiness.

3.5.2.14. Catalase

Catalase is a tetrameric protein found in aerobic organisms. It helps hydrogen peroxide decomposition. This enzyme can be produced by microbial sources such as *A. niger* and *Micrococcus luteus* and from bovine liver. Microorganisms are usually preferred as sources for enzyme production due to their advantages, such as fast growth, easy handling, and genetic tuning for obtaining a desired product. The anaerobic *Bacteroides fragilis* exhibited increased catalase levels in media with haem. Frankenberg et al. isolated catalase from *Enterococcus faecalis*, which completely depends on the haem source without which it cannot synthesize the porphyrin group. A facultative anaerobic catalase-producing *B. maroccanus* resistant to hydrogen peroxides was isolated from textile effluents. A potent catalase-producing bacterium *Pyrobaculum calidifontis* was isolated from hot springs in Los Banos and Calamba, Laguna, Philippines. A thermo-alkaliphilic catalase-positive strain of *B. halodurans* LBK 261 was isolated from the alkaline hot-spring waters of Kenya. A halo (alkali) tolerant catalase-producing *Oceanobacillus oncorhynchi* ssp. *incaldaniensis* was isolated from an algal mat capable of producing catalase at a wide range of pH 6.0–9.5 and salinity of 5–20%. A catalase-positive psychrophile *Bacillus* N2a was isolated from seawater. Other catalase-positive bacteria such as *Rhizobium radiobacter* were isolated from industrial effluent from the beverage industry, *Comamonas testosteroni* and *C. terrigena* from effluent sludge enriched with crude oil along with heavy metals and *Serratia* SYBC08 from hydrogen peroxide sludge. *Psychrobacter piscatorii* T-3, a psychrotolerant bacterium isolated from bleach-rich runoff, has high catalase activity. Fungi and yeast are able to produce catalase. The highest level of catalase activity of 400 mg/g was observed in isolates of *A. niger*. A catalase-positive entomopathogenic fungus *Metarhizium anisopliae* strain Ma10 (CNRCB MaPL10) was isolated from *Geraeus senilis*.

In the fabric industry, catalase is used for removing excess hydrogen peroxide from fabric. This enzyme is mostly used along with other enzymes in the food processing industry. Catalase is often used with glucose oxidases for food preservation. Ough used a glucose oxidase/catalase cocktail for the elimination of oxygen from wine before bottling and evaluated the formation of acetaldehydes. Results showed that color and amount of acetaldehyde were stable if treated properly with enzymes. Catalase is applied in the milk processing industry to eliminate peroxide from milk, to remove glucose from egg whites in the baking industry and in food wrappers to prevent oxidation and control the perishability of food. This enzyme has limited use in cheese production.

3.5.2.15. Peroxidase

Oxidoreductase proteins with iron(III) protoporphyrin IX as the prosthetic group are called peroxidases. They catalyze a variety of inorganic and organic compound oxidations as well as the reduction of peroxides. They are made up of a variety of nonspecific enzymes as well as a unique set of enzymes like glutathione peroxidase, NADH peroxidase, and iodide peroxidase. Their molecular mass ranges from 30,000 to 1,50,000 Da. Microorganisms, animals, and plants all contain peroxidases. They have a role in defense mechanisms against infected or damaged tissues as well as lignification processes in plants.

Phanerochaete chrysosporium is the microorganism with the best-characterized peroxidase-secreting organism. Fungal peroxidases can only be used on an industrial scale due to difficulties arising from protein post-translational modification. Bacterial peroxidases, on the other hand, are more stable and active and can be produced more easily, making them ideal for industrial use. These enzymes are used in conjunction with bacterial laccases to decolorize dyes. Bacterial taxa, including Firmicutes, Proteobacteria, Actinobacteria, and Acidobacteria, have been shown to exhibit peroxidase activity. Actinomycetes are soil bacteria that can grow like fungi, share an ecological niche with them, and produce peroxidases that break down lignin. *Stemomyces viridosporus* T7A produced the first extracellular lignin peroxidase to be secreted.

Peroxidase uses hydrogen peroxide or other peroxides to catalyze a variety of substrates. This enzyme is used in the food industry to enhance the nutritional value of food while also adding flavor, color, and texture. Additional uses include the creation of biosensors, the synthesis of polymers, and the control of environmental contaminants. It can be applied to the treatment of phenolic industrial effluents. The food industry uses thermal inactivation of peroxidases to gauge how well a blanching treatment is working to extend the shelf life of food. Peroxidases have the detrimental effect of giving fruits an unwanted browning and vegetables an off flavor.

3.5.2.16. α-Acetolactate Decarboxylase

The rapid maturation of beer is significantly facilitated by α-acetolactate decarboxylase. This enzyme can be made from recombinant *S. cerevisiae* or from naturally occurring microbes like *Brevibacillus brevis*. Through a two-step process that involves direct substrate decarboxylation to an enol derivative and subsequent protonation to the final product, the enzyme catalytically converts acetolactate to acetoin.

Enzyme-based removal of α-acetolactate and α-aceto-α-hydroxybutyrate assists in overcoming the rate-limiting step of beer maturation. While the maturation of beer without the use of enzymes takes 2 to 12 weeks, the use of α-acetolactate decarboxylase results in maturation within 24 hours depending on the source of the enzyme. Moreover, the off-taste due to the presence of diacetyl in beer is nullified by the action of this enzyme. Studies indicate that both free and encapsulated forms of this enzyme work efficiently in the process, thus aiding the use of immobilized enzymes at reduced costs. Novel inorganic nanoflowers or alginate microbeads immobilized with α-acetolactate decarboxylase are promising strategies with better thermal stability, reusability, and catalytic efficiency.

3.5.2.17. Asparaginase

Asparaginases are a prominent class of pharmaceutical, nutraceutical, and industrially significant enzymes that are widely used by humans among the various enzymes derived from microbes. As the name suggests, asparaginase is an asparagine-depleting enzyme because it catalyzes the breakdown of asparagine into the acid derivatives aspartic acid and NH_3. While asparagine is an essential amino acid for cancerous cells, it is not necessary for humans. The growth of cancerous cells is significantly impacted by asparagine depletion, which is the basis for the enzyme's anticancer properties.

A number of food processing techniques, including baking and frying in oil, turn asparagine into acrylamide, a known carcinogen. The depletion of asparagines by enzymatic treatment has been found effective in reducing the formation of acrylamides from asparagines by 97% among various methods attempting to overcome the acrylamide formation.

3.5.2.18. Debittering Enzymes – Naringinase

Naringinase is mainly responsible for the breakdown of naringin, the principal bitter flavanone glycoside found in citrus fruits. Naringin is broken down to a glycon naringenin and rhamnose as a result of its α-rhamnosidase and β-glucosidase action. Nariginase is produced mostly by fungal isolates, *viz. A. niger*, *Circinella*, *Eurotium*, *Fusarium*, *Penicillium*, *Rhizopus*, and *Trichoderma* and bacteria such as *Bacillus* sp., *Burkholderia cenocepacia*, *Bacteriodes distasonis, Thermomicrobium roseum, Pseudomonas paucimobilis,* etc. Fungal sources of naringinase are found to be more predominantly used than bacterial sources due to increased yield.

Naringinase has a major role in food processing as a debittering enzyme supplemented with fruit juices. Both free and immobilized forms of this enzyme are used to get better results. Immobilization of this enzyme has been done in a variety of substrates such as polyvinyl alcohol cryogels, packaging films, cellulose triacetate nanofiber, graphene, etc. Various food additives, such as biopolymers and sweeteners, can be synthesized using rhamnosidase or naringinase. Yet, another use of naringinase together with β-glucosidase and arabinosidase is to improve the aroma of wine. The use of naringinase is also noted in tomato pulp preparation, kinnow peel waste treatment, and prunin preparation.

Did you get it?

1. How are food enzymes produced from microorganisms?
2. What are the applications of microbial enzymes in the food industry?

3.6 PRODUCTION OF ADDITIVES, COLORANTS, AND FLAVORINGS BY BIOTECHNOLOGY

→Learning Objectives

- Understand the food additives, colorants, and flavorings

Biotechnology has revolutionized the production of food additives, colorants, and flavorings, offering sustainable, efficient, and often healthier alternatives to traditional methods. These compounds play crucial roles in enhancing the sensory attributes, shelf-life, and nutritional profiles of food products. This article explores the biotechnological processes involved in producing food additives, colorants, and flavorings, along with their applications and benefits.

3.6.1 Food Additives

Food additives are substances added to food to preserve flavor or enhance its taste, appearance, or other qualities. Biotechnological methods have enabled the production of additives such as enzymes, antioxidants, and stabilizers with improved efficiency and purity compared to traditional chemical synthesis.

- **Enzymes**:
  - **Production**: Enzymes used in food processing (e.g., amylases, proteases, lipases) are often produced using microbial fermentation. Microorganisms like bacteria and fungi are genetically modified or selected for their ability to produce high yields of specific enzymes.
  - **Applications**: Enzymes are used to improve dough handling in baking, clarify juices, tenderize meats, and enhance flavor development in cheese and alcoholic beverages.
- **Antioxidants**:
  - **Production**: Antioxidants like vitamin C (ascorbic acid) and vitamin E (tocopherols) can be produced through fermentation processes using

genetically engineered microorganisms.

- **Applications**: They are used to prevent oxidative deterioration of fats and oils, extend shelf-life in processed foods, and maintain the color and texture of fruits and vegetables.

- **Stabilizers and Emulsifiers**:
  - **Production**: Biotechnological methods produce stabilizers (e.g., xanthan gum, pectin) and emulsifiers (e.g., lecithin) from microbial fermentation or enzymatic processes.
  - **Applications**: These additives improve texture, stability, and mouthfeel in food products like sauces, dressings, and dairy alternatives.

3.6.2 Food Colorants

Food colorants impart visual appeal to food products and can be obtained from natural sources or synthesized through biotechnological methods. Biotech approaches offer sustainable alternatives to synthetic dyes and provide stable, natural pigments.

- **Natural Pigments**:
  - **Production**: Microorganisms like bacteria, yeasts, and algae are engineered to produce natural pigments such as carotenoids (e.g., beta-carotene, astaxanthin) and anthocyanins.
  - **Applications**: These pigments are used to color beverages, confectionery, and dairy products while meeting consumer demand for clean-label and natural ingredients.
- **Synthetic Dyes**:
  - **Production**: Biotech processes can optimize the production of synthetic dyes through microbial fermentation or enzymatic synthesis, reducing environmental impact compared to chemical methods.
  - **Applications**: Synthetic dyes are used in a wide range of processed foods to achieve vibrant colors that are stable under various processing conditions.

3.6.3 Food Flavorings

Food flavorings enhance taste and aroma profiles, contributing to consumer preference and satisfaction. Biotechnological advancements have expanded the availability of natural flavors and improved the production efficiency of flavor compounds.

- **Natural Flavors**:
  - **Production**: Yeasts and bacteria are engineered to produce natural flavor compounds such as vanillin (from vanilla beans), fruity esters, and savory compounds (e.g., monosodium glutamate).
  - **Applications**: Natural flavors are used in beverages, snacks, and processed foods to replace synthetic additives and meet clean-label trends.
- **Artificial Flavors**:

- **Production**: Biotech processes optimize the synthesis of artificial flavors using microbial fermentation or enzymatic conversion of precursor compounds.
- **Applications**: Artificial flavors replicate specific tastes (e.g., strawberry, chocolate) and are used in a wide range of food products to maintain consistency and consumer preference.

3.7 BIOTECHNOLOGICAL METHODS FOR THE DETECTION OF PATHOGENS AND CONTAMINANTS IN FOODS

→Learning Objectives

- Define biotechnological methods for pathogen detection
- Explain the biotechnological methods for contaminant detection
- Explain quality control of biotechnological foods

Biotechnological methods for the detection of pathogens and contaminants in foods are crucial for ensuring food safety, quality control, and compliance with regulatory standards. These methods leverage advanced techniques in molecular biology, genomics, and bioinformatics to detect and identify microbial contaminants and harmful substances in food products. This article explores the key biotechnological approaches used for pathogen and contaminant detection in food, emphasizing their role in evaluating and controlling the quality of biotechnological foods.

3.7.1 Biotechnological Methods for Pathogen Detection

- **Polymerase Chain Reaction (PCR)**:
  - **Principle**: PCR amplifies specific DNA sequences of pathogens present in food samples, enabling their rapid detection even at low concentrations.
  - **Applications**: Used for detecting bacteria (e.g., Salmonella, Listeria), viruses (e.g., Norovirus, Hepatitis A), and parasites (e.g., Cryptosporidium) in

various food matrices.

- **Real-Time PCR (qPCR)**:
  - **Principle**: qPCR combines PCR amplification with real-time detection of fluorescence signals, allowing quantitative measurement of pathogen DNA/RNA.
  - **Applications**: Provides rapid and sensitive detection of pathogens, essential for foodborne outbreak investigations and monitoring microbial contamination levels.
- **Loop-Mediated Isothermal Amplification (LAMP)**:
  - **Principle**: LAMP amplifies DNA under isothermal conditions, eliminating the need for thermal cycling equipment.
  - **Applications**: Suitable for on-site testing in resource-limited settings, detecting pathogens like E. coli and Salmonella with high specificity and speed.
- **Next-Generation Sequencing (NGS)**:
  - **Principle**: NGS technologies sequence millions of DNA fragments simultaneously, enabling comprehensive analysis of microbial communities and identification of pathogens.
  - **Applications**: Used for microbial profiling, whole-genome sequencing of pathogens, and investigating complex foodborne outbreaks.

3.7.2 Biotechnological Methods for Contaminant Detection

- **Immunoassays**:
  - **Principle**: Immunoassays detect contaminants (e.g., allergens, toxins) based on antigen-antibody interactions, providing rapid and specific detection.
  - **Applications**: Used for allergen testing (e.g., gluten, peanuts), mycotoxin detection (e.g., aflatoxins), and screening for chemical residues in food.
- **Biosensors**:
  - **Principle**: Biosensors combine biological components (e.g., enzymes, antibodies) with transducers to detect contaminants based on specific binding interactions.
  - **Applications**: Enable real-time monitoring of food quality parameters such as pH, temperature, and contaminants (e.g., pesticides, heavy metals).
- **Mass Spectrometry (MS)**:
  - **Principle**: MS analyzes molecular mass and structure of contaminants, providing high sensitivity and specificity.
  - **Applications**: Used for detecting chemical contaminants (e.g., pesticides, veterinary drugs) and identifying unknown substances in food samples.
- **Microarrays**:
  - **Principle**: Microarrays detect multiple contaminants

simultaneously by hybridizing DNA/RNA probes with target sequences.

- **Applications**: Screen for a wide range of pathogens, allergens, and genetic markers associated with foodborne hazards, offering high-throughput analysis.

3.7.3 Quality Control of Biotechnological Foods

Biotechnological methods for pathogen and contaminant detection play a critical role in ensuring the safety and quality of biotechnological foods:

- **Regulatory Compliance**: Facilitate adherence to food safety regulations by monitoring and controlling microbial contamination and chemical hazards.
- **Risk Assessment**: Assess the risk of foodborne illnesses and contamination events, enabling proactive measures to prevent outbreaks and ensure consumer safety.
- **Traceability and Transparency**: Enhance traceability throughout the food supply chain, providing transparency in food production processes and ingredient sourcing.
- **Consumer Confidence**: Build consumer trust by demonstrating rigorous quality control measures and adherence to stringent safety standards in biotechnological food production.

3.8 THE MOST RECENT TREND: LAB-GROWN MEATS

→Learning Objectives

- Explain the benefits of lab-grown meats
- Define challenges and considerations of lab-grown meats

Lab-grown meats, also known as cultured or cultivated meats, represent a transformative innovation in the food industry aimed at addressing sustainability, animal welfare concerns, and the growing demand for protein-rich diets. Lab-grown meats are produced using tissue engineering techniques that replicate the process of meat formation in animals, but in controlled laboratory environments. The process typically involves the following steps:

- **Cell Isolation**: Animal cells, often from muscle tissue (myosatellite cells), are isolated from a live animal through a biopsy procedure that does not harm the animal.
- **Cell Culture**: The isolated cells are then cultured in a nutrient-rich medium that contains essential nutrients, growth factors, and scaffolding materials to support cell growth and differentiation.
- **Tissue Formation**: Cells multiply and differentiate into muscle fibers under controlled conditions, forming 3D structures that resemble natural muscle tissue.
- **Harvesting and Processing**: Once sufficient tissue is grown, it can be harvested, processed, and formed into familiar meat products such as burgers, sausages, or nuggets.

3.8.1 Benefits of Lab-grown Meats

- **Environmental Sustainability**: Lab-grown meats have the potential to significantly reduce the environmental impact of traditional livestock farming, including greenhouse gas emissions, land use, and water consumption.
- **Animal Welfare**: By eliminating the need for raising and slaughtering animals for meat production, lab-grown meats alleviate animal suffering and align with ethical concerns regarding animal welfare.

- **Food Security**: As global populations increase and demand for meat rises, lab-grown meats offer a scalable and efficient solution to meet protein needs without further straining natural resources.
- **Health Considerations**: Lab-grown meats can be engineered to have lower levels of saturated fats and no added antibiotics or hormones, potentially offering healthier meat options compared to conventionally produced meats.

3.8.2 Challenges and Considerations

- **Cost**: Currently, the production of lab-grown meats is expensive due to the high costs of growth media, scaffolding materials, and bioreactor technology. Economies of scale and technological advancements are needed to reduce costs and make lab-grown meats commercially viable.
- **Consumer Acceptance**: Public perception and acceptance of lab-grown meats remain a significant barrier. Consumer attitudes toward food safety, taste, and ethical concerns about "artificial" or "processed" foods need to be addressed through education and transparency.
- **Regulatory Approval**: Regulatory frameworks for lab-grown meats vary globally, and ensuring safety, labeling standards, and compliance with food regulations are essential for market entry and consumer trust.
- **Scale-Up and Production Challenges**: Scaling up production to meet commercial demand while maintaining product quality and consistency presents technical challenges in cell culture, bioreactor design, and tissue engineering.

3.8.3 Future Prospects

Despite challenges, the development of lab-grown meats continues to advance with ongoing research, investment, and collaboration across academic institutions, startups, and established food companies. Future prospects include:

- **Technological Advancements**: Innovations in bioprocessing, cell biology, and tissue engineering are expected to improve production efficiency, reduce costs, and enhance the nutritional profile of lab-grown meats.
- **Market Expansion**: As consumer acceptance grows and regulatory frameworks evolve, lab-grown meats have the potential to become mainstream alternatives to conventional meat products in global markets.
- **Diversification of Products**: Beyond muscle tissues, research is exploring the cultivation of fats, connective tissues, and other components to replicate the sensory experience and nutritional composition of traditional meats more accurately.

In sum, lab-grown meats represent a promising innovation with the potential to revolutionize the food industry by offering sustainable, ethical, and nutritious alternatives to conventional meat production. While overcoming technical, economic, and regulatory challenges is necessary, continued advancements in biotechnology and consumer education are paving the way for a future where lab-grown meats contribute to a more sustainable and humane global food system.

CASE STUDY

BIOTECH INNOVATIONS IN RICE PRODUCTION

Introduction

Rice is a staple food for more than half of the world's population, particularly in Asia. Traditional rice farming faces challenges such as pest infestations, nutrient deficiencies, and climate change impacts. Biotechnology has played a pivotal role in addressing these challenges and enhancing rice production. This case study examines key breakthroughs in rice production through biotech processes, focusing on the development of Golden Rice and drought-resistant rice varieties.

Golden Rice: Addressing Vitamin A Deficiency

Vitamin A deficiency (VAD) is a significant public health issue, especially in developing countries where rice is a major dietary component. VAD can lead to severe visual impairment and increase the risk of disease and death from severe infections.

Golden Rice was developed to combat VAD. This genetically modified rice contains beta-carotene, a precursor of vitamin A, which gives the rice its golden color. The development process involved the introduction of genes from maize and a soil bacterium (Erwinia uredovora) into the rice genome, enabling the rice to produce and accumulate beta-carotene in its grains.

Impact:

- **Health Benefits:** Golden Rice has the potential to significantly reduce VAD-related health issues among populations that rely heavily on rice as their main food source.
- **Nutritional Enhancement:** It provides an essential nutrient that is not naturally present in conventional rice varieties.
- **Regulatory Approval:** After rigorous testing for safety and efficacy, Golden Rice has received regulatory approvals in several countries, including the Philippines, where it has been officially approved for commercial propagation.

Drought-Resistant Rice: Enhancing Climate Resilience

Drought is a major stress factor that severely affects rice production, leading to significant yield losses. Traditional breeding methods have had limited success in developing drought-tolerant rice varieties due to the complex genetic traits involved. Advances in genetic engineering have enabled the development of drought-resistant rice varieties. One notable example is the introduction of the gene DREB1A (Dehydration Responsive Element Binding 1A) from Arabidopsis thaliana into rice plants. This gene regulates the expression of other genes involved in drought tolerance, enhancing the plant's ability to withstand water scarcity.

Impact:

- **Increased Yield Stability:** Drought-resistant rice varieties maintain higher yields under water-limited conditions, ensuring food security for farmers in drought-prone areas.
- **Economic Benefits:** Reduced crop losses translate to improved income stability for farmers, contributing to rural economic development.
- **Sustainability:** These varieties require less irrigation, promoting water conservation and more sustainable agricultural practices.

These advancements underscore the importance of continued research and development in agricultural biotechnology. They highlight how genetic engineering can contribute to global food security and the well-being of vulnerable populations. As regulatory frameworks evolve and public acceptance grows, biotech innovations in rice production are poised to play an increasingly vital role in meeting the challenges of the 21st century.

SUMMARY

- Food biotechnology is a branch of biotechnology that encompasses the application of biological techniques and processes to improve and enhance various aspects of food production, processing, and consumption. This interdisciplinary field combines principles from biology, genetics, microbiology, and other related disciplines to address challenges and opportunities within the food industry.
- Food serves as an interacting medium between various living species because it is a source of nutrients for humans, animals as well as microorganisms. This is a natural consequence of cohabitation. Human and animal food is basically derived from plant and animal sources.
- Pathogenic microorganisms grow in the food utilizing the nutrients in the food and produce toxins which are detrimental to the health of the consumer when such food is consumed. Food also serves as a vector or medium for certain pathogens that cause food infections and diseases.
- Bacteria are microscopic single-celled organisms that thrive in diverse environments. They can live in the soil, in the ocean, and inside the human gut. Humans' relationship with bacteria is complex. Sometimes they lend a helping hand by curdling milk into yogurt or helping with our digestion.
- The microorganisms occurring on and/or in foods are from a practical point of view divided into three groups: molds, yeast, and bacteria. Molds are generally concerned in the spoilage of foods; their use in the food industry is limited (e.g., mold-ripened cheese). Yeasts are the most widely used microorganisms in the food industry due to their ability to ferment sugars to ethanol and carbon dioxide.
- Bacteria that play significant roles in foods are often grouped on the basis of their activity in foods without regards to their systematic classification. Bacteria make up the largest group of microorganisms. People often think of them only as germs and the harm they do. Only a small number of bacteria types are pathogenic (disease-causing). Most are harmless, and many are helpful.
- Fermentation is traditionally a process that enables food preservation, and as such, has been used for centuries until the present. However, nowadays, the main purpose of food fermentation is not to preserve, since other preservation techniques are known, but to produce a wide variety of fermentation products with specific taste, flavor, aroma, and texture.
- Fermentation processes can be classified as spontaneous and induced (e.g., making bread dough by the addition of baking yeast to flour). Fermentation products contain chemical energy, which means that they are not fully oxidized and their complete mineralization requires oxygen. Fermentation is less energetically efficient than oxidative phosphorylation (ATP is produced only by substrate-level phosphorylation).
- Solid-state (substrate) fermentation (SSF) has been defined as the fermentation process occurring in the absence or near-absence of free water. SSF processes generally employ a natural raw material as a carbon and energy source. SSF can also employ an inert material as a solid matrix, which requires supplementing

a nutrient solution containing necessary nutrients as well as a carbon source.

- Koji fermentations are widely practiced. Koji comprises soybeans or grain on which mold grows, and has been used in oriental food preparation for thousands of years. Koji is a source of fungal enzymes, which digest proteins, carbohydrates, and lipids into nutrients that are used by other microorganisms in subsequent fermentations.
- GM foods have the potential to solve many of the world's hunger and malnutrition problems and to help protect and preserve the environment by increasing yield and reducing reliance upon chemical pesticides and herbicides.
- The term GM foods or GMOs is most commonly used to refer to crop plants created for human or animal consumption using the latest molecular biological techniques. These plants have been modified in the laboratory to enhance desired traits such as increased resistance to herbicides or improved nutritional content.
- Enzymes are produced by all living cells as catalysts for specific chemical reactions. In the food industry, enzymes have been used to produce and to increase the quality and diversity of food. Some examples of products that use enzymes are cheese, yogurt, bread syrup, etc.

REVIEW QUESTIONS

1. Develop a comprehensive understanding of food biotechnology, detailing its applications and significance.
2. Investigate the role of microorganisms commonly associated with food, emphasizing their impact on food quality and safety.
3. Explore fermentation biotechnology, discussing its principles and applications in food production.
4. Evaluate the controversies and benefits surrounding genetically modified foods, considering ethical and safety aspects.
5. Define aerobic fermentation and anaerobic fermentation. Provide a brief example of a product produced through each type of fermentation.

MULTIPLE CHOICE QUESTIONS

1. What is the primary objective of food biotechnology?

a. Enhancing flavor

b. Improving nutritional content

c. Increasing shelf life

d. All of the above

2. Which process involves the intentional modification of an organism's genome for food production?

a. Fermentation

b. Cloning
c. Genetic engineering
d. Hybridization

3. **What is the role of recombinant DNA technology in food biotechnology?**
   a. Enhancing food taste
   b. Increasing crop yield
   c. Introducing new food varieties
   d. Improving food safety
4. **Which enzyme is commonly used in the production of genetically modified crops to make them resistant to pests?**
   a. Amylase
   b. Lipase
   c. Protease
   d. *Bacillus thuringiensis* (Bt) toxin
5. **What is the primary function of CRISPR-Cas9 in food biotechnology?**
   a. Enhancing flavor
   b. Genome editing
   c. Increasing shelf life
   d. Accelerating fermentation
6. **Which of the following is an example of a genetically modified organism (GMO) commonly used in food production?**
   a. Quinoa
   b. Soybean
   c. Kale
   d. Chia seeds
7. **What is the significance of biofortification in food biotechnology?**
   a. Enhancing food flavor
   b. Increasing nutritional content
   c. Extending shelf life
   d. Improving food texture
8. **Which technique is employed to preserve food by inhibiting the activity of spoilage-causing microorganisms?**
   a. Pasteurization
   b. Irradiation
   c. Fermentation
   d. Freeze drying

9. **What is the primary purpose of using enzymes in food biotechnology?**
   a. Enhancing food color
   b. Increasing food temperature
   c. Improving food texture
   d. Facilitating biochemical reactions
10. **Which biotechnological method is commonly used to produce high-value proteins, enzymes, and other bioactive compounds in microorganisms?**
   a. Somatic hybridization
   b. Polymerase Chain Reaction (PCR)
   c. Recombinant DNA technology
   d. Tissue culture

Answer to Multiple Choice Questions

1. (d); 2. (c); 3. (c); 4. (d); 5. (b); 6. (b); 7. (b); 8. (b); 9. (b); 10. (c)

REFERENCES

1. Adams, M. R., & Nicolaides, L., (2008). Review of the sensitivity of different food borne pathogens to fermentation. *Food Control, 8*, 227–239.
2. Bartkiene, E., Juodeikiene, G., Vidmantiene, D., Viskelis, P., & Urbonaviciene, D., (2011). Nutritional and quality aspects of wheat sourdough bread using *Lactobacillus luteus* and *L. angustifolius* flours fermented by *Pediococcus acidilactici. International Journal of Food Science and Technology, 46*, 1724–1733.
3. Blandino, A., Al-Aseeri, M. E., Pandie, S. S., Cantero, D., & Webb, C., (2003). Cereal-based fermented foods and beverages. *Food Research International, 36*, 527–543.
4. Bourdichon, F., Casaregola, S., Farrokh, C., Frisvad, J. C., Gerds, M. L., Hammes, W. P., Harnett, J., et al., (2012). Food fermentations: Microorganisms with technological beneficial use. *International Journal of Food Microbiology, 154*, 87–97.
5. Bover-Cid, S., Izquierdo-Pulido, M., & Vidal-Carou, M. C., (2000). Mixed starter cultures to control biogenic amine production in dry fermented sausages. *Journal of Food Protection, 63*, 1556–1562.
6. Brenner, K., You, L., & Arnold, F. H., (2008). Engineering microbial consortia: A new frontier in synthetic biology. *Trends in Biotechnology, 26*, 483–489.
7. Dalié, D. K. D., Deschamps, A. M., & Richard-Forget, F., (2010). Lactic acid bacteria—Potential for control of mould growth and mycotoxins: A review. *Food Control, 21*, 370–380.
8. De Vuyst, L., & Marshall, V. M., (2001). First international symposium on exopolysaccharides from lactic acid bacteria: From fundamentals to applications. *International Dairy Journal, 11*, 659–768.

9. De Vuyst, L., De Vin, F., Vaningelgem, F., & Degeest, B., (2001). Recent developments in the biosynthesis and applications of heteropolysaccharides from lactic acid bacteria. *International Dairy Journal, 11*, 687–707.

10. Escalante, A., Rodríguez, M. E., Martínez, A., López-Munguía, A., Bolívar, F., & Gosset, G., (2004). Characterization of bacterial diversity in pulque, a traditional Mexican alcoholic fermented beverage, as determined by 16S rDNA analysis. *FEMS Microbiology Letter, 235*(2), 273–279.

11. Fernández-Bodega, M. A., Mauriz, E., Gómez, A., & Martín, J. F., (2009). Proteolytic activity, mycotoxins and andrastin A in penicillium roqueforti strains isolated from Cabrales, Valdeón and Bejes-Teresviso local varieties of blue-veined cheeses. *International Journal of Food Microbiology, 136*, 18–25.

12. Gaggia, F., Di Gioia, D., Baffoni, L., & Biavati, B., (2011). The role of protective and probiotic cultures in food and feed and their impact on food safety. *Trends in Food Science and Technology, 22*, S58–S66.

13. GarcíaFontán, M. C., Lorenzo, J. M., Parada, A., Franco, I., & Carballo, J., (2007). Microbiological characteristics of "androlla," a Spanish traditional pork sausage. *Food Microbiology, 24*, 52–58.

14. Gareau, M. G., Sherman, P. M., & Walker, W. A., (2010). Probiotics and the gut microbiota in intestinal health and disease. *Nature Reviews Gastroenterology and Hepatology, 7*, 503–514.

15. Gest, H., (2004). The discovery of microorganisms by Robert Hooke and Antoni van Leeuwenhoek, fellows of the royal society. *Notes and Records of the Royal Society of London, 58*, 187–201.

16. Giraud, F., Giraud, T., Aguileta, G., Fournier, E., Samson, R., Cruaud, C., Lacoste, S., et al., (2010). Microsatellite loci to recognize species for the cheese starter and contaminating strains associated with cheese manufacturing. *International Journal of Food Microbiology, 137*, 204–213.

CHAPTER 4

Plant Biotechnology

LEARNING OBJECTIVES

After studying this chapter, you will be able to:

- Understand the plant tissue culture;
- Learn about transgenic plants;
- Discuss the plant genetic transformation methods.

INTRODUCTORY EXAMPLE

The Role of CRISPR-Cas9 in Plant Biotechnology:

Plant biotechnology has witnessed a revolutionary shift with the advent of CRISPR-Cas9 technology, offering precise and targeted genetic modifications in crops. This powerful tool has opened up new avenues for enhancing climate resilience in crops, ensuring food security in the face of changing environmental conditions.

- **CRISPR-Cas9 Technology:** Clustered Regularly Interspaced Short Palindromic Repeats (CRISPR) and CRISPR-associated protein 9 (Cas9) constitute a revolutionary gene-editing system. Unlike traditional genetic modification methods, CRISPR-Cas9 allows scientists to precisely and selectively alter specific genes within a plant's DNA. This technology enables the development of crops with enhanced traits such as drought tolerance, disease resistance, and improved nutrient utilization.
- **Drought Tolerance in Maize:** With climate change leading to more frequent and severe droughts, ensuring crop resilience to water scarcity is imperative for global food security. Researchers have utilized CRISPR-Cas9 to enhance drought tolerance in crops like maize. By modifying specific genes associated with water retention and stress response mechanisms, scientists have developed maize varieties capable of maintaining productivity even under water-limiting conditions.
- **Disease Resistance in Tomatoes:** Plant diseases pose significant threats to crop yields, impacting food production and economic stability. In the realm of plant biotechnology, CRISPR-Cas9 has been employed to bolster disease resistance in crops like tomatoes. By precisely editing genes responsible for susceptibility to specific pathogens, researchers have developed tomatoes with increased resistance to common diseases. This approach reduces the reliance on chemical pesticides, promoting sustainable and environmentally friendly agricultural practices.
- **Improved Nutrient Utilization in Rice:** Addressing malnutrition is a global challenge, and plant biotechnology plays a pivotal role in enhancing the nutritional content of staple foods. Using CRISPR-Cas9, scientists have targeted genes in rice to improve nutrient utilization. This includes modifying genes associated with nutrient uptake and storage, leading to the development of biofortified rice varieties with increased levels of essential vitamins and minerals. Such advancements contribute to combatting malnutrition, particularly in regions where rice is a dietary staple.
- **Ethical Considerations and Regulatory F**rameworks: The rapid progress in CRISPR-Cas9 technology raises ethical considerations related to genetically modified organisms (GMOs). While the precision of CRISPR-Cas9 minimizes off-target effects, concerns about unintended consequences and environmental impacts persist. Striking a balance between the potential benefits of climate-resilient crops and addressing ethical, environmental, and social considerations remains a challenge for scientists, policymakers, and society at large.

UNIT INTRODUCTION

Plant biotechnology is a field of scientific inquiry and application that focuses on harnessing the unique characteristics of plants to address various challenges in agriculture, medicine, and environmental sustainability. Plant biotechnology utilizes molecular and cellular techniques to manipulate the genetic makeup of plants, aiming to enhance their desirable traits or confer new capabilities. This branch of biotechnology plays a crucial role in ensuring food security, improving crop yields, and developing sustainable agricultural practices.

One fundamental aspect of plant biotechnology is plant cell and tissue culture, which involves the in vitro cultivation of plant cells or tissues. This technique allows researchers and biotechnologists to propagate plants efficiently, enabling mass production of uniform and genetically identical plant material. Micropropagation, a key application of plant tissue culture, has revolutionized the production of disease-free and genetically identical plants on a large scale.

Genetic engineering is another pivotal component of plant biotechnology, involving the manipulation of plant DNA to introduce or enhance specific traits. This technology enables the creation of transgenic plants, which carry foreign genes from unrelated organisms. Transgenic plants have been instrumental in developing crops with improved resistance to pests, diseases, and environmental stressors, contributing significantly to global agriculture. One of the most widely debated applications of plant biotechnology is the development of genetically modified (GM) crops. GM crops, such as insect-resistant Bt cotton and herbicide-tolerant soybeans, have been engineered to withstand environmental challenges, resulting in increased crop yields and reduced dependence on chemical pesticides. However, the adoption of GM crops also raises ethical and environmental concerns, leading to ongoing discussions about their safety and long-term impact. In recent years, plant biotechnology has expanded its horizons to include innovative approaches like genome editing. Technologies like CRISPR/Cas9 allow precise modifications to the plant genome, offering unprecedented control over the expression of specific traits. This advancement holds immense potential for creating crops with enhanced nutritional content, improved resistance to diseases, and increased adaptability to changing environmental conditions.

Furthermore, plant biotechnology extends beyond agriculture into the realm of plant molecular farming. This involves leveraging plants as biofactories to produce valuable pharmaceuticals, industrial enzymes, and other high-value compounds. Plants, such as tobacco and maize, can be engineered to express proteins with therapeutic or commercial applications, offering a cost-effective and scalable alternative to traditional production methods.

4.1. PLANT TISSUE CULTURE

→Learning Objectives

- Assess the advantages and disadvantages of tissue culture;
- Examine the history of plant tissue culture;
- Understand the basics of plant cell and tissue culture;
- Define tissue culture in agriculture;
- Learn about germplasm conservation;
- Understand the genetic transformation;
- Discuss haploid production;
- Explain the techniques of plant tissue culture;
- Define tissue culture in pharmaceuticals.

Plant tissue culture, or PTC, is the process of growing undifferentiated plant cells, tissues, or organs on synthetic media in an aseptic environment with appropriate temperature control. It is a crucial tool for both commercial and fundamental research applications. The foundation of plant tissue culture is the distinct quality of a plant cell, i.e., totipotency. A vegetative cell's totipotency refers to its capacity to proliferate into a whole plant or to divide and differentiate into any kind of specialized cell.

4.1.1. Advantages and Disadvantages of Plant Tissue Culture

Plant tissue culture is a technique used to propagate and grow plants under controlled conditions. Like any scientific method, it comes with its advantages and disadvantages.

- Advantages of Plant Tissue Culture:
    1. **Mass Production of Uniform Plants:** Tissue culture allows for the mass production of genetically identical plants, ensuring uniformity in traits, quality, and characteristics.
    2. **Rapid Clonal Propagation:** It offers a faster and more efficient method of clonal propagation compared to traditional methods like seed germination, resulting in quicker plant multiplication.
    3. **Preservation of Endangered Species:** Tissue culture can be used to conserve and propagate endangered or rare plant species, helping to maintain biodiversity.

4. **Disease-Free Plants:** The process allows the production of disease-free plants by starting with healthy tissues, minimizing the risk of transmitting diseases from one generation to another.
5. **Year-Round Production:** Tissue culture enables year-round production, irrespective of season or environmental conditions, providing a consistent supply of plants.
6. **Selection and Improvement of Traits:** It facilitates the selection and improvement of specific traits, such as resistance to pests, diseases, or environmental stress.
7. **Production of Secondary Metabolites:** Some plant tissues cultured in vitro can produce secondary metabolites of pharmaceutical or industrial importance.

- **Disadvantages of Plant Tissue Culture:**

1. **Cost and Equipment:** The initial setup costs for a tissue culture facility can be high, and maintaining the necessary equipment and sterile conditions can be expensive.
2. **Genetic Uniformity:** While genetic uniformity is an advantage, it can also be a disadvantage in terms of susceptibility to pests and diseases. If a genetically uniform crop is attacked, the entire population may be vulnerable.
3. **Mutation Accumulation:** Due to the rapid cell division in tissue culture, there is a potential for genetic mutations to accumulate, leading to unintended changes in the characteristics of the plant.
4. **Labor-Intensive:** Tissue culture requires skilled personnel to handle the delicate and precise procedures involved, making it labor-intensive.
5. **Risk of Contamination:** Maintaining a sterile environment is crucial, as contamination can result in the loss of cultures or the introduction of unwanted microorganisms.
6. **Environmental Concerns:** Critics argue that the high energy consumption and resource requirements of tissue culture facilities may have environmental impacts.
7. **Somatic Variation:** Somatic variations may occur in cultured tissues, leading to differences between the original plant and the cultured offspring.

Keyword

Plant Cell – It is a eukaryotic cell enclosed by a cell wall, containing a membrane-bound nucleus and other cell organelles.

4.1.2. History of Plant Tissue Culture

Plant tissue culture science originated with the discovery of the cell and the subsequent development of cell theory. The basic structural unit of all living things is the cell, according to Schleiden and Schwann's 1838 proposal. They conceptualized the idea that since each cell possesses the ability to function independently, it should be feasible for them to regenerate into an entire plant in the right conditions. This idea served as the foundation for the first attempt to cultivate isolated single palisade cells from leaves in Knop's salt solution that had been enhanced with sucrose in 1902 by German physiologist Gottlieb Haberlandt. For as long as a month, the cells did not divide but instead grew larger and accumulated starch. Even though he didn't succeed, he is recognized as the father of plant tissue culture because he established the fundamentals of tissue culture technology. Following this, some important discoveries in tissue culture were made, which can be summed up as follows:

- **1902:** Haberlandt proposed the concept of *in vitro* cell culture.
- **1904:** Hannig cultured embryos from several cruciferous species.
- **1922:** Kolte and Robbins successfully cultured root and stem tips, respectively.
- **1926:** Went discovered first plant growth hormone –Indole acetic acid.
- **1934:** White introduced vitamin B as a growth supplement in tissue culture media for tomato root tip.
- **1939:** Gautheret, White, and Nobecourt established endless proliferation of callus cultures.
- **1941:** Overbeek was the first to add coconut milk for cell division in *Datura*.
- **1946:** Ball raised whole plants of *Lupinus* by shoot tip culture.
- **1954:** Muir was first to break callus tissues into single cells.
- **1955:** Skoog and Miller discovered kinetin as a cell division hormone.
- **1957:** Skoog and Miller gave the concept of hormonal control (auxin: cytokinin) of organ formation.
- **1959:** Reinert and Steward regenerated embryos from callus clumps and cell suspension of carrot (*Daucuscarota*).
- **1960:** Cocking was the first to isolate protoplast by enzymatic degradation of cell wall.
- **1960:** Bergmann filtered cell suspension and isolated single cells by plating.
- **1960:** Kanta and Maheshwari developed test tube fertilization technique.
- **1962:** Murashige and Skoog developed an MS medium with a higher salt concentration.
- **1964:** Guha and Maheshwari produced first haploid plants from pollen grains of *Datura* (Anther culture).
- **1966:** Steward demonstrated totipotency by regenerating carrot plants from single cells of tomato.
- **1970:** Power et al. successfully achieved protoplast fusion.
- **1971:** Takebe et al. regenerated first plants from protoplasts.
- **1972:** Carlson produced the first interspecific hybrid of *Nicotianatabacum* by protoplast fusion.

- **1974:** Reinhard introduced biotransformation in plant tissue cultures.
- **1977:** Chilton et al. successfully integrated Ti plasmid DNA from *Agrobacterium tumefaciens* in plants.
- **1978:** Melchers et al. carried out somatic hybridization of tomato and potato, resulting in pomato.
- **1981:** Larkin and Scowcroft introduced the term somaclonal variation.
- **1983:** Pelletier et al. conducted intergeneric cytoplasmic hybridization in Radish and Grape.
- **1984:** Horsh et al. developed transgenic tobacco by transformation with *Agrobacterium*.
- **1987:** Klien et al. developed a biolistic gene transfer method for plant transformation.
- **2005:** Rice genome sequenced under the International Rice Genome Sequencing Project.

4.1.3. Basics of Plant Cell and Tissue Culture

Plant tissues and organs are grown in vitro on synthetic media in an aseptic, controlled environment through the process of plant cell culture. The methodology primarily relies on the notion of totipotentiality of plant cells, which denotes a single cell's capacity to express the entire genome through cell division. Plant cells have totipotent potential, but they also have the ability to change how they grow, behave, and metabolize. These changes are essential to the regeneration of the entire plant. All the nutrients needed for a plant's normal growth and development are present in the plant tissue culture medium. In the case of a solid medium, it is primarily made up of some gelling agents, vitamins, other organic components, carbon sources, macronutrients, and micronutrients. Murashige and Skoog medium (MS medium) is most extensively used for the vegetative propagation of many plant species *in vitro*. The pH of the media is also important because it affects both the growth of plants and the activity of plant growth regulators (PGRs), and was adjusted between 5.4–5.8. Both the solid and liquid medium can be used for culturing. The composition of the medium, particularly the plant hormones and the nitrogen source, has profound effects on the response of the initial explant. The development route of plant cells and tissues in culture medium is largely determined by PGRs, or PGRs. PGRs that are most frequently used are gibberellins, cytokinins, and auxins. The plant species, the tissue or organ being cultivated, and the experiment's goal all influence the kind and concentration of hormones used. In plant tissue culture, auxins and cytokinins are the most commonly used PGRs. The type of culture that is established or regenerated depends on the amount of each. A high auxin concentration typically encourages the formation of roots, whereas a high concentration of cytokinins encourages the regeneration of shoots. A callus is a mass of undifferentiated cells that develops when auxin and cytokinin are in balance.

Maximum root induction and proliferation were observed in *Stevia rebaudiana*, when the medium is supplemented with 0.5 mg/l NAA. Cytokinins generally promote cell division and induce shoot formation and axillary shoot proliferation. A high cytokinin-to-auxin ratio promotes shoot proliferation, while a high auxin-to-cytokinins ratio results in root formation. Shoot initiation and proliferation were found maximum, when the callus of black pepper was shifted to medium supplemented with BA at the concentration of 0.5 mg/l. Gibberellins are used for enhanced growth and to promote cell elongation. Maximum shoot length was observed in *Phalaenopsis* orchids when cultured in a medium containing 0.5 mg/l GA_3.

Keyword

Cytokinins – These are plant-specific chemical messengers (hormones) that play a central role in the regulation of the plant cell cycle and numerous developmental processes.

4.1.4. Tissue Culture in Agriculture

Plant tissue culture is an emerging technology that has a significant impact on industry and agriculture by providing the necessary plants to meet the growing global demand. It has recently made major contributions to the development of agricultural sciences, and as a result, they are now an essential tool in contemporary agriculture.

The pace at which biotechnology is being incorporated into agricultural practices is unprecedented. Plant material that is disease-free and genetically homogeneous can be produced and propagated through tissue culture. In vitro culture of cells and tissues is a helpful technique for somaclonal variation induction. Tissue culture-induced genetic variability may serve as a source of variability for the creation of novel, stable genotypes. It has been encouraging to see biotechnological approaches used in gene transfer studies in tree species, mass micropropagation techniques, and in vitro regeneration. Mature and/or immature zygotic embryo cultures are used in vitro cultures to recover plants derived from intergeneric crosses that do not yield viable seeds. Many improved crop varieties with high yield potential and pest resistance can be made possible through genetic engineering. The technical components of molecular biology and plant tissue culture are essential to genetic transformation technology because of the following:

- Production of improved crop varieties;
- Production of disease-free plants (virus);
- Genetic transformation;
- Production of secondary metabolites;
- Production of varieties tolerant to salinity, drought, and heat stresses.

4.1.5. Germplasm Conservation

In vitro cell and organ culture offers an alternative source for the conservation of endangered genotypes. Germplasm conservation worldwide is increasingly becoming an essential activity due to the high rate of disappearance of plant species and the increased need to safeguard the floristic patrimony of the countries. Tissue culture protocols can be used for preservation of vegetative tissues when the targets for conservation are clones instead of seeds, to keep the genetic background of a crop and to avoid the loss of the conserved patrimony due to natural disasters, whether biotic or abiotic stress. The plant species that do not produce seeds (sterile plants) or that have 'recalcitrant' seeds that cannot be stored for a long period can successfully be preserved via *in vitro* techniques for the maintenance of gene banks.

Cryopreservation plays a vital role in the long-term *in vitro* conservation of essential biological material and genetic resources. It involves the storage of *in vitro* cells or tissues in liquid nitrogen that results in cryo-injury on the exposure of tissues to physical and chemical stresses. Successful cryopreservation is often ascertained by cell and tissue survival and the ability to re-grow or regenerate into complete plants or form new colonies. It is desirable to assess the genetic integrity of recovered germplasm to determine whether it is 'true-to-type' following cryopreservation. The fidelity of recovered plants can be assessed at phenotypic, histological, cytological, biochemical, and molecular levels, although, there are advantages and limitations of the various approaches used to assess genetic stability. Cryobionomics is a new approach to study genetic stability in cryopreserved plant materials. The embryonic tissues can be cryopreserved for future use or for germplasm conservation.

4.1.6. Embryo Culture

Plant tissue culture that is used to cultivate embryos from seeds and ovules in a nutrient medium is known as embryo culture. In an embryo culture, the plant either grows from the embryo directly or indirectly through the development of a callus, which is followed by the growth of shoots and roots. The method was created to produce haploid plants and rare species, and break seed dormancy. It can also be used to test the vitality of seeds. It is a useful method for reducing the length of the dormant period of seeds and cutting the breeding cycle of plants by cultivating removed embryos. With the specific goal of mass multiplication, intra-varietal hybrids of the economically significant energy plant "Jatropha" have been successfully created. In Jucara palm embryo cultures, somatic embryogenesis and plant regeneration have been used to expedite cloning and enhance specific individuals. However, using the embryo culture method can also lead to the conservation of endangered species. Excising embryos from mature seeds is a successful *in vitro* protocol that has been developed recently for the propagation of *Khayagrandifoliola*. The plant has a significant economic value both as medicine and as lumber. Because it provides a means of propagating elite individuals in situations where it is challenging to select and improve natural populations, this technique has significant applications in forestry.

4.1.7. Genetic Transformation

The most recent development in plant cell and tissue culture is genetic transformation, which offers a method for transferring desired traits' genes into host plants and producing transgenic plants. When combined with plant biotechnology and breeding programs, the technique offers a great deal of potential for genetic improvement of different crop plants. It holds promise for introducing traits that are crucial to agronomy, like higher yield, improved quality, and stronger resistance to pests and diseases.

Plants are capable of undergoing genetic transformation through vector-mediated (indirect gene transfer) or vectorless (direct gene transfer) methods. The most popular vector-dependent gene transfer technique for expressing foreign genes in plant cells is *agrobacterium*-mediated genetic transformation. The genetic transformation of root explants was utilized to introduce agronomic traits into plants successfully. Virus-based vectors present a viable means of fast and stable transient protein expression in plant cells, thereby offering an effective means of large-scale recombinant protein production. Recently successful transgenic plants of Jatropha were obtained by direct DNA delivery to mature seed-derived shoot apices via the particle bombardment method. This technology has an important impact on the reduction of toxic substances in seeds, thus overcoming the obstacle of seed utilization in various industrial sector. Regeneration of disease or virus-resistant plants is now achieved by employing genetic transformation techniques. Researchers succeeded in developing transgenic plants of potato resistant to potato virus Y (PVY), which is a major threat to potato crops worldwide. In addition, marker-free transgenic plants of *Petunia hybrida* were produced using the multi-auto-transformation (MAT) vector system.

4.1.8. Protoplast Fusion

A valuable technique for improving crops and plant breeding is somatic hybridization, which creates interspecific and intergeneric hybrids. The process entails fusing the protoplasts of two distinct genomes, choosing the desired somatic hybrid cells, and then regenerating hybrid plants. Protoplast fusion has a growing influence on crop improvement and offers an effective method of transferring genes with desired traits from one species to another. In order to create somatic hybrids with salt tolerance, protoplasts from ditch reed and rice were fused together using an electrofusion process.

In vitro fusion of protoplast opens a way of developing unique hybrid plants by overcoming the barriers of sexual incompatibility. The technique has been applicable in the horticulture industry to create new hybrids with increased fruit yield and better resistance to diseases. Successful viable hybrid plants were obtained when protoplasts from citrus were fused with other related citrinae species. The potential of somatic hybridization in important crop plants is best illustrated by the production of intergeneric hybrid plants among the members of Brassicaceae. To resolve the problem of loss of chromosomes and decreased regeneration capacity, the successful protocol has been established for the production of somatic hybrid plants by using two types of wheat protoplast as recipient and protoplast of *Haynaldiavillosa* as a fusion donor. It is also employed as an important gene source for wheat improvement (Figure 4.1).

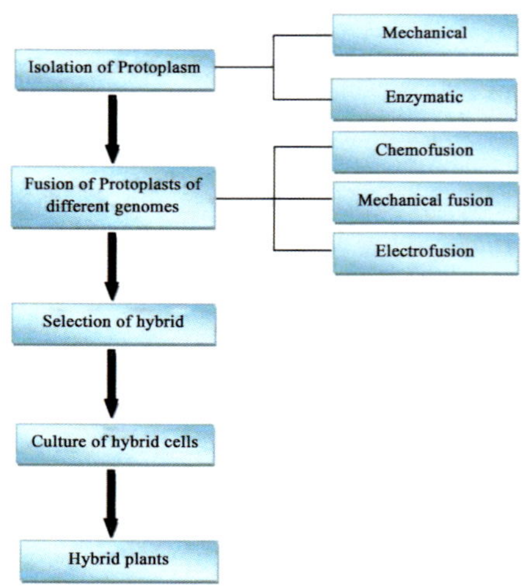

Figure 4.1. *A schematic representation of the production of a hybrid plant via protoplast fusion.*

Source: https://www.researchgate.net/figure/Schematic-representation-of-production-of-hybrid-plant-via-protoplast-fusion_fig1_297777765.

4.1.9. Haploid Production

Instead of using traditional breeding methods, tissue culture techniques allow for the relatively quick production of homozygous plants through protoplast, anther, and microspore cultures. Sterile plants with a single set of chromosomes, known as haploids, can undergo spontaneous or induced chromosome doubling to become homozygous diploids. Plant fertility is restored by chromosomal doubling, which produces double haploids with the potential to become pure breeders of new cultivars. The process of creating haploid plants from immature pollen cells without fertilization is known as androgenesis. Sudherson et al. reported employing pollen grains as the main explants to produce haploid plants of the sturt's desert pea. Because haploidy technology produces inbred lines more quickly and gets around the issues of seed dormancy and non-viability of embryos, it has become an essential component of plant breeding programs. The method is remarkably useful for producing haploid plants that have been induced to be resistant to a variety of biotic and abiotic stresses through genetic transformation.

4.1.10. Techniques of Plant Tissue Culture

4.1.10.1. Micropropagation

The first step in micropropagation is to select plant tissues (explants) from a robust, healthy mother plant. Leaf, apical meristem, bud, and root are among the plant parts that can be used as explants. The following steps, which are depicted in Figure 4.2, can be used to summarize the entire process.

- **Stage 0: Preparation of Donor Plant:** It is possible to introduce any plant tissue *in vitro*. The mother plant should be cultivated *ex vitro* under ideal conditions to reduce contamination in the *in vitro* culture and increase the likelihood of success.
- **Stage I: Initiation Stage:** An explant is surface sterilized and placed in the nutrient medium at this point. It is generally advised to apply fungicide and bactericide products together. The products chosen are determined by the kind of explant that will be introduced. In order to eliminate pollutants from explants while causing the least amount of harm to plant cells, surface sterilization in chemical solutions is a crucial step. The most widely used disinfectants are ethanol, mercuric chloride ($HgCl_2$), calcium hypochlorite, and sodium hypochlorite. Depending on the propagation method, the cultures are incubated in growth chambers under light or dark conditions.
- **Stage II: Multiplication Stage:** Increasing the number of propagules is the goal of this phase. Repeated subcultures are used to multiply the propagule count until the intended (or desired) plant population is reached.
- **Stage III: Rooting Stage:** The rooting stage can take place concurrently in the same culture media that the explants were multiplied in. However, to promote rooting and the development of robust root growth, it may occasionally be necessary to alter the medium, including by modifying the nutritional composition and growth regulator composition.

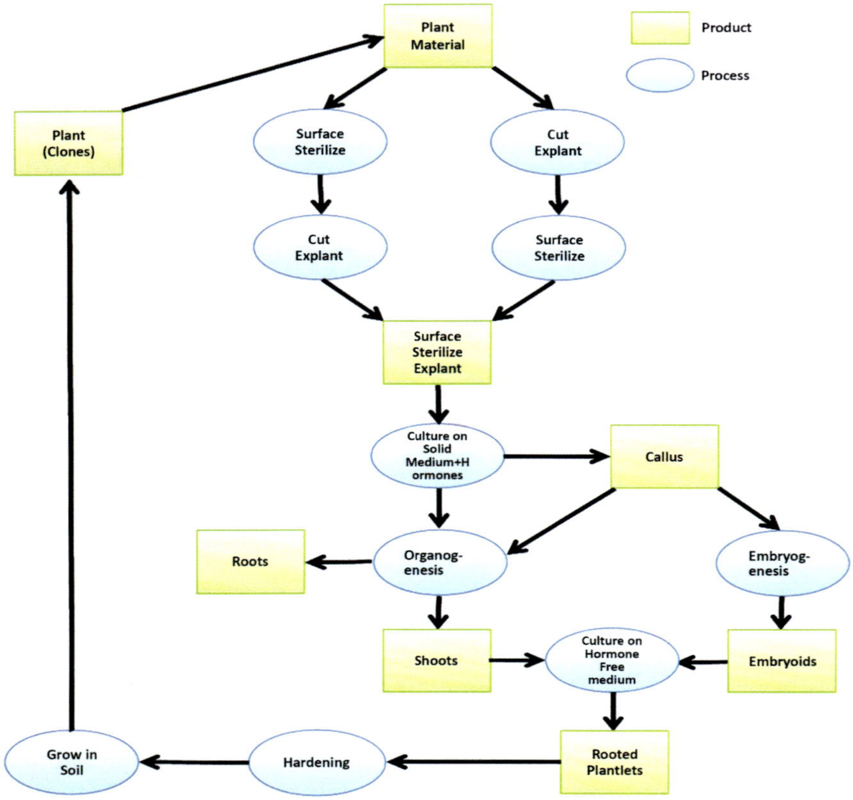

Figure 4.2. *Flow chart summarizing tissue culture experiments.*

Source: https://www.researchgate.net/figure/Flow-chart-summarizing-tissue-culture-experiments_fig2_297777765.

- **Stage IV: Acclimatization Stage:** The *in vitro* plants have reached this point of weaning and hardening. From high to low humidity and from low to high light intensity, hardening occurs gradually. After that, the plants are moved to a suitable substrate (compost, peat, sand, etc.) and progressively solidified in a greenhouse.

4.1.11. Tissue Culture in Pharmaceuticals

Plant cell and tissue cultures present a great potential for the regulated synthesis of a wide range of beneficial secondary metabolites. For the production of important therapeutic secondary metabolites, plant cell cultures combine the advantages of microbial and animal cell cultures with those of whole-plant systems. Biotechnological methods, in particular plant tissue cultures, are found to have potential as a supplement to traditional agriculture in the industrial production of bioactive plant metabolites in the search for substitutes for the production of medicinal compounds from plants. During the past 10 years, a team of microbiologists and plant scientists have investigated the biosynthetic potential of diverse cell cultures across multiple nations. Cell suspension culture: These days, plant cells are cultured on a large scale using cell suspension culture systems so that secondary metabolites can be extracted. The relatively friable part of the callus is transferred into a liquid medium to create a suspension culture, which is then kept alive under the right physical parameters such as temperature, light, agitation, and aeration. Not only can cell cultures produce large quantities of defined standard phytochemicals, but they can also remove interfering compounds from field-grown plants. The advantage of this method is that it can ultimately provide a continuous, reliable source of natural products. The major advantages of the cell cultures include the synthesis of bioactive secondary metabolites, running in a controlled environment, independent from climate and soil conditions. A number of different types of bioreactors have been used for mass cultivation of plant cells. The first commercial application of large-scale cultivation of plant cells was carried out in stirred tank reactors of 200 l and 750 l capacities to produce shikonin by cell culture of *Lithospermumerythrorhizon*. Cell of *Catharanthusroseus, Dioscoreadeltoidea, Digitalis lanata, Panax notoginseng, Taxus wallichiana* and *Podophyllum hexandrum* have been cultured in various bioreactors for the production of secondary plant products.

Keyword

Root Growth – It occurs from the expansion of newly developed cells resulting from the division of meristematic cells located just behind the root cap.

Many cultures of plant cells and tissues are used to produce a variety of medicinally significant alkaloids, anticancer medications, recombinant proteins, and food additives. Developments in the field of using cell cultures to produce pharmaceuticals have enabled the synthesis of a broad range of drugs, including steroids, alkaloids, terpenoids, flavonoids, saponins, and phenolics. Certain ones, like paclitaxel (Taxol) and shikonin, are now commercially available. Around 20 distinct recombinant proteins, including growth factors, cytokines, enzymes, edible vaccines, and antibodies, have been produced in plant cell cultures thus far. Technological developments in immobilization and scale-up methods have led to a significant rise in the uses of plant cell cultures for the synthesis of high-value compounds. Table 4.1 lists a few of the secondary plant products made from different plants' cell suspension cultures.

Table 4.1. *List of Some Secondary Plant Products Produced in Suspension Culture*

Secondary Metabolite	Plant Name
Vasine	Adhatoda vasica
Artemisinin	Artemisia annua
Azadirachtin	Azadirachta indica

Cathin	Bruceaja vanica
Capsiacin	Capsicum annum
Sennosides	Cassia senna
Ajmalicine Secologanin Indole alkaloids Vincristine	Catharanthus roseus
Stilbenes	Cayratia trifoliata
Berberin	Coscinium fenustratum
Sterols	Hyssopus officinalis
Shikonin	Lithospermum erythrorhizon
Ginseng saponin	Panaxnoto ginseng
Podophyllotoxin	Podophyllum hexandrum
Taxane paclitaxel	Taxus chinensis

Source: https://www.google.com/imgres.

4.1.12. Hairy Root Cultures

In the past 20 years, the hairy root system—which is based on inoculation with *Agrobacterium rhizogenes*—has gained popularity as a means of generating secondary metabolites that are produced in plant roots. The synthesis of secondary metabolites can be greatly aided by organized cultures, particularly root cultures. The majority of studies on transformed (hairy) roots have used differentiated cultures rather than cell suspension cultures. Hairy root disease in plants is caused by *A. rhizogenes*.

The neoplastic (cancerous) roots produced by *A. rhizogenes* infection are characterized by the high growth rate, genetic stability, and growth in hormone-free media. High stability and productivity features allow the exploitation of hairy roots as valuable biotechnological tools for the production of plant secondary metabolites. These genetically transformed root cultures can produce levels of secondary metabolites comparable to that of intact plants. Hairy root technology has been strongly improved through increased knowledge of molecular mechanisms underlying their development. Optimizing the composition of nutrients for hairy root cultures is critical to gain a high production of secondary metabolites.

Did you get it?

1. What is plant tissue culture?
2. Explain the history of plant tissue culture.

4.2. TRANSGENIC PLANTS

→Learning Objectives

- Explore the application of transgenic plants;
- Understand the concept of gene constructs;
- Identify vectors for the production of transgenic plants;
- Define transgenic plants for improved quality traits;
- Examine the use of plant biotechnology in floriculture.

Transgenic plants are those whose genomes have undergone modifications due to genetic engineering, either by the insertion of a foreign gene or the deletion of a harmful gene. A foreign gene that has been introduced into a plant may come from a distinct species or even a kingdom. By inserting the *nptII* bacterial antibiotic resistance gene into tobacco, the first transgenic plant was created. Since then, a wide range of transgenic plants with significant agronomic traits, such as pest resistance and drought tolerance, have been developed, ranging from dicots to monocots that are amenable to genetic modifications. This is due to the rapid advancement of plant molecular biology and genetic engineering technology. The primary goal of transgenic plant production is to create crops with superior quality, high yield, and ideal traits. In addition to their benefits for agriculture, the plants can serve as a factory for the production of pharmaceutical proteins.

4.2.1. Application of Transgenic Plants

4.2.1.1. Resistance Against Abiotic Stresses

In order to create stress-resistant plants, a number of genes that offer resistance against environmental stressors like heat, cold, salt, and heavy metals have been found and inserted. By stimulating the synthesis of particular substances known as compatible solutes (proline, betaines), antioxidants (carotenoids, ascorbate), and antioxidant enzymatic machinery, the insertion of these genes into plants aids in the ability of the plant to withstand stress. Abiotic stress causes plants to activate genes that code for the synthesis of particular proteins, such as chaperones, antifreeze proteins, and late embryogenesis abundant (LEA) proteins. These proteins offer defense against abiotic stress-induced desiccation or dehydration.

Did you know?

In 1983, the first transgenic plant was produced when tobacco plants were transformed with a bacterial gene, demonstrating the feasibility of introducing foreign genes into plants.

4.2.1.2. Drought Resistant Plants

When plants experience drought stress, they activate specific genes that produce compounds to safeguard the plant against the damaging impacts of drought or to detoxify harmful substances. Researchers have inserted genes responsible for the production of trehalose and proline, facilitated by enzymes such as trehalose 6-phospho synthase (TPS), trehalose phosphatase, and pyrroline-5-carboxylate synthetase (P5CS), into plants.

Drought-tolerant transgenic rice, wheat, soybean, and cotton have been developed. A gene named P5CS encoding the proline biosynthetic pathway has been introduced in tobacco. The plant show 8–10 fold higher proline content. The genes for the glycine betaine biosynthetic pathway have been introduced into plants. These enzymes convert choline into betaine aldehyde, which gets converted to glycine betaine. Transgenic tobacco-expressing genes for these enzymes [bet A encoding choline dehydrogenase (CDH) and bet B encoding betaine aldehyde dehydrogenase (BADH)] have been developed. Transgenic plants having genes encoding the polyamine biosynthesis enzymes, viz., arginine decarboxylase (ADC), ornithine decarboxylase (ODC), and SAMDC (S-adenosyl methionine decarboxylase) have been developed. Accumulation of polyamines confers tolerance in plants.

In plants, the ERA gene regulates the biosynthesis of ABA (abscisic acid). Through stomatal closure, the down regulation of a gene confers drought tolerance. Plants that carry this gene exhibit strong drought resistance. Corn and canola now contain the ERA gene.

Certain genes that respond to drought are activated in plants to make them drought-tolerant. Through the regulation of downstream stress-responsive genes, a number of regulatory genes, including transcription factors (TFs), are crucial to the regulation of various abiotic stress responses. The dehydration response element (DRE), a cis-acting promoter element, is crucial for controlling gene expression during drought stress. In many plants, including Arabidopsis, the DRE element is the cause of the ABA-independent induction of numerous genes in response to osmotic and cold stress. Through their interaction with a DRE/CRT cis-element present in the promoter region of various abiotic stress-responsive genes, DREBs (dehydration responsive element binding) are important plant transcription factors (TFs) that regulate the expression of many stress-inducible genes and play a critical role in improving the abiotic stress tolerance of plants. DRE-containing genes are activated by the transcription factors DRE1A and DRE2B, which bind to DRE specifically.

4.2.1.3. Resistance of Plants Against Chilling Stress

Innovative approaches made possible by biotechnology can be applied to create transgenic crop plants with increased resistance to cold stress. Numerous genes that respond to freezing stress have been identified and characterized. Plants that are susceptible to chilling stress have lipids that are high in saturated fatty acids. One of the 16 or 18 carbon saturated fatty acids found in lipids has double bonds, which provide resistance against damage at low temperatures. Glycerol-3-phosphate acyltransferases, or GPATs, are thought to be crucial for plants' capacity to tolerate cold temperatures. The GPAT gene from a cold-sensitive pumpkin was inserted into tobacco, resulting in an increase in the degree of fatty acid saturation in the membrane lipids of transgenic tobacco plants.

Plants express the genes that boost the expression of antioxidant enzymes like superoxide dismutase, which enhances resistance to abiotic stress. Plants have been GM to produce osmoprotectants like proline, glycine betaine, and trehalose. These substances shield organelles, aid in the stabilization of macromolecules, and help retain water in the intake. Furthermore, genes like osmotin and Rab proteins have also been inserted into plants. It is believed that cold stress tolerance is mediated by low-temperature inducible genes that have been isolated from plants. The C-repeat binding factor (CBF) genes are involved in signal transduction and adaptation to low temperatures. Some of them have transcription factors called C-repeat binding factor/dehydration-responsive element binding (CBF/DREB1) that control their expression.

4.2.1.4. Resistance of Plants Against Heat Stress

When heat stress occurs in plants, heat-shock proteins (Hsps) and antioxidant enzymes are crucial. In addition to safeguarding and mending damaged proteins and membranes, preserving photosynthesis, and controlling cellular redox state, Hsps are involved in the transduction of stress signals. Plants have been GM to express genes that encode heat shock proteins (HSPs), which provide resistance or tolerance against heat-related stressors. Different Hsps' expression is regulated at the transcriptional and translational levels. Heat-induced transcription has been shown to require the heat-shock element (HSE), a cis-acting DNA sequence.

Plants under HT stress usually accumulate more reactive oxygen species (ROS) in both chloroplasts and mitochondria, which can severely damage DNA and cause cell membrane lipid peroxidation (LPO). Thus, an increased capacity for scavenging and detoxifying the ROS is required to protect the plants against HT. Induction of thermotolerance is ascribed to the maintenance of better membrane thermostability and low levels of ROS accumulation due to improved antioxidant capacity. Thus genes encoding for increased expression of antioxidant enzymes have also expressed in plants. One well-known adaptive mechanism in plants is the accumulation of osmolytes, such as proline, glycine, betaine, and trehalose, which are essential for imparting heat tolerance. Plants with high thermotolerance have been developed through the induction of solute synthesis.

It is thought that membrane lipid saturation plays a significant role in HT tolerance. The HT increases membrane fluidity, which may cause the lipid bilayer to break apart. In addition to providing plants with heat resistance, the silencing of chloroplast-localized fatty acid desaturase has been shown to be beneficial. This method produced transgenic plants that are resistant to high temperatures.

4.2.1.5. Improving Crop Yield and Nutritional Value

Due to limited access to nutrient-dense food, malnutrition is a serious health concern that is particularly prevalent in underdeveloped and developing nations. One of the more successful approaches to solving this issue is the genetic engineering of staple crops. Numerous crops have been successfully modified to date for increased nutritional value and better yield. A technique used in agriculture to improve the nutritional value of crops is called biofortification. A well-known example is the golden rice, a GM variety of *Oryza sativa* used to biosynthesize beta-carotene. Two beta-carotene synthesis genes, phytoene synthase (*psy*) and lycopene β-cyclase (β-*lcy*), which are derived from *Narcissus pseudonarcissus*, were added to create golden rice. Together with a bacterial phytoene desaturase (crt*I*, from *Erwinia uredovora*), these genes were driven by the endosperm-specific glutelin promoter.

4.2.1.6 Pest and Insect Resistance GM plants

Genetically modified (GM) plants engineered for pest and insect resistance represent a significant advancement in agricultural biotechnology. These plants are designed to produce specific proteins that protect them from pest infestations, reducing the need for chemical pesticides and enhancing crop yields.

GM plants achieve pest and insect resistance through several mechanisms:

- **Bt Toxin Production**:
  - The most common method involves the insertion of genes from the bacterium *Bacillus thuringiensis* (Bt) into the plant's genome. Bt produces crystal proteins (Cry proteins) that are toxic to specific insect pests.
  - When insects ingest Bt proteins, the toxins bind to receptors in their gut lining, causing cell lysis and death. This method targets pests such as the European corn borer, cotton bollworm, and various lepidopteran insects.
- **Protease Inhibitors**:
  - Plants can be engineered to produce protease inhibitors, which interfere with the digestive processes of insect pests.
  - These inhibitors prevent pests from digesting proteins efficiently, leading to starvation and death.
- **RNA Interference (RNAi)**:
  - This technique involves the use of double-stranded RNA molecules to silence specific genes in pests, disrupting vital physiological processes.
  - RNAi can be tailored to target genes that are crucial for pest survival, reproduction, or development.

Benefits of Pest and Insect Resistance GM Plants

- **Reduced Pesticide Use**:
  - One of the primary advantages is the significant reduction in chemical pesticide applications. This leads to lower production costs, reduced environmental contamination, and decreased exposure of non-target organisms, including beneficial insects and humans, to harmful chemicals.
- **Increased Crop Yields**:
  - By protecting plants from pest damage, GM crops can achieve higher yields. This is particularly important in

regions where pest infestations are prevalent and can cause substantial crop losses.

- **Environmental Sustainability**:
  - The reduced need for chemical pesticides contributes to more sustainable agricultural practices. It helps in preserving soil health, reducing water pollution, and promoting biodiversity by protecting beneficial organisms.
- **Economic Benefits**:
  - Farmers can benefit economically from higher yields and lower costs associated with pesticide use. This can lead to increased profitability and improved livelihoods, especially for smallholder farmers in developing countries.
- **Improved Crop Quality**:
  - GM plants often exhibit improved quality traits, such as better grain integrity and reduced mycotoxin contamination, as healthier plants are less susceptible to fungal infections that can occur through pest damage.

Examples of Pest and Insect Resistant GM Plants

- **Bt Corn**:
  - Bt corn is engineered to produce Cry proteins from *Bacillus thuringiensis*, making it resistant to the European corn borer and other pests. Bt corn has been widely adopted in countries like the United States, Brazil, and South Africa, leading to significant reductions in chemical pesticide use.
- **Bt Cotton**:
  - Bt cotton produces Bt toxins that protect against the cotton bollworm and other lepidopteran pests. The adoption of Bt cotton has been particularly successful in India and China, resulting in increased yields and reduced pesticide applications.
- **Bt Brinjal (Eggplant)**:
  - Bt brinjal has been developed to resist the brinjal fruit and shoot borer, a major pest in South Asia. Bangladesh was the first country to approve and cultivate Bt brinjal, leading to significant reductions in pesticide use and higher crop yields.
- **RNAi Potatoes**:
  - Potatoes engineered using RNAi technology target genes essential for the survival of the Colorado potato beetle, a notorious pest. These GM potatoes provide effective pest control with minimal environmental impact.

4.2.1.7. Transgenic Plants as Bioreactors for Recombinant Proteins

The first human recombinant protein was produced in 1989 using plants as a biofactory. Mammalian expression systems yield low and costly product yields from recombinant proteins, whereas bacterial systems are unable to perform post-translational modification during the formation of complex proteins. This led to a change in production techniques to plant cell systems, which offer more affordable and superior alternative sources for the synthesis of recombinant proteins. Transgenic plants generate a variety of recombinant proteins, such as proteins, antibodies, metabolites or catabolites, and vaccines. Transgenic plants, including potatoes, bananas, algae, and tobacco, have been shown to produce antibodies and vaccines against cholera, malaria, and gastrointestinal tract diseases. Additionally, rice and wheat seeds have been successfully used to express an anticancer antibody that recognizes the cells of lung, breast, and colon cancer. Even so, commercializing plant-produced antibodies and vaccines has proven to be challenging. As of now, the United States Department of Agriculture has only approved one plant-produced vaccine for Newcastle disease, with multiple vaccines undergoing clinical trials.

4.2.2. Gene Constructs

A promoter region, a gene coding region, and a terminator/stop region make up a basic functional gene construct. Furthermore, depending on the purpose of the study, specific gene constructs might include unique sequences like enhancer, silencer, or reporter sequences. Transgene construction is always the first step in plant transformation. All transgene constructs share the same elements, apart from the presence of selectable markers and the desired gene. A suitable gene construct is essential to the creation of a successful transgenic line.

4.2.2.1. A Typical Plant Gene

Plants typically have two types of genes: structural and regulatory. Regulatory genes typically have their own promoter, enhancer, or silencer region and are found 5⊠ upstream of other genes. Contrarily, the sequence of structural genes starts with the catabolite activator protein (cap) site and continues with a terminator, exons, introns, leader sequence, and polyadenylation site (poly-A tail). These components are in charge of DNA transcription. After transcription, the pre-mRNA is converted to mature mRNA by RNA splicing, which removes the introns (a non-coding region). When ribosomal subunits attach to the promoter of this mature mRNA, translation is triggered, and the mRNA is transported to the cytoplasm. The ribosome then proceeds downstream to the subsequent codon, forming a peptide chain with the aid of tRNAs, and translation stops when it reaches the chain terminator (stop codon, TAA/TAG). Translation then starts at the start codon (ATG).

4.2.2.2. Promoters/Enhancers

Usually, a gene's 5⊠ upstream region is home to the promoter region. Promoters have a well-established role in controlling the expression of genes, much like an on/off switch. The promoter sequence in DNA transcription is identified by transcription factors. These transcription factors bind to the promoter's consensus region to enlist the RNA polymerase. DNA transcription starts with the formation of the RNA polymerase transcription complex.

Constitutive promoters, tissue-specific promoters, and inducible promoters are the three primary categories into which the promoters fall. Throughout the majority of the developmental stages, constitutive promoters are active and directly contribute to the maintenance of a moderate and steady level of gene expression. Gene expression can be limited to specific tissues or developmental stages by means of tissue-specific promoters. Environmental stimuli have a significant impact on the gene expressions linked to the inducible promoters, enabling the external regulation of gene expression. Selected promoters for plant transformation are displayed in Table 4.2.

Table 4.2. *Examples of Promoters Used in Plant Transformation*

Promoter	Source	Activity
CaMV35S	*Cauliflower mosaic virus*	Constitutive
Ubiquitin RUBQ1, RUBQ2 and rubi3	Rice	Constitutive
Ubiquitin Gmubi3	Soybean	Constitutive
SCR, SRK	*Brassica rapa*	Pollen and stigma specific
Exo70C2	*Arabidopsis*	Pollen- and root-specific
LMW Glu, HMW Glu-1D1	Wheat	Seed specific
Expansin PcExp2	Sour cherry	Ripened fruits
Potato class I patatin	Potato	Tuber/storage organ specific
NtHSP3A	Tobacco	Stress inducible

Source: Lee-Yoon, L. Transgenic Plants: Gene Constructs, Vector, and Transformation Method. doi: 10.5772/intechopen.79369.

Enhancers are short (50–1500 bp) sections of a gene that activator proteins can attach to and recognize. By binding to the enhancer and forming an enhancer-bound transcription factor complex, these proteins—also known as transcription factors—interact with the mediator complex (TFIID) and ultimately aid in the recruitment of RNA polymerase II. The enhancer becomes more accessible to the transcription proteins when the enhancer-bound transcription factor complex makes a loop toward the intervening sequence and makes contact with the promoter region. On the other hand, silencers serve as enhancers' exact opposite. Repressor transcription factors bind to silencing regions on their genomes. It is known that these repressors inhibit a gene's transcription. The intended regulation of gene expression informs the selection of an appropriate enhancer and promoter in plant genetic engineering. When a transgene causes mild toxicity to the target plant, gene expression remains at a basal level. Conversely, increased levels of gene expression make it easier to identify and track transgenes, which are typically underexpressed in nature.

4.2.2.3. Reporter Genes

Reporter genes are those that are linked to regulatory sequences or to genes of interest to enable the localization of expressed proteins and the detection of transgene expression. Reporter gene sequences encode proteins or protein products that have been catalyzed and are then detectable using tools or basic assays. Conversely, nontransgenic plant development is inhibited by selectable marker genes that confer resistance against specific chemical agents, such as herbicidal-resistant genes, antibiotic genes, and anti-metabolic genes. Green fluorescent protein (GFP), chloramphenicol acetyltransferase (CAT), beta-galactosidase (LacZ), luciferase (Luc), and beta-glucuronidase (GUS) are common reporter genes used to track plant transgene expression. The expression of the tagged protein can be regulated, transgene localization can be detected, and transformed cells can be distinguished from nontransformed cells thanks to these reporter genes. There are also dual reporter systems like Luc/Luc and GUS/Luc available for improved protein separation detection. Highly sensitive, stable, and dependable reporter genes should be able to withstand repeated measurements across a variety of cell types and tissues. Even though current reporter systems have been thoroughly studied, the perfect reporter genes with all the desired characteristics are still unobtainable. Every reporter system exhibits unique positive and negative characteristics. As a result, depending on the purpose of the study, careful thought must be given when selecting a suitable reporter gene.

4.2.2.4. Problem Posed by Antibiotic Resistance Reporter Genes

Although not yet flawless, current plant transformation methods are fairly effective. There isn't a method that can guarantee 100% transformation efficiency. Markers are required to differentiate between the transformed and nontransformed plant cells. The main markers used in transformant selection to effectively remove non-transformants are antibiotic-resistant genes or herbicides. Three factors determine how effective an antibiotic resistance system is: (i) the resistance gene must be expressed in transformed cells; (ii) the explant used for transformation must completely inhibit the growth of nontransformed cells; and (iii) the selective agent used must be effective. A selection of the antibiotics used in transgenic plant screening is displayed in Table 4.3.

Table 4.3. *Selective Antibiotics are Used for the Screening of Transgenic Plants*

Antibiotics	Mechanism of Action	General Working Concentration (μg/mL)	Selection
Kanamycin	Inhibiting ribosomal translocation and eliciting miscoding	50	*nptII*
Hygromycin B	Inhibit protein synthesis	20–200	*hph*
Streptomycin	Inhibit protein synthesis	100	*spt*
Spectinomycin	Inhibit protein synthesis	100	*aadA*
Phleomycin	DNA breakage	10	*ble*
Bleomycin	DNA breakage	10	*ble*

Source: Lee-Yoon, L. Transgenic Plants: Gene Constructs, Vector, and Transformation Method. doi: 10.5772/intechopen.79369.

The first successful transgenic plant identification has come from antibiotic screening. However, using antibiotics almost always results in problems with the environment and the safety of GM food. This is mostly because of the worry about gene pollution, which occurs when an antibiotic gene from a GM plant finds its way into the environment through microorganisms. It is well known that bacteria can take up and incorporate foreign DNA fragments into their genome. It's possible that the bacteria around the GM plant absorbed DNA fragments from the transgenic plant and, as a result, became resistant to the antibiotics. Furthermore, gene escape is a possibility because pollen can spread the antibiotic resistance gene to nearby plants. Additionally, eating the transgenic plant containing the antibiotic resistance gene could cause the genes to be transferred to the probacteria that live in the stomachs. For this reason, in transgenic whole plant screening, the antibiotic resistance marker genes are typically disregarded.

An attempt was made to swap out the antibiotic marker gene with a potential marker gene, like reporter genes. Reporter genes, like GFP, are widely used as biosensors and are not considered hazardous to the environment. It may be possible to stop the needless emergence of antibiotic resistance in the environment by engineering plants with these reporter genes. Conversely, the elimination of the transgenic plant's antibiotic resistance gene can also prevent gene escape. Gene markers from transgenic plants may be removed with the help of cutting-edge genome editing techniques like CRISPR/Cas9 and TALEN. These cutting-edge instruments are still in the shaping stages.

Remember

The development and use of GM crops raise ethical and societal considerations, including safety and socio-economic impacts.

4.2.3. Vectors for the Production of Transgenic Plants

The target gene is delivered to a target cell via a vector so that it can be expressed and replicated there. Three parts make up a common vector: a selectable marker, a multicloning or recombination site, and an origin of replication. The origin of replication is an AT-rich region on the vector that binds to a protein complex to start the replication process. This unwinds the vector, allowing polymerases to help replicate it. The multicloning site is a region that has several distinct sequences, also referred to as restriction sites, that can be broken by a particular restriction enzyme to enable the insertion of the desired gene. Site-specific recombination between two plasmids is made possible by the recombination site. Selectable markers are genetic markers that serve the purpose of validating the insertion of the vector into *Agrobacterium sp.*, as described in the gene construct section. Plant viral-based vectors and Ti plasmid-based vectors are frequently used in plant transformation.

4.2.3.1. Plasmid Vectors

1. **Ti Plasmid:** The most widely used vector for creating transgenic plants is the Ti plasmid. Depending on the Ti plasmid class, the estimated size of the plasmid ranges from 200 to 800 kbp. The virulence region, the opine catabolism region, and the transfer DNA (T-DNA) region comprise the three primary regions of the Ti plasmid. The T-DNA segment that is inserted into the plant's genome has a size of roughly 24 kbp. Repeat sequences on either end of this region are commonly referred to as the left and right borders. The crucial area required for the transfer of DNA that causes tumorigenesis is the right border. On the other hand, the virulence region is in charge of encoding the vir genes, which facilitate the transfer of T-DNA. The biosynthesis of opine and phytohormones (auxin and cytokinin) is also encoded by the T-DNA sequence. The three oncogenes found in T-DNA—the auxin biosynthesis gene, cytokinin, and opioid—are primarily responsible for plant tumor formation, which results in crown gall disease. The unchecked proliferation of plant cells is caused by the synthesized growth hormones, which exacerbate the condition by promoting the formation of crown galls.

   Opines are the main carbon source utilized by the *A. tumefaciens* which are not naturally synthesized from plant metabolism. Therefore, *A. tumefaciens* will develop its own biosynthetic machinery for the production of nutrients by genetically modifying the host cells. The opine catabolism region encodes the genes for proteins involved in opines catabolism. The origin of DNA replication allows stable maintenance of the Ti plasmid in the bacterium. For plant transformation, the Ti plasmid is usually disarmed, with the tumor-inducing genes removed and replaced with the reporter genes together with the gene of interest.

   With the addition of the desired genes and selectable markers, the already large Ti plasmid would only grow larger. Large plasmids have low copy numbers in nature and are difficult to handle. But in the end, this disadvantage prompted the creation of a co-integrative system that, when combined with the binary vector system, resolved the issue with large-sized plasmids.

2. **The Co-Integrative Vector:** It is developed through homologous recombination between an intermediate vector and disarmed Ti plasmid. The intermediate vector is normally the *E. coli* plasmid harboring the gene of interest. Both the intermediate vector and disarmed Ti vector consist of some common sequences, which allow the homologous recombination of the two plasmids to occur. The recombination will result in a large co-integrative vector containing the merged *E. coli* plasmid and disarmed Ti plasmid. This co-integrative vector will later be introduced back into the *Agrobacterium* for transgenic plant transformation. However,

the enormous size of the plasmid as a result of the recombination may prove an ominous challenge to be manipulated. Thus, the use of this vector had been discontinued since the binary vector system was introduced.

3. **The Binary Vector:** When scientists discovered that T-DNA could function on its own without attaching to the Ti plasmid, they created the binary vector system, a two-plasmid system. The helper vector and mini vector are the two plasmids used in the binary system. The term "mini vector" describes a smaller plasmid that contains both the replication origin of *E. coli* and T-DNA and *Agrobacterium Tumefaciens*, which permits the cloning of the plasmid in *E. coli* and *A. tumefaciens*. The wild-type Ti plasmid without the T-DNA region is referred to as the helper vector. Since the wild-type Ti plasmid offers the template for every gene required for gene transfer and integration, it is also referred to as a helper plasmid. The Agrobacterium is treated to both of these helper and mini vectors simultaneously, and the resultant transformed Agrobacterium will be utilized in plant transformation.

4.2.3.2. Plant Virus Vectors

Viruses are intracellular obligate parasites that require molecular machinery from a specific host to replicate. Viruses have not been found to infect plants through the use of transmission vectors such as aphids, insects, nematodes, and fungi. These viruses have been modified and are used as alternative sources for plant transformation; common plant viruses used in transgenic plant production include the *Cauliflower mosaic virus* (CaMV), *Tobacco mosaic virus* (TMV), *Alfafa mosaic virus* (AMV), *Potato virus X* (PVX), and *Cowpea mosaic virus* (CPMV). The wild-type plant viral vectors have been improved and modified to accommodate their use with *Agrobacteria* as well as the plant host for an increased efficiency level through two approaches. The first approach would be designing virus vectors that are similar to wild types carrying the gene of interest, which are capable of infecting plants. The creation of a "deconstruct" virus, which is accomplished by deleting undesirable viral genes—like the one that expresses coat protein, for example—and substituting them with useful genes, like reporter genes or antibiotic resistance genes, makes transgenic screening easier.

4.2.4. Transgenic Plants for Improved Quality Traits

Transgenic crops have been developed for increased yield and quality, such as biomass and harvest index. Transgenic crop plants have been used to modify the nutritional value, edible oil composition, and grain quality of cereals. There is tobacco that is transgenic and overexpresses PhyA. PhyA responds to far-red light by elongating its stems. It has been proposed that a high concentration of PhyA suppresses the shade avoidance response.

The process of creating transgenic plants with a high chlorophyll content involves overexpressing the gene coding for the bacterial hemoglobin molecule (VHb). High growth rates, altered metabolic pathways, and increased productivity are all displayed by these plants. Studies on these plants have shown an 80–100% increase in biomass production.

Transgenic crops for the production of nutraceuticals have been developed. Neutraceuticals are compounds isolated from foods that act as health supplements. Efforts have been made to increase the oil and protein content of oil crops such as soybean, palm, rapeseed, and sunflower.

Edible oils can have their ratio of unsaturated to saturated fatty acids changed through a transgenic technique. 'The fatty acid content of rapeseed rises when the gene encoding sn-2 acyltransferase is transferred. Crop plants with higher levels of glucosinolate have higher nutritional values. Following the transgenic approach, the stearic acid content in soybean and safflower has increased and the linoleic acid content in flax has decreased. The ratio of saturated to unsaturated fatty acids is changed by stearoyl-ACP desaturase (ACP) activity in seeds. The initial desaturation step in the biosynthesis of seed oil is catalyzed by the enzyme. Stearoyl-ACP is changed into oleoyl ACP by it. Brassica stearoyl-ACP desaturase (ACP) seed-specific antisense gene constructs have been used to create transgenic plants. Reduced desaturase activity leads to higher stearate levels in seeds, and polyunsaturated fatty acids (PUFAs) make up 30% of the seed oil. The nutritional and commercial value of the crop plant is increased by the transfer of genes that cause the PUFA content of rapeseed to rise. These crops have shown a 60 percent higher fatty acid content.

Keyword

Soybean (Glycine max) – It is a versatile legume that is widely cultivated for its edible beans and various industrial uses.

By altering the amino acid composition and seed starch quality, transgenic plants have been cultivated to produce higher-quality seeds. An attempt has been made to enhance the essential amino acid composition of grains, cereals, oilseeds, and legumes. Transgenic "Brassica" and soybean (**Glycine max**) rich in lysine have been created. By modifying the biosynthesis pathway that turns aspartic acid into lysine, threonine, and methionine, the lysine content of the seed grains could be increased. Increasing the amount of albumin protein in legumes and prolamins in cereals will yield methionine-rich seeds. In general, crops like potatoes, corn, and rice have high levels of amylopectin and amylase. Thus, plants that produce resistant starch have had their amylopectin-to-amylase ratio changed. Potatoes with a transgenic composition high in amylase have been created.

4.2.4.1. Golden Rice

Transgenic crops like golden rice and golden mustard rich in vitamin A have been developed. Golden rice has been developed by inserting the genes from Narcissus and Erwinia. The developed rice variety possesses a high content of β carotene and carotenoids, which are the precursors of vitamin A. The work was initiated by the International Rice Research Institute (IRRI). *A. tumefaciens*-mediated transformation of embryogenic rice calli with the introduction of the phytoene synthase (psy1) gene from Zea mays (Zmpsy 1) and the carotene desaturase I (crtI) gene from *Pantoea ananatis* resulted in GR2E rice. High levels of provitamin A (β-carotene) in rice endosperm get converted to vitamin A when metabolized by the human body. The Golden Rice Project was introduced in 1999 by Ingo Potrykus and Peter Beyer. Similarly, golden mustard was developed by introducing a gene that elevated the β carotene levels. Transgenic crops rich in iron content have also been developed. The gene targeted in these plants is ferritin, which provides iron for synthesizing iron proteins such as ferredoxin and cytochromes.

Figure 4.3. *Photograph showing golden rice (left) and normal rice (right).*

Source: https://www.researchgate.net/figure/Golden-Rice-right-and-normal-rice-The-yellow-color-is-the-consequence-of-the-presence_fig1_259434854.

The synthesis of ferritin is regulated at the transcriptional level. The animals can use ferritin as a rich source of iron if the content is increased in the cereal grain. For this, the genes for ferritin protein were isolated from the soybean, *Phaseolus,* and transferred into the rice. As a result, the plants produced seeds rich in ferritin. The experiments were conducted in Japan in 1999 and Europe in 2001. Later on, the International Rice Research Institute developed a rice variety rich in Fe and Zn by crossing these two varieties (Figure 4.3).

4.2.4.2. Flavr Savr Tomato

The tomato fruit ripening process involves the enzyme polygalacturonase (PG), whose expression is regulated by antisense RNA. This enzyme is widely distributed in ripe tomato fruit and has been implicated in the production of ethylene, which softens fruit skin. Pectin in cell walls is dissolved by polygalacturonase. They have developed tomatoes that express antisense RNA against PG. By introducing a reverse-orientation copy of the gene (also known as an "antisense" copy), the accumulation of PG in ripening tomatoes can be suppressed. When the PG antisense gene was inserted, the tomato produced less PG. PG could not be produced by the transgenic tomato.

The immediate precursor of ethylene is 1-aminocyclopropane 1 carboxylic acid (ACC), which is broken down by a gene called ACC deaminase. Fruit ripening is delayed in tomatoes that express the antisense ACC synthase or ACC oxidase gene (Figure 4.4). Flavr Savr tomatoes were identified by their expression of the ACC synthase gene. The fruits taste better because they contain higher concentrations of sugar. The Flavr Savr tomatoes keep well for a long time. Fruits like melon and tomatoes have been used to develop this technique.

Tomatoes with higher sucrose and lower starch content could be created by introducing the sucrose phosphate synthase or isopentenyl transferase (ipt) genes from *A. tumefaciens*.

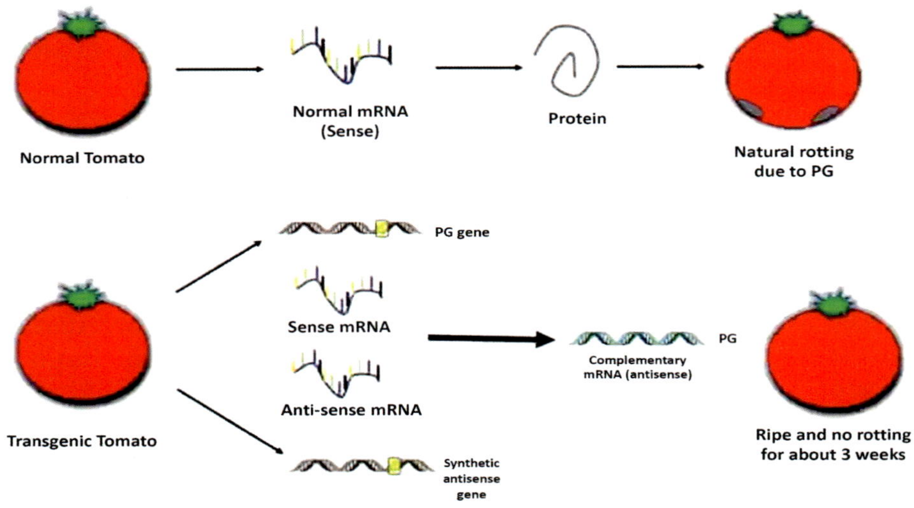

Figure 4.4. *Comparison between normal and Flavr Savr tomato.*

Source: https://sciwri.club/archives/8310.

Did you know?

The first GM crop, the Flavr Savr tomato, was commercialized in 1994. It was engineered for improved shelf life.

4.2.5. Use of Plant Biotechnology in Floriculture

Biotechnology has been used to create flowering plants with different flower colors. Anthocyanidins, a class of compounds, and their glycosylated derivatives, anthocyanins, are what give flowers their color. Whereas anthocyanins are red and blue, anthocyanidins are white and yellow in color. The genes responsible for the synthesis of anthocyanidins cause color changes in flowers. The intensity of the color of flowers has also been enhanced by transcriptional activators of a gene involved in the anthocyanin biosynthesis pathway. Gene editing has produced color-changing flowers in the rose, tulip, and carnation, which produce delphinidin (3'5' hydroxylated anthocyanins). Flowers turn blue when the gene encoding for delphinidins is expressed.

The' flavonoid genes in these plants have also been altered; in the case of petunia, chrysanthemum, gerbera, and roses, the chalcone synthase enzyme creates white, variegated flowers. Furthermore, the 'color of the flower was modified via the chalcone synthase gene. For instance, when this gene is expressed, Petunia flowers start to produce pink and white petals. Plants have been genetically engineered to produce more flavonoids by utilizing RNA antisense to block the flavonoid pathway at a particular step. For instance, Petunia has been shown to express the dihydroflavanol-4-reductase gene (DFR). The orange compound pelargonidin is produced when the gene is expressed.

The life period of flowers can be increased by roductiong genes aco and acFs encoding for enzymes ACC synthase and ACC oxidase involved in the biosynthesis of ethylene. The genes encoding enzymes such as ACC synthase and ACC oxidase have been cloned in many species, including carnation. As a result, the flowers produced exhibit longer shelf life.

Besides these, the genes for monoterpenes responsible for inducing fragrance in plants have also been introduced in other plants. It has also been done by manipulating gene S-linalool synthase coding for monoterpenes.

Did you get it?

1. Describe in brief how plant biotechnology is applied to develop plants resistant to abiotic and biotic stresses.
2. Discuss in brief the concept of developing transgenic plants for improved quality traits.

4.3. PLANT GENETIC TRANSFORMATION METHODS

→Learning Objectives

- Identify the PGT methods.

The natural process of gene transfer from one organism to another, known as genetic transformation, produces the diversity of biological traits. Transforming a gene of interest into a particular plant and achieving the desired expression has become a common strategy known as plant genetic transformation.

Plant genetic transformation is the process of incorporating genes into a plant's nuclear genome to produce the desired expression. To successfully implement gene transfer techniques for crop improvement, promoters are needed in addition to a selectable gene marker. Gene transfer requires a gene from one plant to another. Therefore, a non-coding DNA region upstream 'coding sequence of a gene that is necessary for the transcription of that gene is called a promoter. Furthermore, they are essential to the regulation and expression of plant genes.

Why Do We Use Plant Transformation Technology?

1. To Develop a Pest Resistant Variety:
    - It is now possible to introduce foreign genes that impart resistance to insects into the plant genome with the advent of genetic engineering techniques based on recombinant DNA technology.
    - Combined with plant tissue culture, recombinant DNA technology has helped to develop innovative methods for the economic management of various types of biotic stresses, including insect pests.
    - In reducing the losses incurred by biotic stresses, including insect pests,

such innovations will be of tremendous benefit.

- Transgenic plants that display insecticidal Bt proteins alone or in combination with proteins conferring herbicide resistance are revolutionizing agriculture.

2. For Enhancement of the Plant Variety Quality:
    - In improving plant productivity and improving the quality of plant products, plant transformations/transgenics have tremendous utility.
    - Improvements may be related to improving the nutritional value of the plant or to improving the functional properties of the production or consumption process.
    - In order to block the development of certain metabolites, transgenic plants can be used by regulating the over-expression or inhibition (antisense expression) of some of the essential enzymes as shown.

4.3.1. PGT Methods

Two approaches to plant genetic transformation are used, depending on the species of the plant. There are two types of methods: biological and physical.

4.3.1.1. Physical Gene Transfer Methods

1. Electroporation:
    - In this method, the electric pulse of high field strength is used to form pores in the cell membrane.
    - If DNA is present at an appropriate concentration in the buffer solution, it will be taken up via these pores.
    - Plant materials are incubated in a buffer solution that contains DNA and exposed to intense, high voltage electric pulse.

Advantages:

- It is possible to transform both intact cells and tissue.
- The transformation efficiency depends on the materials of the plant.

Disadvantages:

- DNA is obtained by 40 to 50 percent of incubated cells.
- Near to 50 percent of the cells transformed will survive.

2. Biolistic Gene Gun/Particle Bombardment:
    - It is also termed microprojectile bombardment.
    - For delivery into cells, foreign DNA is coated with high-velocity gold or tungsten particles.
    - This method is popularly used for its ability to transfer foreign DNA into mammalian cells and microorganisms.

Advantages:

- All plant species can be transformed using this process.
- The protocol for transformation is relatively simple.

Disadvantages:

- Difficulty in acquiring single-copy transgenic events.

- High costs for microcarriers and supplies.
- Intracellular target is irregular or random (cytoplasm, vacuole, nucleus, plasmid, etc.)
- Transfer DNA is not protected.

3. Microinjection Method:
    - A direct physical approach involving the mechanical introduction of the desirable DNA into a target cell is microinjection.
    - The microinjection technique involves transferring the gene into the cytoplasm or nucleus of a plant cell or protoplast through a micropipette.
    - The most important application of this is the introduction of DNA into animal oocytes and embryos, either in the study of transient expression or in the generation of transgenic animals.
    - The key disadvantages of microinjection are that it is long, costly, and needs to be carried out by trained and certified workers.

Advantages:

- This method does not need a protoplast.
- The device is easy and inexpensive.
- Methods may be helpful for the transfer of genes into cereals that do not easily regenerate from cultured cells.
- Technically straightforward.

Limitations:

- Less specific
- Less efficient
- The transformation frequency is low.

4. Liposome-Mediated Transformation:
    - Liposome-mediated transformation includes liposome adhesion to the surface of the protoplast, its fusion at the attachment site, and the release of plasmids inside the cell.
    - Liposomes are lipid spheres that are used to bring molecules into cells.
    - These are artificial vesicles that can behave as delivery vehicles for exogenous materials, including transgenes.
    - They are regarded as spheres of lipid bilayers encircling the molecule to be transported and encourage transport after fusing with the cell membrane.
    - Cationic lipids are those with a positive charge that are used for nucleic acid transfer.
    - Liposomes can interact more readily with the negatively charged cell membrane than uncharged liposomes.
    - Fusion between the cationic liposome and the cell surface results in the transmission of DNA directly through the plasma membrane.

Advantages:

- High reproducibility degree.
- Stability for the long term.
- Low toxicity level.
- Nucleic acid protection from degradation

4.3.1.2. Chemical Mediated Gene Transfer Methods

1. Polyethylene Glycol (PEG)-Mediated Gene Transfer:
   - PEG disrupts the plasma membrane of protoplasts in the presence of divalent cations (using Ca2+) and makes it permeable to naked DNA.
   - PEG disrupts the plasma membrane of protoplasts in the presence of divalent cations (using Ca2+) and makes it permeable to naked DNA.
   - The protoplastic culture is taken into a tube and 40 percent PEG 4000 (w/v) dissolved in mannitol is added to this tube and calcium nitrate is added slowly and incubated for a few minutes.

Keyword

Cell Surface – It is the locus for many important biochemical functions of cells and for the interactions of cells with one another and with their environment.

Advantages:

- A large number of protoplasts can be transformed concurrently.
- A wide variety of plant species can be used successfully.

Limitations:

- The DNA is prone to rearrangement and degradation.
- Random integration of foreign DNA into the genome may lead to unfavorable characteristics.
- A challenging task is the regeneration of plants from converted protoplasts.

2. Calcium Phosphate Co-Precipitation:
   - The DNA is permitted to mix with calcium chloride solution and isotonic phosphate buffer to form a DNA-calcium phosphate precipitate.
   - The cells are transformed when the actively dividing cells in the culture are exposed to this precipitate for several hours.
   - The efficacy of this technique depends on the high DNA concentration and the protection of the complex precipitate.
   - The addition of dimethyl sulfoxide (DMSO) enhances transformation efficiency.
3. DEAE-Dextran Mediated Transfer:
   - With a high molecular weight polymer diethyl amino ethyl(DEAE)dextran, the desirable DNA can be complexed and transferred.
   - This approach's main drawback is that it does not generate stable transformants.

4.3.1.3. Biological Method of Gene Transfer

1. ***Agrobacterium*-Mediated Gene Transfer in Plants:**
   - *A. tumefaciens* is a widely occurring soil bacterium that is responsible for causing crown gall, and has the capability to integrate new genetic material into the plant cell.
   - T DNA is a term given for the genetic material that is situated on a Ti plasmid.
   - A Ti plasmid is a common circular fragment of DNA found in almost all bacteria.
   - Agrobacterium-mediated transformation is highly efficient and hence is the most commonly used method for plant genetic engineering.
   - In the course of transformation, various components of Ti plasmid play a role in the effective transfer of the gene of interest into the plant cells. They are:
   - T-DNA border sequences that are responsible for demarcating the T-DNA to be transferred to the plant genome.
   - The vir genes that are necessary for the transfer of T-DNA region to the plant but they cannot be transferred by themselves.
   - Modified T-DNA region where the genes that are responsible for the formation of crown gall are eliminated and replaced with genes of interest.
2. **Steps for *Agrobacterium*-Mediated Plant Transformation Process:**
   - The *Agrobacterium*-mediated transformation process comprises several steps. They are:
   - Isolation of the gene of interest from the source organism.
   - Development of a functional transgenic hybrid involving the gene of interest, promoters to drive expression, codon modification, if required to enhance successful protein production, and marker genes to favor tracking of the introduced genes in the host plant.
   - Insertion of the transgene into the Ti-plasmid.
   - Integration of the T-DNA-containing plasmid into *Agrobacterium*.
   - The transformed *Agrobacterium* is mixed with plant cells to permit the transfer of T-DNA into plant chromosomes.
   - Regeneration of the new plantlets that are GM.
   - Examination for characteristic performance or expression of transgene at lab, greenhouse, and field level.

Did you get it?

1. Why do we use plant transformation technology? Explain it?
2. Explain the genetic transformation methods.

FOCUS ON CAREERS

- **Plant Geneticist:**

  A plant geneticist is a scientist who studies plant DNA to improve the 'characteristics of the plant for agricultural and research purposes. Plant geneticists often focus on a particular specialty, such as a plant type, purpose, or industrial role. They are usually interested in the ways plants develop and reproduce and they often follow best practices in scientific methodology. Plant geneticists can work in a wide variety of settings, including colleges and universities, government agencies, and private businesses and corporations.

- **What Does a Plant Geneticist Do?**

  Plant geneticists study the features of plants and crops and use their understanding of DNA to help create new characteristics and varieties of plants. They might focus on a specific type of plant or industry, depending on the context of their specific role. They often work on plant features such as size, drought, and pest resistance, nutritional value, and appearance. A plant geneticist often works on many stages of the research process, from planning their research methodology to implementing studies and evaluating their results. Professionals in this role also frequently collaborate with other scientists to help achieve their research objectives.

- **Plant Geneticist Skills:**

  The skills you will need to begin a career and succeed as a plant geneticist include transferable career skills as well as specialized scientific skills that are unique to this role. Here are some skills you might need in your plant geneticist career:

  - Agricultural expertise;
  - Computer science knowledge;
  - Scientific methodology;
  - Attention to detail;
  - Public speaking;
  - Clear written communication;
  - Leadership;
  - Team management;
  - Mathematical ability;
  - Critical thinking.

- **Responsibilities of Plant Geneticist:**

  - Discovering what genes are responsible for certain traits in plants and crops (e.g., size, taste, nutritional value, resistance to diseases).

- Collecting a wide range of data from plant trials (tests on plants conducted to determine their level of health).
- Finding how to improve crops and plants for a future where it may be increasingly difficult to grow crops in natural environments.
- Writing reports for academic journals and keeping tabs on what research is being done in your field already.
- Teaching plant genetics or related fields in universities to students at different levels of higher education.
- Making estimates on the financial costs of research and figuring out ways to fund it.
- Cross-growing plants for further tests.
- Keeping note of all research completed and results of tests.

SUMMARY

- Plant biotechnology is a field of scientific inquiry and application that focuses on harnessing the unique characteristics of plants to address various challenges in agriculture, medicine, and environmental sustainability. Plant biotechnology utilizes molecular and cellular techniques to manipulate the genetic makeup of plants, aiming to enhance their desirable traits or confer new capabilities.
- Plant tissue culture (PTC) refers to the cultivation of undifferentiated plant cells, tissues or organs on synthetic media under aseptic environment and suitable controlled physical conditions. It is an important tool for both basic research as well as commercial applications.
- In plant cell culture, plant tissues and organs are grown *in vitro* on artificial media under an aseptic and controlled environment. The technique depends mainly on the concept of totipotentiality of plant cells which refers to the ability of a single cell to express the full genome by cell division.
- PGR's play an essential role in determining the development pathway of plant cells and tissues in a culture medium. The auxins, cytokinins, and gibberellins are the most commonly used PGRs.
- Biotechnology has been introduced into agricultural practice at a rate without precedent. Tissue culture allows the production and propagation of genetically homogeneous, disease-free plant material. Cell and tissue *in vitro* culture is a useful tool for the induction of somaclonal variation.
- Embryo culture is a type of plant tissue culture that is used to grow embryos from seeds and ovules in a nutrient medium. In embryo culture, the plant develops directly from the embryo or indirectly through the formation of callus and then subsequent formation of shoots and roots.
- Genetic transformation is the most recent aspect of plant cell and tissue culture that provides the means of transfer of genes with desirable traits into host plants and recovery of transgenic plants.
- Plant cell and tissue cultures hold great promise for the controlled production of a myriad of useful secondary metabolites. Plant cell cultures combine the merits of whole-plant systems with those of microbial and animal cell cultures for the production of valuable therapeutic secondary metabolites.
- Transgenic plants are plants that have had their genomes modified through genetic engineering techniques, either by the addition of a foreign gene or the removal of a certain detrimental gene.
- Plants exposed to drought stress activate genes coding for the production of certain compounds that protect the plant from the harmful effects of drought or detoxification enzymes. The genes that encode for trehalose and proline biosynthesis via enzymes trehalose 6-phospho synthase (TPS), trehalose phosphatase, pyrroline-5-carboxylate synthetase (P5CS) have been introduced in plants.

- A typical plant gene consists of regulatory and structural genes. Regulatory genes are usually located at the 5′ upstream of a gene, with its own promoter, enhancer, or silencer region.
- The genes that increase the synthesis of flavonoids by blocking the flavonoid pathway at a specific step (using RNA antisense) have been introduced into plants. For example, another gene named the dihydroflavanol-4-reductase gene (DFR) has been expressed in Petunia. The expression of the gene produces an orange compound, pelargonidin.
- Genetic transformation is a natural process in which a gene is transferred from one organism to another, resulting in biological trait diversity. Plant genetic transformation has become a popular strategy for transforming a gene of interest into a specific plant and obtaining the desired expression.

REVIEW QUESTIONS

1. What are the general steps of plant tissue culture?
2. Describe various techniques of plant tissue culture.
3. Write a note on golden rice.
4. What is Flavr Savor tomato? How is it developed?
5. How has plant biotechnology been applied in floriculture?
6. Describe the various applications of transgenic plants.
7. Explain the vectors for the production of transgenic plants.

MULTIPLE CHOICE QUESTIONS

1. Plant biotechnology involves:

a. production of valuable products in plants
b. rapid clonal multiplication of desired genotypes
c. production of virus-free plants
d. all of these

2. All are plant-derived alkaloids except

a. menthol
b. nicotine
c. quinine
d. codeine

3. The modification of exogenous compounds by plant cells is called

a. Biotransformation
b. bioconversion

c. both a and b
d. biophytomodification

4. **Which technology is commonly used for introducing foreign genes into plants?**
   a. CRISPR-Cas9
   b. Restriction enzymes
   c. Agrobacterium-mediated transformation
   d. Polymerase Chain Reaction (PCR)
5. **The first commercially available genetically modified crop was:**
   a. Bt Cotton
   b. Roundup Ready Soybeans
   c. Flavr Savr Tomato
   d. Golden Rice
6. **What is the role of Bt in Bt crops?**
   a. Enhancing color and flavor
   b. Improving yield
   c. Providing resistance to pests
   d. Increasing nutritional content
7. **Which enzyme is commonly associated with the CRISPR-Cas9 system?**
   a. Ligase
   b. Helicase
   c. Restriction enzyme
   d. Cas9
8. **Which regulatory framework assesses the safety of genetically modified organisms (GMOs) before their release?**
   a. FDA (Food and Drug Administration)
   b. EPA (Environmental Protection Agency)
   c. USDA (United States Department of Agriculture)
   d. GMOFA (Genetically Modified Organism Regulatory Authority)
9. **In plant biotechnology, what is the purpose of RNA interference (RNAi)?**
   a. Enhancing protein synthesis
   b. Silencing specific genes
   c. Inducing mutations
   d. Accelerating photosynthesis
10. **What is a potential application of synthetic biology in plant biotechnology?**
    a. Enhancing plant growth using chemical fertilizers
    b. Creating artificial plants

c. Controlling plant diseases with traditional methods

d. Engineering plants with novel traits

Answer to Multiple Choice Questions

1. (d); 2. (a); 3. (c); 4. (c); 5. (c); 6. (c); 7. (d); 8. (a); 9. (b); 10. (d)

REFERENCES

1. Atkinson, N. J., & Urwin, P. E., (2012). The interaction of plant biotic and abiotic stresses: From genes to the field. *Journal of Experimental Botany, 63*(10), 3523–3543.
2. Galun, E., (2007). *Plant Patterning: Structural and Molecular Genetic Aspects* (p. 333). World Scientific Publishing Company.=. ISBN 9789812704085.
3. Jones, H. D., & Sparks, C. A., (2009). Stable transformation of plants. *Methods Mol. Biol., 513*, 111–130.
4. Kauth, P. J., Vendrame, W. A., & Kane, M. E., (2006). In vitro seed culture and seedling development of *Calopogon tuberosus*. *Plant Cell, Tissue, and Organ Culture*, *85*(1), 91–102.
5. Keshavareddy, G., Kumar, A. R. V., & Ramu, S. V., (2018). Methods of plant transformation: A review. *International Journal of Current Microbiology and Applied Sciences, 7*(7), 2656–2668.
6. Kikkert, J. R., Vidal, J. R., & Reisch, B. I., (2005). Stable transformation of plant cells by particle bombardment/biolistics. *Methods Mol. Biol., 286*, 61–78.
7. Krenek, P., Samajova, O., Luptovciak, I., Doskocilova, A., Komis, G., & Samaj, J., (2015). Transient plant transformation mediated by *Agrobacterium tumefaciens*: Principles, methods, and applications. *Biotechnol. Adv., 33*(6 Pt 2), 1024–1042.
8. Leibfried, A., To, J. P., Busch, W., Stehling, S., Kehle, A., Demar, M., & Lohmann, J. U., (2005). WUSCHEL controls meristem function by direct regulation of cytokinin-inducible response regulators. *Nature*, *438*(7071), 1172–1175. doi: 10.1038/nature04270.
9. Umezawa, T., Fujita, M., Fujita, Y., Yamaguchi-Shinozaki, K., & Shinozaki, K., (2006). Engineering drought tolerance in plants: Discovering and tailoring genes to unlock the future. *Current Opinion in Biotechnology, 17*(2), 113–122.

CHAPTER 5

Animal Biotechnology

LEARNING OBJECTIVES

After studying this chapter, you will be able to:

- Explore the fundamentals of animal tissue culture;
- Understand the animal-cell culture media;
- Explain the transgenic animals.

INTRODUCTORY EXAMPLE

In Vitro Fertilization (IVF) in Mammals:

In vitro fertilization (IVF) is a reproductive technology extensively employed in mammals to address fertility issues and facilitate conception. The process typically begins with ovarian stimulation in female mammals, where hormones such as follicle stimulating hormone and leutinizing hormone are administered to encourage the development of multiple eggs within the ovaries. This stimulation aims to enhance the chances of successful fertilization.

Following ovarian stimulation, the matured eggs are retrieved from the ovaries of a female through a minor surgical procedure known as follicular aspiration or egg retrieval. Simultaneously, sperm is collected from the male partner or a donor. Various methods, including testicular sperm extraction or epididymal sperm aspiration, may be employed for sperm retrieval. The next crucial step involves the actual fertilization of the eggs. This can be achieved through conventional insemination, where sperm is introduced to the eggs in a laboratory dish, or through a more advanced technique called intracytoplasmic sperm injection, where a single sperm is directly injected into an egg.

Once fertilization occurs, the resulting embryos undergo a period of incubation and development in a carefully controlled laboratory environment. This culture period allows for the selection of the healthiest embryos for transfer back into the female's uterus. Embryo transfer typically occurs 2 to 5 days after fertilization, and the number of embryos transferred is often a consideration based on factors such as the female's age and health.

The success of IVF is contingent upon the implantation of the transferred embryo in the uterus. Pregnancy is monitored through regular blood tests to measure hormone levels, providing insights into the progression of the pregnancy. Additionally, ultrasound examinations may be employed to visualize and monitor the developing fetus.

It is noteworthy that while the basic principles of IVF are shared across mammalian species, specific procedures and techniques may vary, influenced by the unique reproductive characteristics of each species. Ethical considerations, legal regulations, and the application of IVF for research purposes further contribute to the dynamic landscape of assisted reproductive technologies in mammals.

UNIT INTRODUCTION

Animal biotechnology is the combination of science and engineering which is very useful in transforming living organisms. The main objective of animal biotechnology is to make products, to improve animals, and to develop microorganisms for specific agricultural uses. Creating transgenic animals is the best example of animal biotechnology.

In the agricultural sector, animal biotechnology is employed to enhance livestock through selective breeding and genetic modification. This involves introducing specific genes that confer desirable traits, such as disease resistance, improved growth rates, and increased production of valuable proteins like milk or wool. By using transgenic animals, scientists aim to create livestock with enhanced traits that can contribute to a more sustainable and efficient agricultural industry.

In the field of medicine, animal biotechnology is instrumental in producing therapeutic proteins and other biopharmaceuticals. Transgenic animals are genetically modified (GM) to produce human proteins in their milk, blood, or other tissues, providing a cost-effective and scalable method for large-scale production of medicines. Additionally, animal biotechnology contributes to the development of animal models for studying human diseases, enabling researchers to better understand disease mechanisms and test potential treatments.

Conservation efforts also benefit from animal biotechnology, particularly in the preservation of endangered species. Scientists use genetic techniques to assist in breeding programs, ensuring the genetic diversity and survival of rare and endangered animals. Moreover, transgenic animals can play a role in ecological restoration by introducing genes that enhance adaptability to changing environments or by serving as biological sensors for environmental monitoring.

Ethical considerations play a significant role in the development and application of animal biotechnology. Researchers must carefully navigate issues related to animal welfare, genetic modification, and environmental impact. Striking a balance between scientific advancements and ethical responsibility is crucial to ensuring the responsible and sustainable application of animal biotechnology.

5.1. ANIMAL TISSUE CULTURE

→Learning Objectives

- Understand the development of animal cell culture;
- Grasp the concept of cell culture;
- Explain the characteristics of cell cultures;
- Learn about the reconstruction of three-dimensional structures;
- Explain the types of cell culture;
- Explore the concept of a cell line;
- Understand the growth cycle;
- Define cell viability.

Human, animal, or insect cells are cultivated in a conducive artificial environment through a process known as cell culture. The cells could come from strains of established cells, multicellular eukaryotes, or preexisting cell lines. Although the idea of keeping live cell lines apart from their original tissue source was discovered in the 19th century, animal cell culture became a widely used laboratory technique in the mid-1900s. One of the main instruments employed in the life sciences today for research with commercialization and economic value potential is animal cell culture. Science has advanced our understanding of cell growth and differentiation, growth factor identification, and the mechanisms underlying the normal functions of different cell types thanks to the development of basic culture media, which have made it possible for scientists to work with a wide variety of cells under controlled conditions. Additionally, high cell density bioreactor and culture conditions have been studied using new technologies.

Mass culture of animal cell lines is a fundamental requirement for many biotechnology products (e.g., vaccines against viruses). More complex proteins that are glycosylated, or modified with carbohydrates, currently need to be produced in animal cells, even though many simpler proteins are being produced using rDNA in bacterial cultures. The current focus of cell culture research is on determining how culture conditions affect recombinant protein biological activity by measuring viability, productivity, and stability of post-translational modifications like glycosylation. Anticancer drugs, enzymes, immunobiologicals (interleukins, lymphokines, monoclonal antibodies (mABs), and hormones are among the biologicals made in animal cell cultures using recombinant DNA (rDNA) technology.

5.1.1. Development of Animal Cell Culture

Mammalian cell cultures were first established in the early 1900s. The cultures were initially established to investigate normal physiological processes like nerve development as well as the evolution of cell cultures. The first nerve fiber growth in vitro was demonstrated by Ross Harrison in 1907. On the other hand, animal cell culture was first carried out commercially in the 1950s. Large-scale cell cultures were required as a result of the significant polio epidemics of the 1940s and 1950s and the ensuing need for viral vaccines. Table 5.1 lists the polio vaccine, which was created from a deactivated virus, as one of the first manufactured goods derived from animal cells in culture.

Table 5.1. *Milestones in Cell Cultures and Microfluidics*

1878	Claude Bernard	Established that a physiological state of the cell similar to the live cell can be maintained even after the death of the organism.
1907	Harrison	Cell entrapment and frog embryo nerve fiber growth in vitro.
1912	Alexis Carriel	Initiated tissue culture of chick embryo heart cells using embryo extracts as cultural media passaged for a reported period of 34 years.
1913	Steinhardt, Israeli, and Lambert	Grew vaccinia virus in fragments of guinea pig corneal tissue.
1916	Rous and Jone	Used trypsin to suspend attached cells in culture.
1927	Carrel and Rivera	First viral vaccine against chicken pox.
1948	Sanford, Earle, and Likely	Single-cell culture using microscale glass capillaries
1949	Enders, Weller, and Robbins	Poliovirus grown on human embryonic cells in culture.
1952	Gey	Establishment of a continuous cell line from a human cervical carcinoma (HeLa cells).
1955	Eagle	Established nutrient requirements of cells in culture and defined culture media for growth.
1956	Little Field	HAT (hypoxanthin, aminopterin, thymidine) medium introduced for cell selection.
1961	Hayflick and Moorhead	Studied human fibroblasts (WI-38) and showed a finite lifespan in culture.

1965	Ham	First serum-free HAMS's media.
1975	Kohler and Milstein	First hybridoma secreting a monoclonal antibody.
1977	Genetech	First recombinant human protein: somatostatin.
1979	Terry, Jerman, and Angell	Gas chromatography system using silicon-etched microchannels
1985	Collen	Recombinant tissue plasminogen activator (tPA) in mammalian cells. Human growth hormone produced from recombinant bacteria was accepted for therapeutic use.
1986	FDA approval	First monoclonal antibody was approved by the FDA for use in humans (Orthoclone OKT3).
1986	Genetech	First recombinant protein (interferon alpha-2a) was commercialized.
1989	Amgen	Erythropoietin (EPO) recombinant protein produced in CHO cells was available commercially.
1992	Manz, Harrison, Verpoorte, Fettinger, Paulus, Ludi, and Widmer	Micro-machined glass chip electrophoresis for separating molecules
1992	FDA approval	First genetically engineered recombinant blood-clotting factor used in the treatment of hemophilia A.
1996	Wilmut	Production of transgenic sheep (Dolly) through nuclear transfer technique.
1997	Hadd, Raymond, Halliwell, Jacobson, and Ramsey	Microchip device for enzyme assay
1998	Duffy, McDonald, Schueller, and Whitesides	Creation of microfluidic systems using PDMS (polydimethylsiloxane)
2000	Panaro, Yuen, Sakazume, Fortina, Kricka, Wilding	Lab-on-chip system (bioanalyzer)
2002	Cloneaid	Claimed to produce a cloned human baby named EVE.
2003	Zheng	Protein crystallization chip

2004	FDA approval	First antiangiogenic monoclonal antibody that inhibits the growth of blood vessels or angiogenesis (for cancer therapy).
2005	Sanford, Earle, and Likely	Single-cell culture in open microfluidic systems
2005	Birch	Reported antibody titers at an industrial scale of 5 g/L and more.
2009	Nathalie Cartier-Lacave	Combined gene therapy with blood stem cell therapy, which may be a useful tool for treating fatal brain disease.
2011	Melanie Welham, David Tosh	1 M molecule treatment causes stem cells to turn into precursors of liver cells.
2012	Willison and Klug	Single-cell and single-molecule proteomics utilizing nanospace technology
2012	Maria Blasco	First gene therapy successful against aging-associated decline in mice.

Source: Anju, V., Megha, V., & Anchal, S. Animal Tissue Culture Principles and Applications. doi: 10.1016/B978-0-12-811710-1.00012-4.

5.1.2. Concept of Cell Culture

Tissue culture is the in vitro culture and propagation of single cells, tissues, or entire organs in a suitable lab setting. When supplemented with a medium containing nutrients and growth factors, many animal cells can be made to grow outside of their original organ or tissue under specific circumstances. The culture conditions may not accurately reflect in vivo conditions in terms of temperature, pH, CO_2, O_2, osmolality, and nutrition when it comes to in vitro cell growth. Furthermore, sophisticated incubation conditions, a consistent supply of nutrients for growth, and sterility are required for the cultured cells. The medium itself plays a significant role in determining how cells grow in culture. Animal cells are currently cultivated in either natural or artificial media, depending on the requirements of the experiment. The most crucial and crucial stage in the culture of animal tissue is the selection of the culture medium. Depending on the kind of cells that must be cultivated for cell growth differentiation or the manufacture of specific pharmaceutical products, this will vary. Furthermore, there are now serum-containing and serum-free media that provide the cell culture with different levels of benefit. The growth of cell lines requires sterile conditions.

Nowadays, a vast variety of tissues and organisms' cells are cultivated in laboratories. In the past, studying growth, growth requirements, the cell cycle, and the cell itself was the main goal of cell culture. Currently, homogenous cultures made from primary cell cultures are helpful resources for researching the biology and origins of the cells. Artificial tissues can be produced with the help of organotypic and histotypic cultures, which replicate the corresponding organs and tissues.

5.1.3. Characteristics of Cell Cultures

Animal cell cultures are distinct from microbial cultures in that they exhibit particular traits. Animal cells are characterized by their slow growth rate, need for solid substrata to support anchorage, fragility due to the absence of a cell wall, and sensitivity to various physiochemical conditions like pH and CO_2 levels. The following are some of the basic bioprocess variables.

5.1.3.1. Temperature

Since temperature directly affects growth and production processes, it is one of the most fundamental variables. Incubators with thermostats can be used to regulate temperature on a small scale. Large-scale cell cultures grown in bioreactors, however, necessitate more delicate temperature control. The techniques used by various bioreactors to keep the cell culture at a certain temperature vary. A heat blanket and water jacket equipped with a temperature sensor regulate the temperature inside a bioreactor.

5.1.3.2. pH

You can adjust the pH of the culture medium by adding an acidic (HCl) or alkali (NaOH, KOH) solution. The culture's pH can be kept constant by adding CO_2 gas to the bioreactor, using naturally occurring buffering solutes, or buffering with sodium bicarbonate. The most often used electrode in the bioreactor is an electrochemical pH electrode made of silver chloride.

5.1.3.3. Oxygen

The most basic element that must be consistently added to the cell culture medium is dissolved oxygen. In aerobic cultures, it is eaten along with a carbon source. One way to introduce dissolved oxygen to the medium is by diffusion through a liquid surface or membranes.

Did you know?

The discovery and use of antibiotics, such as penicillin and streptomycin, in the 1940s and 1950s helped control contamination and improve the longevity of cell cultures.

5.1.3.4. How Are Cell Cultures Obtained?

Cell cultures are obtained through a series of steps that involve isolating cells from their original tissue and providing them with an environment conducive to growth and proliferation. The process typically starts with the collection of a tissue sample from an organism. This sample can be derived from various sources such as organs, blood, or tumors.

- **Tissue Collection and Dissection**: The tissue is first dissected into smaller pieces to increase the surface area and facilitate enzyme access.
- **Enzymatic Digestion**: Enzymes such as trypsin or collagenase are used to dissociate the tissue into individual cells by breaking down the extracellular matrix that holds the cells together.
- **Cell Suspension**: The resulting cell suspension is then filtered to remove any remaining tissue fragments and centrifuged to concentrate the cells.
- **Seeding and Culturing**: The isolated cells are seeded into culture vessels containing a nutrient-rich medium, which supplies the necessary growth factors, vitamins, and minerals. The culture environment is typically

controlled for temperature, pH, and gas composition (CO2 and O2 levels).

- **Incubation and Maintenance**: The cells are incubated under optimal conditions to promote cell attachment, growth, and division. The culture is regularly monitored and the medium is replenished as needed to maintain cell health and viability.

Through these steps, primary cell cultures can be established, which can then be sub-cultured or immortalized for continuous cell line generation.

5.1.3.4. Cell Passage and Use of Trypsin

The technique of subculturing cells to generate a large number of new cells from pre-existing ones is called passage. Subculturing prevents the senescence linked to extended high cell density and yields a more homogenous cell line. Transferring a small number of cells into each new vessel is the process of splitting cells. Cells can be propagated, described, and stored following subculturing. Proteins must be used to separate adherent cell cultures from the tissue culture flask or dish surface. Cells and the surface are joined tightly by proteins that the cells secrete. Trypsin and EDTA combined are used to break down proteins at particular locations. Trypsin is a proteolytic or protein-degrading enzyme that breaks down peptide bonds to hydrolyze peptides that have been digested by pepsin. Trypsin's effectiveness is increased when EDTA sequesters specific metal ions that have the ability to inhibit trypsin activity. Figure 5.1 illustrates the trypsinization process and the method for removing adherent cells.

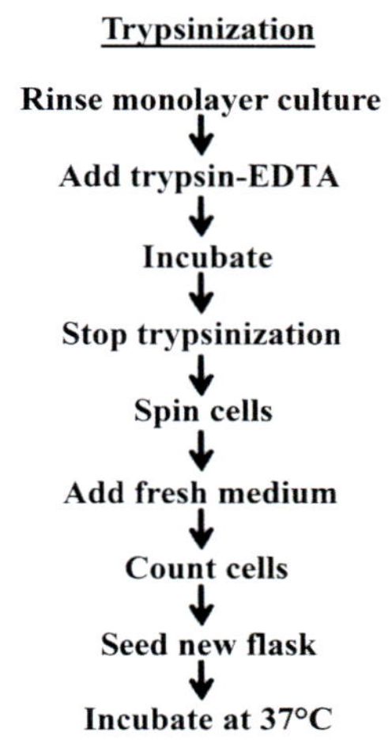

Figure 5.1. *Trypsinization of adherent cells.*

Source: Anju, V., Megha, V., & Anchal, S. Animal Tissue Culture Principles and Applications. doi: 10.1016/B978-0-12-811710-1.00012-4.

5.1.4. Reconstruction of Three-Dimensional Structures

Reconstruction of three-dimensional (3D) structures in cell culture involves cultivating cells in a manner that mimics the natural architecture and environment of tissues in the body. Unlike traditional two-dimensional (2D) cultures where cells grow in a flat monolayer, 3D cultures provide a more physiologically relevant context, enhancing cell-cell and cell-matrix interactions.

- **Scaffold-Based Techniques**: Cells are seeded onto biocompatible scaffolds made from materials like collagen, gelatin, or synthetic polymers. These scaffolds provide a supportive framework that allows cells to adhere, proliferate, and form complex structures similar to those found in vivo.
- **Spheroid and Organoid Cultures**: Cells are allowed to self-assemble

into spheroids or organoids. Spheroids are simple, spherical clusters of cells, while organoids are miniaturized, functional versions of organs, often derived from stem cells. These structures can replicate key features of their respective tissues, including cellular diversity and functionality.

- **Bioprinting**: Advanced techniques like bioprinting involve the precise layering of cells and biomaterials to create intricate 3D structures. This method uses computer-aided design to fabricate tissues and organ models with high accuracy.
- **Hydrogel Encapsulation**: Cells are encapsulated in hydrogels, which provide a hydrated environment resembling the extracellular matrix. Hydrogels support nutrient diffusion and waste removal, promoting cellular activities within the 3D structure.

Employing these techniques, researchers can create more accurate models of human tissues, improving the study of disease mechanisms, drug testing, and regenerative medicine applications. There are two main categories into which these cultures can be separated.

5.1.4.1. Histotypic Culture

By growing cell cultures to high cell densities and using the right extracellular matrix (ECM) and soluble factors, it is possible to achieve cell-cell interactions that resemble tissue-like densities. This can be accomplished in one of three ways: (a) by growing the cells in a relatively large reservoir that has enough medium and a filter fitted over the cells; (b) by growing the cells in high concentrations on agarose or agar or as stirred aggregates (spheroids); and (c) by growing the cells on the outside of hollow fibers that have medium pumped through them from a reservoir after the cells are seeded on the outside.

5.1.4.2. Organotypic Culture

Cells of differentiable lineages are recombined to simulate both homotypic and heterotypic cell interactions. Co-culturing mammary gland fibroblast and epithelial cell clones enables the cells to differentiate functionally in the right hormonal environment, resulting in the production of milk proteins.

5.1.5. Types of Cell Culture

5.1.5.1. Primary Cell Culture

These cells are extracted directly from tissues and organs through enzymatic digestion and chemical or mechanical disintegration. These cells are cultured in appropriate glass or plastic jars using sophisticated culture media. Despite their heterogeneity and typically slow growth rate, these cultures are still preferred over cell lines because they are more like the cell types found in the tissues from which they originate. Numerous cell types can be distinguished by their morphological structure in culture: (i) epithelium cells, which are polygonal in shape and appear flattened when adhered to a substrate to form a continuous thin layer (i.e., monolayer on solid surfaces); (ii) epitheloid type, round in shape, devoid of sheets like epithelial cells, and not adhering to the substrate; (iii) fibroblast type, angular in shape, elongated, and forming an open network of cells instead of densely packed cells, bipolar or multipolar, and adhering to the substrate; and (iv) connective tissue type, derived from fibrous tissue, cartilage, and bone, and typified by a profusion of fibrous and amorphous extracellular materials.

5.1.5.1.1. Advantages and Disadvantages of Primary Cell Culture

The most effective experimental models for in vivo research are these cultures. They exhibit traits not found in cultured cells and have the same karyotype as the parent. They do, however, have a short lifespan and are hard to come by. Another significant drawback is the possibility of bacterial and viral contamination. Depending on the kind of cells in the culture, the primary cell culture can also be divided into two types.

5.1.5.1.2. Anchorage-Dependent/Adherent Cells

These cells require a stable, nontoxic, and biologically inert surface for attachment and growth and are difficult to grow as cell suspensions. Mouse fibroblast STO cells are anchorage cells.

5.1.5.1.3. Anchorage-Independent/ Suspension Cells

These cells do not require a solid surface for attachment or growth. Cells can be grown continuously in liquid media. The source of cells is the governing factor for suspension cells. Blood cells are vascular in nature and are suspended in plasma, and these cells can be very easily established in suspension cultures.

5.1.5.2. Secondary Cell Culture

The availability of fresh nutrients at regular intervals allows primary cell cultures that are passaged or subcultured and grown in fresh medium for an extended period of time to form secondary cultures, which are longer-lasting than the cells of primary cell cultures. The adherent cells are broken down enzymatically during the passaging or subculturing process. The necessary number of cells is then cleaned and re-suspended in the proper volumes of growth media. Secondary cell cultures are the best option because they are easily cultivated and widely accessible; they have been applied to virology, immunity, and toxicology studies.

5.1.5.2.1. Advantages and Disadvantages of Secondary Cell Culture

This kind of culture can be altered to grow continuously and is helpful for obtaining a large population of similar cells. The cellular properties of these cell cultures are preserved. The primary drawback of this system is that the propensity of the cells to differentiate into aberrant cells over time in culture.

5.1.6. Cell Line

0The primary culture, when subcultured, becomes a cell-line or -strain that can be finite or continuous, depending on its lifespan in culture. They are grouped into two types on the basis of the lifespan of the culture.

5.1.6.1. Finite Cell Lines

Cell lines with a limited number of cell generations and growth are called finite cell lines. The cells are slow growing (24–96 hours). These cells are characterized by anchorage dependence and density limitation.

5.1.6.2. Indefinite Cell Lines

Cell lines derived from cancerous or in vitro transformed cells are immortal and can be cultured as monolayers or suspensions. With a generation time of 12–14 hours, these cells divide quickly and can be subcultured indefinitely. Because of their different chromosomal counts, the cell lines can show signs of aneuploidy or heteroploidy. Immortalized cell lines are modified cells that have different growth characteristics. An illustration of an immortal cell line is HeLa cells. These are human papillomavirus 18 (HPV18)-transformed epithelial cells taken from fatal cervical carcinoma. It is simple to work with and maintain infinite cell lines. On the other hand, these cell lines tend to evolve over time.

Keyword

Transformed Cells – These occur when cells take in foreign DNA and express the genes encoded on that DNA, giving the cell new characteristics.

5.1.6.3. Commonly Used Cell Lines

Mammalian cell culture is now a requirement for the industrial production of biologically active compounds. Numerous cell lines used to produce therapeutic proteins, vaccines, pharmaceuticals, and anti-cancer agents, have developed as a result of advances in animal cell culture technology. Human, animal, or insect cells can all be used to create cell lines. Because they grow more quickly, cell lines that can grow in suspension are favored. Chinese hamster ovary (CHO) is the most widely used mammalian cell line. A multitude of general factors need to be taken into account when choosing a cell line, including growth characteristics, population doubling time, growth fraction, saturation density, plating efficiency, and the capacity to grow in suspension. Some of the often-used cell lines are displayed in Table 5.2.

Table 5.2. Commonly Used Cell Lines and Their Origins

Cell Line	Origin	Organism
H1, H9	Embryonic stem cells	Human
HEK-293	Embryonic kidney transformed with adenovirus	Human
HeLa	Epithelial cell	Human
HL 60	Human promyelocytic leukemia cells	Human
MCF-7	Breast cancer	Human
A549	Lung cancer	Human
A1 to A5-E	Amnion	Human
ND-E	Esophagus	Human
CHO	Ovary	Chinese hamster
Vero	Kidney epithelial cell	African green monkey

Cos-7	Kidney cells	African green monkey
3T3	Fibroblast	Mouse
BHK21	Fibroblast	Syrian hamster
MDCK	Epithelial cell	Dog
E14.1	Embryonic stem cells (mouse)	Mouse
COS	Kidney	Monkey
DT40	Lymphoma cell	Chick
S2	Macrophage-like cells	Drosophila
GH3	Pituitary tumor	Rat
L6	Myoblast	Rat
Sf9 and Sf21	Ovaries	Fall Army worm (*Spodoptera frugiperda*)
ZF4 and AB9 cells	Embryonic fibroblast cells	Zebrafish

Source: Anju, V., Megha, V., & Anchal, S. Animal Tissue Culture Principles and Applications. doi: 10.1016/B978-0-12-811710-1.00012-4.

Advantages of continuous cell lines:

- Continuous cell lines show faster cell growth and achieve higher cell densities in culture.
- Serum-free and protein-free media for widely used cell lines may be available in the market.
- The cell lines can be potentially cultured in suspension in large-scale bioreactors.

The major disadvantages of these cultures are chromosomal instability, phenotypic variation in relation to the donor tissue, and a change in specific and characteristic tissue markers.

5.1.7. Growth Cycle

The cultured cells exhibit a distinct growth pattern that includes an exponential or log phase, lag phase, and plateau phase. During the log phase and plateau phase, the doubling time of the cell population can be computed. This is important because it can be used to measure how the cells react to various culture conditions, such as variations in nutrient concentration and the impact of harmful or hormonal components. The population doubling time, which is impacted by nongrowing and dying cells, indicates the rates of cell division within the culture.

5.1.7.1. Phases of the Growth Cycle

The population doubling time, lag time, and saturation density of a particular cell line can be established and characterized for a specific cell type. A growth curve consists of a normal culture and can be divided into a lag phase, a log phase, and a plateau phase.

1. **Lag Phase:** The cell population needs time to recover during this initial growth phase of the subculture and reseeding. Before rapid growth, the number of cells is relatively constant. In this stage, the cell spreads out, adheres to the substrate, and replenishes the portions of the glycocalyx lost during trypsinization. The cytoskeleton resurfaces during the spreading process, and it is most likely a necessary component of the process.
2. **Log Phase:** Because of constant cell division, the number of cells and the population of cells are growing exponentially during this time. The initial seeding density, the cell growth rate, and the density at which density inhibits cell proliferation affect how long the log phase lasts. This phase, in which the population is at its most uniform, and the growth fraction and viability are high (typically 90–100%), represents the most reproducible form of the culture. On the other hand, the cells in the cell cycle might not be evenly distributed, and the cell culture might not be synchronized.
3. **Plateau Phase:** The culture reaches confluency at the end of the log phase, when growth rates decrease and, in certain situations, cell fatigue causes the proliferation of cells to stop. The growth surface is occupied, and the cells are in contact with each other and the surrounding cells. At this point, the culture moves into the stationary phase, and the growth fraction decreases to a range of 0% to 10%. A relative increase in the synthesis of specialized versus structural proteins may also occur, as well as modifications to the composition and charge of the cell surface.

Remember

Be aware of and comply with local and international regulations governing the use of biotechnologies in animals. This includes ethical review processes and safety assessments.

5.1.8. Monitoring Cell Growth

An extensive range of cell-based assays can be performed on animal cell cultures to examine morphology, protein expression, cell growth, differentiation, apoptosis, and toxicity in various settings. Product yields can be raised with appropriate cell growth monitoring. The maximum cell growth in a batch reactor depends on several factors. Monitoring the health and growth stage of cells in culture, as well as minute variations in pH, temperature, humidity, O_2, CO_2, dissolved nutrients, etc., is facilitated by routine observation of the cells may influence the growth of cells. Constantly tracking the growth rate also yields a record indicating when the cells have reached their maximum density in a specific amount of time.

5.1.9. Cell Viability

A reliable indicator of the health of the cell culture is the quantity of viable cells present. Cell viability is assessed by erythrosin B and trypan blue using the breakdown of cellular membrane integrity. Both of these dyes are taken up and retained by dead cells, that do not have an intact membrane, but they are unable to pass through the membrane when it is intact. Trypan blue is not as good as erythrosin B stain, which yields results that are more accurate and have fewer false positives and negatives.

5.1.9.1. Cytotoxicity

The basic functions of cells are impacted by the harmful chemicals present in the culture medium. The cells may die as a result of the cytotoxic effect, or their metabolism may change. In many experimental scenarios involving in vitro and in vivo studies, methods to quickly and accurately assess viable cell number and cell proliferation are a crucial requirement.

Determining the cell number can be helpful in assessing growth factor activity, toxic compound concentration, drug screening, exposure duration, colony size changes, carcinogenic effects of chemicals, and solvent effects (such as ethanol, propylene, etc.)

The assays to measure viable cells (viability assays) are as follows:

- [3-(4,5-dimethylthiazol-2-yl)-2,5-diphenyltetrazolium bromide] (MTT)/MTS/ resazurin assay.
- Protease marker assay.
- ATP assay.

The MTT assay allows simple, accurate, and reliable counting of metabolically active cells based on the conversion of pale yellow tetrazolium MTT. Nicotinamide adenine dinucleotide in metabolically active viable cells reduces tetrazolium compounds into brightly colored formazan products or reduces resazurin into fluorescent resorufin (Figure 5.2). MTT and resazurin assays are widely used, as they are inexpensive and can be used with all cell types.

The protease marker assay utilizes the cell-permeant protease substrate glycylphenylalanyl-aminofluorocoumarin (GF-AFC). The substrate, which lacks an aminoterminal blocking moiety, is processed by aminopeptidases within the cytoplasm to release AFC. The amount of AFC released is proportional to the viable cell number.

This assay has better sensitivity than resuzurin, and the cells remain viable; thus, multiplexing is possible. The ATP assay is the most sensitive cell viability assay. It is measured using the beetle luciferase reaction to generate light. The MTT assay and procedure is given in Figure 5.3.

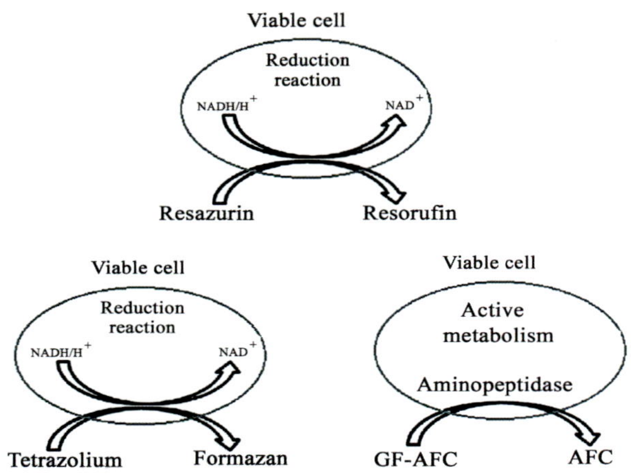

Figure 5.2. *Schematic summary of biochemical events in different viability assays.*

Source: Anju, V., Megha, V., & Anchal, S. Animal Tissue Culture Principles and Applications. doi: 10.1016/B978-0-12-811710-1.00012-4.

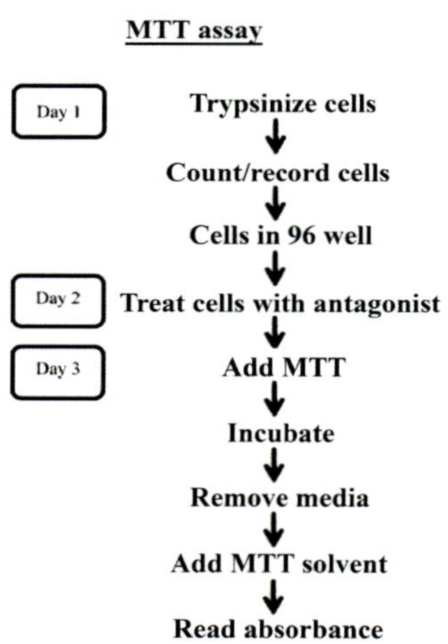

Figure 5.3. *MTT assay.*

Source: Anju, V., Megha, V., & Anchal, S. Animal Tissue Culture Principles and Applications. doi: 10.1016/B978-0-12-811710-1.00012-4.

Assays to detect dead cells are as follows:

- Lactate dehydrogenase (LDH) release;
- Protease release;
- DNA staining.

The outer membranes of the living cells in culture are unbroken. A cell is deemed "dead" when its membrane integrity is lost. Vital dyes that can only enter dead cells can be used to stain the cytoplasm or nucleus, or marker enzymes that leak out of dead cells into the culture medium can be used to measure the activity of dead cell leakage. Within every type of cell is an enzyme called LDH. It is responsible for catalyzing the conversion of lactate to pyruvate when coenzyme NAD^+ is present. The damaged cells release LDH quickly. Cell death is evaluated based on the quantity of released LDH (Figure 5.4). Although this assay is commonly used, its sensitivity is limited due to the 9-hour half-life of LDH at 37°C.

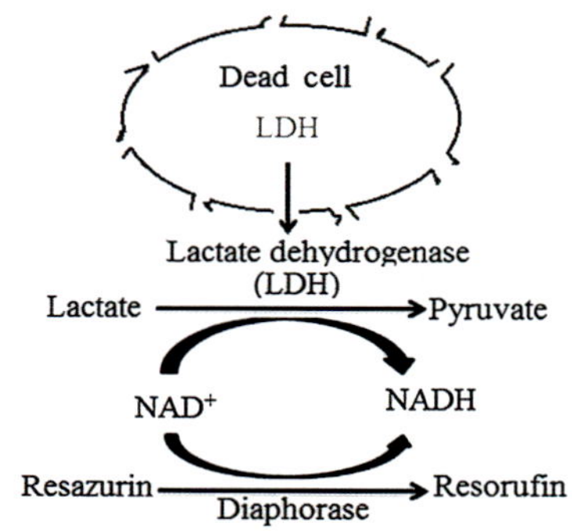

Figure 5.4. *Principle of the LDH release assay.*

Source: Anju, V., Megha, V., & Anchal, S. Animal Tissue Culture Principles and Applications. doi: 10.1016/B978-0-12-811710-1.00012-4.

The basis for the protease release assay is the intracellular release of proteases from the dead or compromised cell into the culture medium. Aminoluciferin, a substrate for luciferase, is released when the released proteases cleave the substrate (Figure 5.5) and causes a "glowtype" signal to be produced.

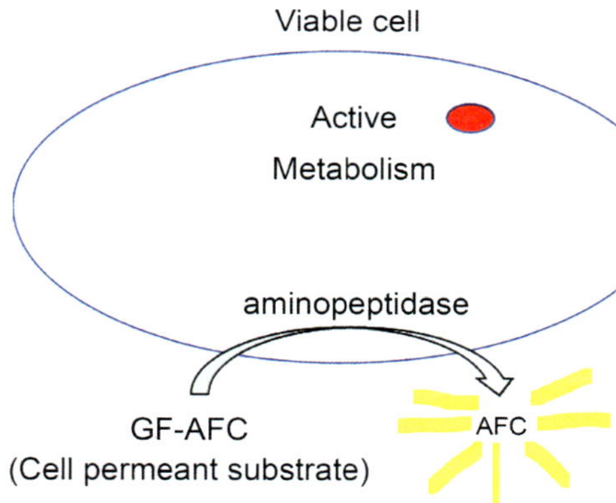

Figure 5.5. *Principle of the luminescent protease release assay.*

Source: Anju, V., Megha, V., & Anchal, S. Animal Tissue Culture Principles and Applications. doi: 10.1016/B978-0-12-811710-1.00012-4.

5.1.9.2. Hayflick Phenomenon

The number of divisions a normal cell population undergoes prior to entering the senescence phase is known as the Hayflick limit or Hayflick's phenomenon. In 1974, Macfarlane Burnet first used the term "c limit.." A population of normal human fetal cells divides in culture 40–60 times before stopping, as shown by Hayflick and Moorhead (1961). It seems that aging and the maximum number of passages are related. Telomere length has an impact on this phenomenon. Telomeres on DNA of a cell shorten as a result of repeated mitosis. Human telomere shortening, which is associated with aging, eventually prevents cells from dividing. This explains why cells from elderly people passage at a lower rate.

Did you get it?

1. Discuss the key aspects of the development of animal cell culture.
2. Identify and describe the characteristics of cell cultures.

5.2. ANIMAL-CELL CULTURE MEDIA

→Learning Objectives

- Identify the points of caution when using culture media.

The technology of cell culture has had an enormous impact on human society. For instance, cell culture technology has been crucial to recent advances in biology. Furthermore, practical technologies based on cell culture have been developed in multiple domains, such as evaluating the toxicity and efficacy of novel drugs, producing vaccines and biopharmaceuticals, and assisting with assisted reproduction. With the recent technological breakthrough of reprogramming somatic cells, scientists from all over the world are vying for the top spot in the field of regenerative medicine advancements. Cell culture technology is also seen as a basis for future advancement and widespread use in this field.

It's unlikely that anyone would contest the idea that the most crucial component of cell culture technology is the culture medium. A medium facilitates both cellular functions and cell survival and proliferation. As a result, the quality of the medium direct impacts the outcomes of assisted reproductive technology research as well as the rate at which biopharmaceuticals are produced and treated. Therefore, choosing a medium that is appropriate for their goals is crucial for researchers working with cell cultures. Sometimes, it is best for researchers to alter a medium themselves. Additionally, to determine the root of any issues with their experiments, researchers need to be aware of the characteristics of the medium they are working with.

Currently, synthetic media can be divided into multiple categories according to the kind of supplements added, such as chemically defined media, serum-containing media, serum-free media, and protein-free media (Tables 5.3 and 5.4). Serum-containing media inherently contain a variety of serum-derived substances, the concentrations of which can vary from batch to batch and cloud the composition of the medium. This condition increases the risk of microbial contamination and reduces the reproducibility of the culture results. However, because serum contains many active substances required for animal cell survival and growth, serum-containing media can be easily designed and used effectively for a wide range of cell types. Serum-free media, in contrast, have a defined composition, resulting in a high reproducibility of results, and the cultivation process can be validated. In addition, target cells can be grown selectively in an intermingled cell population if the culture conditions are configured to benefit them. Among the serum-free media, subgroups of protein-free media (which do not contain any protein at all) and chemically defined media (which do not contain any undefined ingredient) provide additional stability and reproducibility for culture systems, facilitating the identification of the cellular secretions and reducing the risk of microbial contamination. However, the serum-free media are difficult to design: only specific cell types have been cultivated this way to date.

Keyword

Cell Culture – It refers to the removal of cells from an animal or plant and their subsequent growth in a favorable artificial environment.

Table 5.3. Categories of Animal-Cell Culture Media

Category	Definition	Type	Example
Natural media	Consisting of natural biological substances, such as plasma, serum, and embryo extract.	Coagulant or clots	Plasma separated from heparinized blood, serum, and fibrinogen.
		Tissue extracts	Extracts of chicken embryos, liver, and spleen and bone marrow extract.
		Biological fluids	Plasma, serum, lymph, amniotic fluid, and pleural fluid.
Synthetic media	Composed of a basal medium and supplements, such as serum, growth factors, and hormones.	Serum-containing media	Human, bovine, equine, or other serum is used as a supplement.
		Serum-free media	Crude protein fractions, such as bovine serum albumin or α- or β-globulin, are used as supplements.
		Xeno-free media	Human-source components, such as human serum albumin, are used as supplements, but animal components are not allowed as supplements.
		Protein-free media	Undefined components, such as peptide fractions (protein hydrolysates) are used as supplements.
		Chemically defined media	Undefined components, such as crude protein fractions, hydrolysates, and tissue extracts, are not appropriate as supplements, but highly purified components, such as recombinant proteins are appropriate supplements.

Source: Tatsuma, Y., & Yuta, A., (2017). Animal-Cell Culture Media. doi: 10.1002/rmb2.12024.

Table 5.4. *Types and Characteristics of Basal Media*

Category	Name	Features
Connaught medical research laboratories (CMRL) media	Medium 199	Developed to cultivate chicken embryonic cells under protein-free conditions, it is prepared by the sequential addition of amino acids, vitamins (including fat-soluble vitamins), and nucleic-acid precursors. Its composition is extremely complex because the components that are thought to be necessary on theoretical grounds, including inactive components, are added to the medium. Often used for organ culture.
	CMRL1066	Created by amending Medium 199 in order to culture mouse L cells under protein-free conditions, the modifications included increased levels of reducing substances (cysteine, glutathione, and ascorbic acid), the elimination of fat-soluble vitamins, changes in the types of nucleic-acid precursors, and the addition of coenzymes.
Eagle media	Basal medium Eagle (BME)	Supplemented with the minimal components that are necessary for mouse L cells and human HeLa cells to reach the index of proliferative capacity and including 13 amino acids and eight vitamins, it is unsuitable for cells whose cultures require many components because of its simple composition.
	Minimum essential medium (MEM)	BME is modified according to the cellular need for amino acids, so the concentrations of the majority of amino acids are twofold, as compared to BME. Non-essential amino acids, which cells can biosynthesize, were not included in the original MEM formulation; however, researchers can add non-essential amino acids to reduce the biosynthetic load.
	Dulbecco's modified MEM (DMEM)	Modified to have fourfold the concentrations of amino acids and vitamins that are present in BME and developed to study the plaque-forming ability of the polyomavirus in mouse embryonic cells. Various modifications have been made since, with supplementation, for example, of the non-essential amino acids, glycine, and serine, iron, and pyruvate. The glucose concentration can also be increased to 25 mmol L^{-1} in order to accommodate cells with high nutritional requirements. In the event that pH changes are suspected due to metabolites, sodium bicarbonate is doubled in concentration, equilibrated, and then used at 10% CO_2.
	α-MEM	MEM modified for hybrid-cell-line research on mice and hamsters, its composition is MEM supplemented with non-essential amino acids and vitamins (ascorbic acid, biotin, and cyanocobalamin), pyruvate, lipoic acid, and nucleosides.
	Iscove's modified DMEM (IMDM)	Supplemented with several non-essential amino acids and vitamins that are not present in DMEM (cyanocobalamin and biotin) and additional selenite, pyruvate, and HEPES. Transferrin, bovine serum albumin, and soybean lipids are added as serum substitutes. With its high concentrations of amino acids and vitamins, IMDM is suitable for high-density cultures and cultures of rapidly proliferating cells.

Tissue Culture Section of the National Cancer Institute (NCTC) media	NCTC109	Developed to culture L cells under protein-free conditions, its amino acid composition is based on the results of a componential analysis of the compounds that have been ultrafiltered from horse serum and chicken embryonic-tissue extract. Its composition is quite complex, with coenzymes, nucleobases, reducing agents, and a wealth of vitamins added, as well (i.e., A, C, D, E, and K, in addition to the B-group vitamins). Cysteine was included in the original composition, but after it was found to negatively affect cells, a version was developed from which cysteine was removed (NCTC135).
Ham media	Ham's F-10	Enables colony formation by a single Chinese hamster ovary (CHO) cell under serum-free conditions, developed by adding two kinds of purified serum proteins (serum albumin and fetuin) instead of serum and by examining in detail the types and concentrations of amino acids and trace elements This medium was the first to contain the trace elements, copper, and zinc (iron having been included in other media already). The CHO cells have an inferior proliferative capacity in this medium alone, compared to the one with serum added. Furthermore, the culture of the cell lines other than CHO necessitates the addition of serum.
	Ham's F-12	The serum albumin and fetuin (used in Ham's F-10) are replaced by two compounds with definite chemical composition: linoleic acid and putrescine, which enables colony formation by a single CHO cell under protein-free conditions. Often cited as the world's first chemically defined medium, the levels of several amino acids are higher than in Ham's F-10, while those of the vitamins (except choline and inositol) and potassium phosphate are reduced. Its composition must be modified (e.g., by reducing the zinc concentration) for a protein-free culture of cells other than CHO cells. MCDB301, in which 20 trace elements were added, was later developed after it became obvious that trace elements contaminating water or raw materials are necessary for a protein-free culture of CHO cells in Ham's F-12.
	Kaighn's modified Ham's F-12	The concentrations of the amino acids, pyruvate, biotin, calcium, magnesium, putrescine, and phenol red are increased with respect to those in Ham's F-12, among other compositional modifications, in order to support the proliferation and differentiation of primary cultured cells.
Roswell Park Memorial Institute (RPMI) media	RPMI 1640	Based on the 5A Medium (developed by McCoy et al., 1959) and modified for the long-term culture of peripheral blood lymphocytes, it is characterized by low levels of calcium and magnesium and high levels of phosphate. Multiple media were developed on the path to this medium (eg, RPMI 1629, 1630, and 1634). It is widely used as a medium for suspension cultures; for example, of white blood cells, lymphocytes, and hybridomas.

Molecular, Cellular, and Development Biology (MCDB) media	For example, MCDB202 MCDB301 MCDB153 MCDB110 MCDB402 MCDB170 MCDB131	A series of Ham's F-12-based media was developed in order to grow specific cell types in serum-free culture. The composition of each MCDB medium is optimized to promote the growth of a specific cell type. In particular, MCDB202 is a medium for the serum-free culture of chicken embryo fibroblasts; MCDB301 is for the serum-free culture of CHO cells; MCDB153 is for human keratinocytes; MCDB110 is for human fibroblasts; MCDB402 is for mouse fibroblasts; MCDB170 is for the mammary epithelium; and MCDB131 is a medium for the serum-free culture of human vascular endothelial cells.
Mixed media	DMEM/F-12	Constitutes a 50:50 mixture of component-rich Ham's F12 medium and the nutrient-rich DMEM medium and compatible with the requirements of a variety of cell types, it is most often used as a basal medium for serum-free culture.
	RPMI 1640/ DMEM/F-12 (RDF)	Developed for the serum-free culture of hybridomas, it is a mixture of RPMI 1640, DMEM, and Ham's F-12 at 2:1:1 and is typically used as a serum-free medium with added insulin, transferrin, ethanolamine, and selenite.
Other media	Waymouth's MB752/1	Developed with as simple a composition as feasible, so that mouse L929 cells could be cultured without the addition of serum and other proteins, it is composed of a total of 40 components, including glucose, inorganic salts, amino acids, vitamins, purine bases, and hypoxanthine. It is characterized by high concentrations of glucose, histidine, lysine, glutamine, choline, and thiamine.
	Trowell's T-8	Designed for the long-term culture of adult rat liver epithelial cells. With its comparatively simple composition, it contains no non-essential amino acid and hardly any vitamins, but is characterized by high glucose and insulin concentrations. It is used for short-term organ culture.
	Leibovitz's L-15	The buffering capacity is mediated by phosphates and free basic amino acids instead of sodium bicarbonate, so that the pH of a culture is maintained in ambient air without the use of a CO_2 incubator. Instead of glucose, pyruvate (and galactose) is added at a high concentration in order to control pH drops due to the lactic acid that is produced during glucose metabolism and to promote the release of CO_2 from the respiratory chain. Another major characteristic is that each of its constituent amino acids is added at approximately its respective maximal concentration. Once used for cell and tissue transport and primary cultures, its popularity has declined as researchers began to use HEPES as a buffering agent and as they realized, moreover, that a certain amount of sodium bicarbonate is necessary for optimal cell proliferation.
	Fischer's Medium	Contains a high concentration of folate because this medium was developed by using folate-dependent L5178Y lymphoma cells.

Source: Tatsuma, Y., & Yuta, A., (2017). Animal-Cell Culture Media. doi: 10.1002/rmb2.12024.

Popular media formulations for cell culture are currently offered for sale. As a result, some researchers are utilizing culture media without fully comprehending their history or specific components, including the purpose behind their creation, the kinds of cells for which they are appropriate, and their precise composition. This review article provides a brief overview of the evolution of animal cell culture media, along with commentary on the various types of media currently in common use, their respective functions, potential drawbacks, and issues related to their application.

5.2.1. Points of Caution When Using Culture Media

5.2.1.1. Setting Up a Serum-Free Culture

Two things need to be considered when using a serum-free culture system. First, during subculturing, accidental clones of cells in a culture vessel can be chosen because serum-free media frequently encourages the proliferation of a specific cell clone or subtype of cell more frequently than serum-containing media. Second, because serum contains the toxin-neutralizing activity, impurities in a serum-free medium, whether avoidable or inevitable, may have stronger effects on the cultured cells than can impurities in a serum-containing medium. Additional safety measures for serum-free cultures include reducing trypsin concentration and choosing adhesion factors carefully.

5.2.1.2. pH Changes

The pH of the culture medium that is used for animal-cell cultures is generally maintained by the equilibrium relationship between sodium bicarbonate ($NaHCO_3$) in the culture medium and CO_2 in the incubator (Figure 5.6). Many laboratories set the CO_2 concentration in the incubator to 5%-6% by convention, but technically, it should be adjusted according to the concentration of the $NaHCO_3$ in the culture medium. For example, 5% CO_2 is appropriate for a medium with 26 mmol L^{-1} $NaHCO_3$ added (e.g., MEM medium), 2% CO_2 for a medium with 14 mmol L^{-1} added (e.g., Ham's F-12 medium), and 10% CO_2 for a medium with 44 mmol L^{-1} $NaHCO_3$ added (e.g., DMEM). These numbers are theoretical values that are calculated using the Henderson–Hasselbalch equation (Figure 5.6): in practice, checking the pH of a culture medium after it reaches equilibrium and then making minute adjustments to the CO_2 concentration are preferable.

Because of its potent buffering ability, 4-(2-hydroxyethyl)-1-piperazineethanesulfonic acid (HEPES) may have buffering effects in a medium. When handling cultured cells outside of a CO_2 incubator, the pH of the medium containing HEPES is more stable than that of the medium containing only bicarbonate. It may also be advantageous for cultures with a high cell density, such as those where an accumulation of metabolites like lactate can cause the pH to drop quickly. Serum-free media don't have serum's buffering power, so HEPES's pH-buffering function might be more important. However, HEPES may be detrimental to some cell types, including ES cells and chick embryo epiphyseal chondrocytes. Furthermore, it has been shown that when HEPES-containing media are exposed to visible light, their levels of cytotoxic products—primarily hydrogen peroxide—increase.

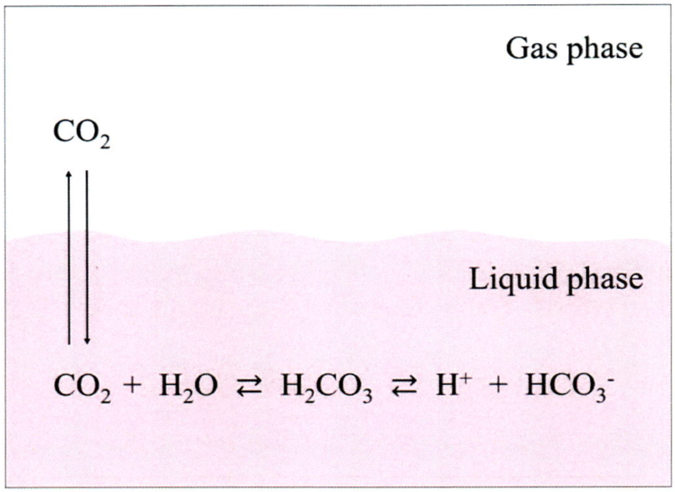

Figure 5.6. *The pH control mechanism of culture media, based on the bicarbonate buffer system and the Henderson–Hasselbalch equation.*

Source: Tatsuma, Y., & Yuta, A., (2017). Animal-Cell Culture Media. doi: 10.1002/rmb2.12024.

Keyword

Oxidative Stress – It is an imbalance between free radicals and antioxidants in your body. This can cause damage to organs and tissues and result in various diseases.

When added to a medium as a pH indicator, phenol red can occasionally influence cell proliferation and exhibit estrogenic activity. Therefore, when working with estrogen-sensitive cells, such as mammary gland cells, researchers should not overlook testing whether such properties of phenol red affect their cells and empirical results.

5.2.1.3. Oxidative Stress

Experiments involving culture, in general, and particularly those conducted in environments with elevated oxygen concentrations, like organ culture experiments, may subject cells to **oxidative stress**, which has an adverse effect on the tissues and cells in culture. Antioxidant-active supplements, such as vitamins C and E, glutathione, selenite, β-mercaptoethanol, dithiothreitol, or lipoic acid, are beneficial, particularly when serum (which is rich in antioxidants) is not added to the medium. Furthermore, free-floating iron and copper ions encourage the creation of reactive oxygen species. Therefore, the best course of action is to complex these ions with suitable carriers (such as transferrin, albumin, or chelating agents) and supply them to the cells while they are isolated from reactive systems. Free iron ions, which are easily hydroxylated and precipitate in an aqueous solution, also require caution.

5.2.1.4. Nutrient Requirements

Cells cannot function properly unless there are adequate amounts of nutrients in the medium. Depending on their rate of proliferation and metabolic activity, certain cell types require greater amounts of nutrients than others. The choice of a medium should consider these cell properties. For instance, glucose was initially intended to be present in DMEM at 5.6 mmol L^{-1}. Now, different suppliers offer a modified version of DMEM with a high glucose concentration of 25 mmol L^{-1}, which can be used for cells that require higher amounts of glucose. The accumulation of metabolites, such as lactate, and a drop in pH are two cautions when using this high-glucose medium for actively proliferating cells. It is recommended that researchers either use HEPES to give the medium a stronger pH-buffering capacity or replace the medium at appropriate intervals.

One of the essential amino acids, glutamine, serves as both a biosynthetic material for proteins and nucleic acids and an energy source for mammalian cells grown in culture. Compared to other amino acids, cells in culture require ~3–40 times more glutamine. Adding glutamine could be beneficial when growing cells that require a lot of nutrients. On the other hand, glutamine breaks down quickly in culture media and produces toxic ammonia. As a result, scientists might think about incorporating glutamine into the medium right before use or using derivatives of glutamine that are resistant to degradation, like glycyl-L-glutamine or L-alanyl-L-glutamine.

The cell is capable of biosynthesizing non-essential amino acids (NEAAs). For this reason, certain basal media, like BME and MEM, do not contain NEAAs. However, in the cultured state, not all cell types are able to generate adequate amounts of NEAAs. Consequently, adding NEAAs to the medium may guarantee better culture conditions. The addition of NEAAs to the medium may help cells even if they are able to biosynthesize them at the required rate, as it reduces the cells' biosynthetic load.

Did you get it?

1. Define the categories of animal-cell culture media
2. Explain the types and characteristics of basal media

5.2.2 3D Spheroid Culture, Cell Bioprinting and Organ-On-A Chip

In recent years, advancements in tissue engineering and cell biology have led to the development of sophisticated techniques for cultivating cells in three-dimensional (3D) environments. These methods, including 3D spheroid culture, cell bioprinting, and organ-on-a-chip technology, aim to replicate the complex structures and functions of human tissues more accurately than traditional two-dimensional (2D) cell cultures. Each approach offers unique advantages and applications in biomedical research, drug discovery, and personalized medicine.

3D Spheroid Culture

3D spheroid culture involves the spontaneous aggregation of cells into spherical clusters when cultured under specific conditions. Unlike 2D cultures, which lack the architectural complexity of native tissues, spheroids more closely mimic in vivo environments. They enable cells to interact in three dimensions, facilitating cell-cell and cell-matrix interactions crucial for maintaining tissue-like functions.

Formation and Applications: Spheroids can form through various methods, including hanging drop techniques, spinner flasks, or on scaffold surfaces. These methods promote the aggregation of cells into compact structures resembling microtissues or small organs. Spheroids have been used extensively in cancer research to study tumor biology, drug response, and metastasis. Their 3D nature allows researchers to assess drug penetration, toxicity, and efficacy more accurately than 2D models.

Advantages:

- **Physiological Relevance:** Spheroids mimic tissue architecture and microenvironment, enhancing the relevance of experimental results.
- **High-throughput Screening:** They enable rapid screening of drug candidates and toxicity testing in a more biologically relevant context.

Cell Bioprinting

Cell bioprinting is a cutting-edge technology that allows precise placement of cells, biomaterials, and biochemical factors to create 3D structures layer by layer. Similar to 3D printing, bioprinting uses computer-aided design (CAD) models to fabricate complex tissue constructs with spatial control over cell distribution and scaffold architecture.

Techniques and Materials: Bioprinting employs various techniques such as inkjet, extrusion, and laser-assisted printing, each suited to different types of biomaterials and cell types. Bioinks composed of cells encapsulated in hydrogels or polymers serve as the building blocks for tissue constructs. These bioinks provide structural support and create an environment conducive to cell viability and function.

Applications: This technology holds promise for creating personalized tissue implants, modeling disease processes, and studying tissue development and regeneration. Bioprinted tissues and organs-on-a-chip offer platforms for drug testing, disease modeling, and potential replacements for damaged tissues.

Advantages:

- **Precision:** Allows precise control over the spatial arrangement of cells and biomaterials.
- **Customization:** Enables the creation of patient-specific tissues and organs for regenerative medicine applications.

Organ-on-a-Chip

Organ-on-a-chip (OOC) platforms are microfluidic devices designed to simulate the structure and function of human organs. These systems integrate living cells with microengineering technologies to replicate physiological functions and responses under controlled conditions.

Design and Functionality: OOC devices typically consist of microchannels lined with cells derived from specific organs or tissues. They incorporate fluid flow to mimic blood circulation and physiological fluid dynamics, creating environments that closely resemble human organs in vivo.

Applications: OOC technology is revolutionizing drug development and personalized medicine by providing more accurate models of human physiology and disease. Researchers use OOCs to study organ interactions, drug metabolism, toxicity, and disease mechanisms in a controlled and reproducible manner.

Advantages:

- **High Fidelity:** Mimics the complex microenvironment and functions of human organs.
- **Reduced Animal Testing:** Offers an ethical alternative to traditional animal models for drug testing and disease research.

These technologies represents significant advancements in biomedical research and therapeutic development. These approaches provide researchers with more physiologically relevant models to study disease mechanisms, evaluate drug candidates, and advance personalized medicine initiatives. As these technologies continue to evolve, their impact on understanding human biology and improving healthcare outcomes is expected to grow, paving the way for innovative treatments and therapies tailored to individual patient needs.

5.3. TRANSGENIC ANIMALS

→Learning Objectives

- Explore the main methods used to produce transgenic animals;
- Identify the expression vectors used to obtain transgenic animals;
- Examine various applications of transgenic animals;
- Understand the potential role of transgenic animals in agriculture.

An animal that has had a foreign gene purposefully added to its genome is said to be transgenic. Recombinant DNA technology is used in the introduction of the foreign gene. Apart from the gene itself, the DNA typically consists of additional sequences that allow the gene to be integrated into the host's DNA and express itself correctly in the host cells. It has been possible to create transgenic sheep and goats that express foreign proteins in their milk. Human proteins can now be synthesized by transgenic chickens in the "white" of their eggs. In the long run, these animals should prove to be important sources of proteins for use in human medicine.

5.3.1. The Main Methods Used to Produce Transgenic Animals

Modern techniques such as intra-pronuclear zygotic DNA microinjection (MI) and somatic cell nuclear transfer (NT) enable the production of recombinant product (RP)-producing livestock that carry the necessary transgene in every cell of their organism and pass it on to their progeny. These days, the most widely employed technique is DNA microinjection into a zygote's male pronucleus (Figure 5.7). Linear DNA can integrate into the genome of living organisms or cell lines as soon as it enters the nucleus. DNA typically integrates into heterochromatin and transcriptionally inactive gene-poor regions. One genomic site can integrate one, multiple, or even hundreds of copies of the injected construct. This technology, which was first tested on mice, is still a dependable way to produce TAs. This process was employed to create the initial agricultural TAs. However, MI is now used mainly to produce transgenic mice, rabbits, and pigs. This is due to the inefficient method due to the low frequency of incorporation of recombinant DNA into the genome and the availability of zygotes at the two pronuclear stages. The result depends on carrying out a large number of surgical procedures, which entails the need to keep a substantial number (200–300 heads) of experimental livestock and perform skilled animal handling. Furthermore, the only way to determine the expression level of the integrated transgene is to examine the original TAs and their offspring. The reproductive cycle in large animals (including the time before they reach physiological maturity and the need to obtain females producing recombinant proteins in milk from the original transgenic males) is approximately 0.9/2.3 years for goat females/males, 1.0/2.3 years for pigs, and 2.3/4.5 years for cows. These limitations increase the cost of obtaining the original TAs and the time required to organize the work.

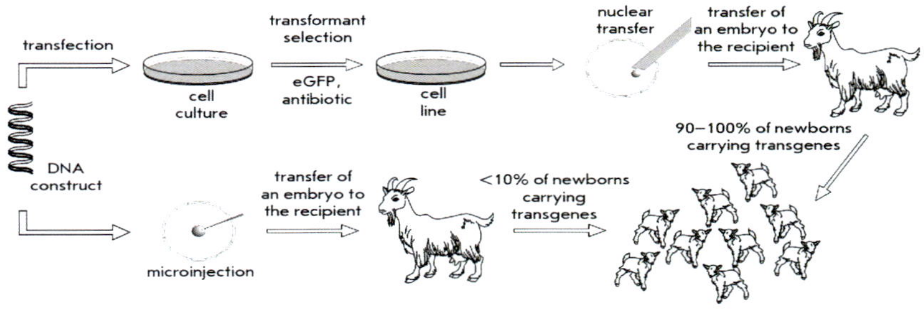

Figure 5.7. *Scheme for producing transgenic animals using the methods of nuclear transfer (upper panel) and intranuclear microinjection of DNA (lower panel).*

Source: ncbi.nlm.nih.gov/pmc/articles/PMC3612824/figure/F1/.

A somatic mammary gland cell was nuclear transferred (NT) into an oocyte in 1997 to create a sheep clone. This success made the development of simpler and less expensive processes for creating agricultural TAs possible (Figure 5.7), as the majority of the modifications in this instance are transferred from a farm to a lab, where somatic cell transfection is executed, and clones that exhibit the transgene's integration into the genome are chosen. The enucleated oocyte is then implanted into female recipients after the somatic cell nucleus has been injected into it. Normally, NT is carried out using fibroblast cells. Most large farm animals generated recently have been acquired by NT. The FDA and EMEA will find it more difficult to approve the recombinant proteins that are produced in this instance because the transfected cells were chosen based on marker genes for antibiotic resistance. To boost the effectiveness of this kind of selection, fluorescent proteins—like the enhanced green fluorescent protein (eGFP)—are frequently added as a second selective agent. Furthermore, selection markers are eliminated from the genomes of the chosen cell lines using systems based on site-specific recombinases.

The low rate of in utero embryo survival and the unwell health of the newborn animals are among the negative effects of the NT technique. This is ascribed, among other things, to insufficient somatic nucleus reprogramming, which impairs the expression of multiple genes necessary for the appropriate development of embryogenesis. Furthermore, obtaining the right oocytes and activating them call for significant time and financial commitments. As a result, the AgroResearch Company (New Zealand), one of the pioneers in using NT for the production of agricultural TAs, has rejected the technique. The business is currently creating different processes to create agricultural TAs.

An alternative to MI and NT techniques could be the use of embryonic stem (ES) cell-based site-specific transgenesis technology. Using this technique, a transgene is inserted into the ES cells' genome, clones that have the appropriate integration of the necessary number of copies are chosen, and finally, transgenic ES cells are placed inside the blastocyst cavity and transplanted into a female recipient. Upon transplanting these cells into the ovaries of adult mice, the transgene can be present in up to 30% of the newborn mice. Nonsurgical techniques, frequently employed in animal husbandry, can be used for any kind of animal handling. A small experimental herd is necessary for the production of transgenes because only a small number of blastocysts are needed. But so far, this technique has only been refined for mice and rats; ES cell lines for agricultural animals are still unavailable. A comparable method entails transforming stem cells, which are the progenitor cells of sperm, and then transferring those cells into the seminiferous tubules of male infertile individuals. The other methods for obtaining TAs are used relatively rarely. Thus, TAs can be effectively produced using retroviruses containing the required transgene. To achieve this objective, the zygotes lacking protective coating are cultured in a medium supplemented with lentiviral particles, followed by transplantation into female recipients. The integration of one to several copies of the transgene occurs depending on the lentiviral titer; almost 100% of the offspring can be transgenic in this case. The advantages of this method include efficient production of any species of TAs and the opportunity to produce TAs carrying only one transgene copy, which is sometimes necessary for scientific purposes. The main drawbacks of the method include the inability to use the introns present in the gene construct and the limitation of the transgene length (approximately 8000 bp), which is determined by the size of the viral particle. As a result, it is very difficult to achieve a high level of transgene expression using this method.

Using vectors based on mobile genetic elements—which transposase integrates into the genome—is a promising way to obtain TAs. The zygote is coinjected with the transgene surrounded by terminal repeats of the transposon and the gene encoding transposase. A single copy of the transgene is integrated into one or more locations within the animal's genome as a result of the reaction that transposase catalyzes. This method has produced large farm animals (e.g., pigs). In this instance, the transgene integration efficiency can reach 50% and depends on the transposon type, transgene length, concentration, and DNA injection site. But as of yet, no information has been found about the target gene's expression levels in the TAs made with this technique. It is important to specifically mention the set of techniques that involve inserting a replication-defective adenovirus containing the target protein gene into an organism's organs or tissues. This method causes the organ or tissue under consideration to produce recombinant proteins in a short-term, nonhereditary manner.

Did you know?

In 1985, researchers created the first transgenic cows, followed by the production of transgenic sheep in 1997. These achievements paved the way for the potential use of transgenic livestock in agriculture for the production of pharmaceutical proteins in milk.

5.3.2. The Expression Vectors Used to Obtain Transgenic Animals

Regulating regions of genes whose protein products make up the majority of milk are present in the expression vectors used to produce recombinant proteins in milk. The most well-known instances of the latter type are the goat β-casein gene, the α-lactalbumin and α-S1-casein genes of cows, the regulatory regions of the lactoglobulin sheep gene, and the acidic protein gene of rodents (mouse, rat, and rabbit). A long 5'-region (1–7 kb) found in expression vectors usually comprises a promoter, tissue-specific enhancers that boost expression in mammary glands, the first non-coding exons, and introns positioned in between (Figure 5.8(A)). Gene transcription can be improved by regulatory elements found in the first introns of the genes. The 3'-untranslated region (UTR) of a gene, whose size can vary from 0 point 5 to 10 kb and even more, is also included in the expression vector. The final non-coding exons and introns, a polyadenylation site, and the surrounding sequences, which may improve transcription termination, are usually found in the 3'-UTR. A vector's 5'- and 3'-UTRs may be associated with one or more distinct genes.

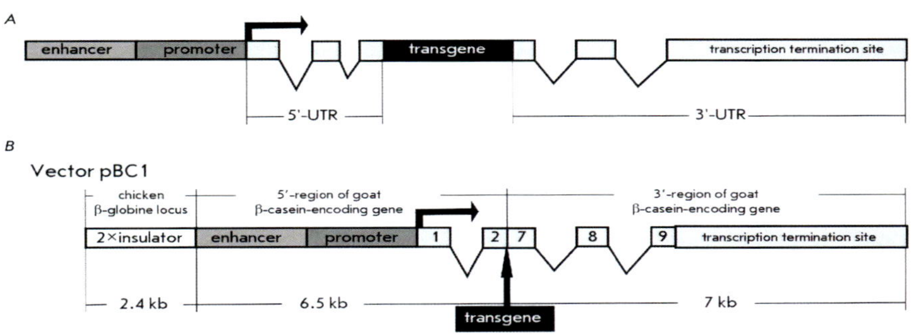

Figure 5.8. *Vectors used for the production of recombinant proteins. (A) Structure of a vector used to produce recombinant proteins in transgenic animals and cell lines. (B) Structure of the pBC1 vector used for the production of recombinant proteins in the milk of transgenic animals.*

Source: ncbi.nlm.nih.gov/pmc/articles/PMC3612824/figure/F2/.

Among the promoters used for expression in mammary glands, the β-casein gene promoter is one of the most efficient and is used to produce target proteins in the mammary glands of mice, goats, and cows (Table 5.5). The most popular commercial vector for the production of recombinant proteins in the milk of TAs – pBC1 (Invitrogen) – was produced using the aforementioned promoter (Figure 5.8(B)). This vector allowed one to obtain most of the transgenic goat lines characterized by a high level of target protein production. The regulatory region of the β-casein gene, of length 6.2 kb and consisting of a promoter and a hormone-dependent enhancer that stimulates the promoter only in the mammary gland cells, is

Remember

Transgenic animals can be engineered for improved agricultural productivity, such as enhanced growth rates, increased milk or meat production, and resistance to diseases.

used in this vector. The structure of the vector also includes a 7.8-kb-long 3'-region of the β-casein gene, which ensures efficient transcription termination. The latter is required for the formation of a stable mRN A encoding the target protein and for the prevention of the transcription of the adjacent genomic regions capable of causing the formation of repressed chromatin by RN A interference. To accumulate the target protein in milk, the coding region of the gene must contain the signal peptide sequence required for secretion. This sequence can be obtained from any gene encoding the secreted protein.

Table 5.5. *Comparison of Production of Human Milk Recombinant Proteins (RPs) in Transgenic Animals (TAs) Using Various Variants of the Gene Construct*

RP/ Construct	Regulatory Elements	Transgenic Animal/Method of Production	Maximum Level of RP Production (mg/mL)
Lactoferrin			
Native gene	Bacmid	Cow/NT	3.4
Native gene	WAP-gene (21 kb) (mouse)	Mouse/MI	30
–”–	β-casein promoter (goat) + insulator	Mouse/MI	160
–”–	β-casein promoter (goat) + insulator	Goat/MI	10.8
–”–	αS1-casein promoter (cow)	Cow/MI	3
cDNA	β-casein promoter (goat) + insulator	Goat/MI	0.7
cDNA	αS1-casein promoter (cow)	Mouse/MI	0.036
–”–	β-casein promoter (goat) + insulator	Mouse/MI	4
Lysozyme			
cDNA	β-casein promoter (goat) + insulator	Cow/NT	0.026
–”–	β-casein promoter (goat) + insulator	Pig/NT	0.00032
–”–	α-S1- casein (cow)	Goat/MI	0.27
–”–	α-S1- casein (cow)	Mouse/MI	0.00071
α-Lactalbumin			
Native gene	Bacmid	Cow/NT	1.55

Source: ncbi.nlm.nih.gov/pmc/articles/PMC3612824/table/T1/?report=objectonly.

Depending on the aim, either a complete gene with introns or its cDNA or a mini-gene containing only some of the introns is inserted into the vector. The use of a gene with an unmodified exon–intron structure allows one to obtain much higher levels of target protein production in TAs in comparison with the use of cDNA. For instance, the human lactoferrin gene was expressed using the same vector, pBC1, in several independent studies (Figure 5.8(B)). The concentration of recombinant lactoferrin did not exceed 4 mg/mL of mouse milk and 0.7 mg/mL of transgenic goat milk if cDNA was used to produce the TAs (Table 5.5). Transgenic mice were produced using the native lactoferrin gene with introns (50 kb long). The concentration of recombinant lactoferrin in their milk was as high as 160 mg/mL (Table 5.5). Transgenic goats carrying one copy of the construct but expressing up to 10 mg/ml of the recombinant human lactoferrin in their milk have also been produced. The difference in the expression of recombinant lactoferrin using the αS1-casein promoter of cows was similar (Table 5.5). This example demonstrates that the presence of introns in the coding region of the transgene results in a two-fold increase in the amount of the target protein in milk.

Effective transgene expression largely depends on the location of the construct's integration into the genome. Gene-poor regions are typified by frequent breaks in DNA, and this is where injected DNA usually gets incorporated. These areas' chromatin usually has an adverse effect on the transgene integrated nearby's expression. Furthermore, multiple integrations of the construct are usually found on the same genomic site, which may result in transcriptional repression because repetitive sequences form heterochromatin.

A number of regulatory elements are used to protect the transgene expression from repression and to maintain the direct relationship between the number of copies and the level of transgene expression in mammalian cell cultures: A/T-rich regions of DNA, which bind to the nuclear matrix fraction (known as MAR/SARelements); regulatory elements (UC OE) that activate the promoters of the "household" genes; STAR-elements that can block the spread of heterochromatin; and insulators. Among the mentioned regulatory elements, only insulators are used in vector constructs to produce TAs. Insulators are the regulatory elements that block the interaction between an enhancer and a promoter, if located between them. Moreover, some insulators can act as a boundary between the transcriptionally active chromatin and heterochromatin. The insulator from a cluster of chicken β-globin genes (HS4 insulator) is one of the most intensely studied vertebrate insulators. It is 1200 bp long and is located at the 5'-end of the β-globin locus. A 250-bp-long core region characterized by full insulator activity has been found in it. This segment contains the binding site for the CTC F protein, the only characterized vertebrate insulator protein. The CTC F protein is responsible for the ability of the HS4 insulator to block enhancers. It also assists the USF1 and USF2 proteins (the boundary between active chromatin and heterochromatin) to bind to the insulator. The HS4 insulator sequence also binds to the BGP1/Vezf1 protein, which protects the GC-rich sequences of an insulator against methylation, which leads to a disruption of the binding of insulator proteins to DNA and, as a result, to inactivation of the insulator. According to the existing model, BGP1/Vezf1 also terminates the weak transcription initiated in the heterochromatic region, which can play an important role in protecting the β-globin locus against the propagation of inactive chromatin. The pBC1 vector constructed by Invitrogen (USA) for the production of TAs contains two 1.2-kb-long copies of the HS4 insulator at the 5'-end of the vector (Figure 5.8(B)). A thorough analysis of the effect of the HS4 insulator on the transcriptional activity of several promoters, including the goat β-casein promoter and rabbit WAP-promoter, revealed that the insulator significantly increases the level of transgene expression and the number of transgenic lines that are characterized by a significant production of the target protein. Meanwhile, the HS4 insulator neither affects the variability of the transgene expression and its ectopic expression in other tissues of the organism nor ensures a direct correlation between the number of transgene copies and its level of expression. Thus, the HS4 insulator acts as a universal transcription regulator that can be used to increase the activity of weak promoters. However, it does not allow efficient transgene expression exclusively in a mammary gland, which is important for the production of many target proteins that can adversely affect the health of TAs.

An alternative to insulators and other regulatory elements in the transgene construction could be to enlarge the regulatory sequences. Many loci encoding milk protein genes have extended 5'- and 3'-regions that may include tissue-specific enhancers and insulators that shield the gene from neighboring influences. For example, a 50 kb construct was created by substituting a structural portion of the human lactoferrin gene (29 kb) for the 3 kb coding region of the mouse WAP gene, which is 24 kb in length. It has been possible to generate transgenic mice with mammary glands that exhibit high tissue-specific transgene expression and whose milk can produce up to 30 mg/mL of recombinant human lactoferrin (Table 5.5).

> **Keyword**
>
> **Lactoferrin** – It is a multifunctional protein of the transferrin family. Lactoferrin is a globular glycoprotein with a molecular mass of about 80 kDa that is widely represented in various secretory fluids, such as milk, saliva, tears, and nasal secretions.

Another method for obtaining TAs that efficiently produce target proteins is the integration of large DNA segments (up to 250 kb) into the genome. Vectors based on bacterial artificial chromosomes (bacmids), which enable cloning of sequences up to 400 kb long, can be used to prepare these extended gene constructs. The regulatory regions of tissue-specific genes can occupy large genomic regions and be a part of the neighboring genes. For example, several enhancers that stimulate the pig WAP gene are found 140 kb away from the gene they regulate and are separated from it by the other genes. When large DNA fragments are used, it is highly likely that all of the regulatory elements of this gene are included in the transgene. The use of this approach is assumed to result in specific transgene expression exclusively in the mammary gland and that the influence of the surrounding chromatin on transgene expression is minimized. This approach allows one to obtain TAs whose level of transgene expression closely corresponds to the expression of the endogenous counterpart. For instance, transgenic cows expressing the genes of human lactoferrin and human α-lactalbumin have been produced (Table 5.5). In general, this method allows one to achieve stable transgene expression at a level similar to that of native genes. Thus, the levels of lactoferrin and alpha-lactalbumin production in transgenic cows were 3.4 and 1.55 mg/mL, respectively (Table 5.5). The problem is associated with the other genes, which are a part of the bacmid structure and whose expression can adversely affect the health of the TAs. It should also be mentioned that the use of a bacmid does not completely suppress the effect of the genomic surrounding: the expression is partially dependent on the genome integration site. In this case, there is no direct observable relationship between the number of bacmid copies and the expression level. This can be attributed to the fact that the initiated RN A interference adversely affects the gene expression in a bacmid.

5.3.3. Applications of Transgenic Animals

Transgenic animals, which have had foreign genes inserted into their genome, have various applications across different fields. Here are some key applications of transgenic animals:

1. Biomedical Research:
   - i. **Disease Modeling:** Transgenic animals are used to model human diseases, allowing researchers to study disease progression, develop, and test new therapies, and understand the underlying genetic factors involved in various disorders.
   - ii. **Drug Development:** Transgenic animals can be engineered to produce specific human proteins, enabling the testing of potential drugs and vaccines for safety and efficacy.
2. Agriculture:
   - i. **Livestock Improvement:** Transgenic animals are created to express desirable traits such as increased resistance to diseases, improved growth rates, and enhanced production of valuable proteins like milk or wool.
   - **Disease Resistance:** Genetic modification can confer resistance to specific diseases in livestock, reducing the need for antibiotics and improving overall animal health.
3. Biotechnology and Industry:
   - i. **Biopharmaceutical Production:** Transgenic animals can be used to produce therapeutic proteins and other biopharmaceuticals in their milk, blood, or other tissues, providing a cost-effective method for large-scale production.
   - ii. **Transgenic Microorganisms:** Some transgenic animals are engineered to produce useful substances, such as enzymes or proteins, in their milk or other secretions.
4. Conservation and Ecology:
   - i. **Endangered Species Conservation:** Transgenic animals can contribute to the preservation of endangered species by serving as hosts for embryos or by introducing genes that enhance their survival in changing environments.
   - ii. **Control of Invasive Species:** Genetic modification can be employed to control or eradicate invasive species by introducing genes that limit their reproduction or survival.
5. Research in Neurobiology and Behavior:
   - i. **Study of Brain Function:**

Transgenic animals with specific genes inserted or modified can help researchers understand brain function, neurodevelopment, and behavior.

ii. **Optogenetics:** GM animals expressing light-sensitive proteins can be used in optogenetic studies to manipulate and control specific neurons, providing insights into brain circuits and behavior.

6. Environmental Monitoring:
   i. **Biological Sensors:** Transgenic animals can be engineered to act as biological sensors by expressing specific proteins in response to environmental pollutants or contaminants, helping in environmental monitoring.
7. Xenotransplantation:
   i. **Organ Transplants:** Transgenic pigs, for example, have been GM to produce organs with reduced immunogenicity, potentially making them suitable for xenotransplantation, which involves transplanting organs from animals to humans.
8. Production of Specific Nutrients:
   i. **Nutrient-Enriched Products:** Transgenic animals can be designed to produce nutrient-enriched products, such as milk with higher levels of specific vitamins or proteins, providing potential health benefits.

It's important to note that ethical considerations, regulatory frameworks, and public acceptance play crucial roles in the development and deployment of transgenic animals in various applications.

5.3.4. Potential Role of Transgenic Animals in Agriculture

Salmon that expresses growth hormone for commercial use (AquAdvantage) is currently getting close to FDA approval; the results indicate that the product is safe for both humans and the environment. The economic impact of transgenic salmon is linked to a nearly twofold increase in growth, which considerably lowers the cultivation costs. Thus, it is reasonable to assume that authorization for the commercial application of different TAs will be acquired soon. These TAs can be used to accomplish a number of significant goals, including: 1) creating modified milk that contains human recombinant proteins; 2) changing the milk's composition to increase the production of dairy products more efficiently; 3) enhancing the traits of farm animals (such as quick growth and recycling); and 4) enhancing the animals' resistance to prion, viral, and bacterial infections.

The importance of artificial infant feeding and newborn nutrition is growing. Breast milk differs greatly from goat and cow milk in terms of composition. Therefore, the concentrations of lysozyme (0.03–3 mg/mL), lactoferrin (2.0–5.8 mg/mL), and lactalbumin (1.8–3.1 mg/mL) in human milk are significantly higher. These proteins strengthen intestinal epithelium structure, boost immunity, positively impact intestinal microbiota, and shield the organism from infections. In contrast, cow milk has much lower concentrations of these proteins: 0.03–0.49 mg/mL for lactoferrin, 0.05–0.22

mg/mL for lysozyme, and 1.47 mg/mL for lactalbumin. Because animal milk is used to prepare artificial feeding mixtures that are made from hydrolysates and lack functional proteins, these mixtures do not offer the best nutrition for infants.

To produce modified milk, transgenic cows expressing human lactalbumin (1.5 mg/mL), lysozyme (0.03 mg/mL), or recombinant lactoferrin (3.4 mg/mL) have been obtained in China. The next stage of production is to produce milk that has the ideal concentrations of all three humanrecombinant proteins. In the United States, transgenic goats have been produced whose milk contains recombinant human lysozyme at a concentration of 0.27 mg/mL, or 67% of the lysozyme concentration in breast milk. It has been shown that the health of young pigs and goats is improved by pasteurized milk containing human lysozyme. A grant from the Brazilian government was given to the University of California, Davis (USA) and the Institute of Biomedicine of the Federal University of Ceara (Brazil) to investigate the potential of using milk containing recombinant human lysozyme to treat diarrhea in children from low-income households. It is also planned to produce transgenic goats that express recombinant lactoferrin in their milk, which will enable the production of milk that concurrently contains two human proteins.

The government-owned agro research company in New Zealand produces transgenic cows with the aim of increasing efficiency in cheese production. Caseins, the most valuable proteins, comprise approximately 80% of milk proteins. The casein fraction in cow milk consists of α-S1-, α-S2-, β- and k-caseins, encoded by a single copy of each gene. Caseins aggregate into large micelles. The micellar structure and its stability may vary depending on the ratio of caseins, which affects the physical and chemical properties of the milk. Cheese is made via the aggregation of casein micelles, which retain water and fat by forming a protein network. An increased content of β- and k-caseins leads to a reduction in the micellar size and increases thermal stability, which is necessary for cheese production. To increase the amount of β- and k-caseins in milk, transgenic cows with additional copies of the gene were produced. The endogenous β-casein gene of cows with its regulatory sequences was used. Since the k-casein gene is characterized by a relatively low expression level, a chimeric gene containing the regulatory region of the β-casein gene and the coding region of the k-casein was used for the production of transgenic cows. The transgenic cows were eventually produced; their milk was characterized by a 20% increase in the level of β-casein expression and a twofold increase in k-casein synthesis. This result clearly demonstrates the fact that the milk content can be altered by transgenesis, which can increase efficiency in the multi-billion dollar cheese production industry.

A challenge in swine breeding is the high piglet mortality rate, which is linked to insufficient levels of α-lactalbumin in milk. To address the issue, we created transgenic pigs with the α-lactalbumin gene from cows, increasing the lactose content in their milk. As a result, the mortality rate of the piglets, given the altered milk, dropped noticeably. The high phosphorus levels in swine excrement pollute the surrounding environment, which is another issue with swine breeding. The solution to this issue was to create transgenic pigs with a bacterially inserted phytase-encoding gene. Phosphates were consequently reduced by 75% in the feces of the transgenic pigs.

Resistance to diseases is another extremely important aspect in the application of transgenesis in agriculture. Thus, the losses inflicted by mastitis (inflammation of the mammary gland caused by bacterial infection) in cattle exceed 1.7 billion dollars a year in the USA alone. Mastitis is typically caused by staphylococci. Lysostaphin, a powerful peptidoglycan hydrolase secreted by *Staphylococcus simulans,* exhibits a bactericidal effect against staphylococci, causing mastitis. Transgenic cows whose milk contains lysostaphin at a concentration of 0.014 mg/mL have been obtained. It was demonstrated that such cows are characterized by increased resistance to staphylococcal infections.

Bovine spongiform encephalopathy (BSE, also known as mad cow disease) is the most lethal disease affecting cattle in countries of the Northern Hemisphere. Removal of the prion protein gene that causes the disease was proposed as an approach to combat it. Transgenic cows lacking the gene (and, thus, resistant to BSE) have been produced. It is obvious that the use of such cows can reduce the incidence and spread of the disease epidemics.

These illustrations show how promising the application of TAs in agriculture is. Fears about the safety of transgenic food products among the general public are the main obstacle to the widespread distribution of TAs. This leads to the imposition of more onerous regulatory requirements, which makes obtaining authorization to use TAs challenging. The FDA approved a process for reviewing applications for using TAs in 2009, following more than 10 years of development. Both the public and the government see TAs as a means of addressing the issue of food security and raising living standards, and the approval process for new products is more straightforward in developing nations. As a result, nations like China, Brazil, and Argentina are currently carrying out the majority of the projects pertaining to the applications of TAs in agriculture.

Did you get it?

1. Outline the methods employed to produce transgenic animals.
2. Explain the role of expression vectors in obtaining transgenic animals.

ROLE MODEL

RUDOLF JAENISCH

The first person to create a transgenic animal

Rudolf Jaenisch is a biology researcher who was the first person to create a transgenic animal. This procedure sparked further developments within the fields of biological and genetic engineering, as this process is used to add specific genetic traits to an organism via artificial inclusion. He is currently a biology professor at the Massachusetts Institute of Technology.

Jaenisch was born on April 20th, 1942 in Międzygórze, Lower Silesian Voivodeship in Poland. He is from a family of German physicians, and like his father and grandfather, was expected to attend medical school. Despite attending medical school and enjoying it, he didn't feel comfortable with the ways in which it was taught and soon lost interest. Seeking to find something he was interested in, Jaenisch began to attend the Max Planck Institute of Biochemistry at Martinsried in Germany to do biological research while continuing his M.D. degree.

Jaenisch immensely enjoyed research in bacteriophages and found that he had a passion for research. Joining one of the leading professors in the development of phages, P. H. Hofschneider, allowed Jaenisch to join his team of researchers. He went on to complete his medical school thesis on phage replication and expression within bacteria. Eventually, Jaenisch went on to graduate in 1967 with his M.D.

After completing two additional years of experimentation at the Max Planck Institute at the University of Munich, he chose to travel to the United States to pursue postdoctoral training. Searching for a mentor in his work, he found geneticist Arnold Levine, who had recently set up his first laboratory at Princeton University. As a result, Levine is credited for being a major inspiration to Jaenisch's research later on in his career. Jaenisch was Levine's first postdoc, and had a significant amount of trust within him, and three weeks into the program, Levine traveled to Europe and allowed Jaenisch to run the lab. It was surprising to Jaenisch that Levine would allow him to run the lab by himself at such an early stage, but as a result, he not only built experience as an independent researcher, but also noted the time he spent with Levine was key to his overall career development. In 1974, during his time spent at Levine's laboratory,

Jaenisch began to research mammalian cells infected with the simian virus (SV40), a pathogen known to cause tumors within connective tissue. With the help of graduate students, Jaenisch continued to research the effects and replication of the SV40 virus, trying to find deeper meaning in the spread of SV40. During his research, he stumbled upon an article by developmental geneticist Beatrice Mintz at the Fox Chase Cancer Center on generating chimeric mice. From reading the article, Jaenisch had suddenly questioned if early developing embryos would be affected by the cancerous effects of the SV40.

Wanting to propose the idea of injecting the SV40 into mice embryo, Janeisch, under the account that his mentor was in Europe at the time, traveled to the Fox Chase Cancer Center to contact Mintz and propose the idea. After much skepticism and careful planning, Jaenisch was given the opportunity to practice the experiment at the Fox Chase Cancer Center, and, while at the grace of Levine, continue half the research in his laboratory. In an interview with Jaenisch on the creation of the first transgenic organism, he credits Mintz as one of the major influences that led to its development.

During his time researching with Mintz and Levine, Jaenisch accepted a position at the Salk Institute for Biological Studies in California, where he began to collaborate with fellow researcher Paul Berg in detecting SV40 within the viral DNA of mice. In 1974, using a technique known as nick translation which takes a piece of radioactive DNA and subsequently marks it into an organism in order to trace its development, found that when the SV40 was injected into the embryo, it had fully incorporated itself into the tissue of the mouse without causing cancer. Jaenisch eventually became ecstatic, as he discovered that the SV40 had fully incorporated itself into the genome of the mouse embryo; as a result, the first transgenic organism was born from the merging of mouse and SV40 DNA.

In 1977 Jaenisch went on to the Heinrich Pette Institute in Hamburg, Germany to study viral infection on mice. The research led him to investigate DNA methylation, which played a role in silencing gene activity during early development. After seven more years, he traveled back to the United States to work at the Whitehead Institute for Biomedical Research in Cambridge, Massachusetts, where he studied epigenetics and attempted to understand how DNA methylation controls gene expression. From there, in 1996, following the birth of the cloned Dolly the Sheep, Jaenisch set out to try to understand how a fully differentiated nucleus from an adult cell that had been placed inside an enucleated egg could be changed by the cytoplasm in the egg. Eventually, in 2007, he became one of the first researchers to identify master gene regulators that effectively reprogram adult cells. Today, he currently serves as a biology professor at the Massachusetts Institute of Technology.

SUMMARY

- Animal biotechnology is the combination of science and engineering which is very useful to transform living organisms. The main goal of animal biotechnology is to produce goods, enhance animals, and cultivate microorganisms for particular agricultural purposes. The creation of transgenic animals exemplifies animal biotechnology.
- Human, animal, or insect cells are grown in a suitable artificial setting through a process called cell culture. The cells could come from strains of established cells, multicellular eukaryotes, or preexisting cell lines.
- Tissue culture involves the in vitro culture and propagation of single cells, tissues, or entire organs in a controlled laboratory environment. When supplemented with a medium containing nutrients and growth factors, many animal cells can be made to grow outside of their original organ or tissue under specific circumstances.
- Animal cell cultures are distinct from microbial cultures in that they exhibit particular traits. Animal cells are characterized by their slow growth rate, need for solid substrata in order to support anchorage, fragility due to the absence of a cell wall, and sensitivity to various physiochemical conditions like pH and CO_2 levels.
- The availability of fresh nutrients at regular intervals allows primary cell cultures that are passaged or subcultured and grown in fresh medium for an extended period to form secondary cultures, which are longer-lasting than the cells of primary cell cultures.
- Mammalian cell culture is now a requirement for the industrial production of biologically active compounds. Numerous cell lines that are used to produce therapeutic proteins, vaccines, pharmaceuticals, and anti-cancer agents have developed as a result of advances in animal cell culture technology.
- A reliable indicator of the health of the cell culture is the quantity of viable cells present. Cell viability is assessed by erythrosin B and trypan blue using the breakdown of cellular membrane integrity.
- Practical technologies based on cell culture have been developed in multiple domains, such as evaluating the toxicity and efficacy of novel drugs, producing vaccines and biopharmaceuticals, and assisting with assisted reproduction.
- The pH of the culture medium that is used for animal-cell cultures generally is maintained by the equilibrium relationship between sodium bicarbonate ($NaHCO_3$) in the culture medium and CO_2 in the incubator.
- Modern techniques such as intra-pronuclear zygotic DNA microinjection (MI) and somatic cell nuclear transfer (NT) enable the production of recombinant protein-producing livestock that carry the necessary transgene in every cell of their organism and pass it on to their progeny.

REVIEW QUESTIONS

1. Categorize and explain different types of cell culture techniques, highlighting their applications and significance.
2. Discuss the factors influencing the growth cycle of cells in culture, and how it impacts experimental outcomes.
3. Define transgenic animals and briefly explain the process of creating them.
4. Explore the factors influencing the selection of specific culture media for different cell types in animal-cell culture.
5. Discuss the role of nutrient requirements in designing effective culture media for animal cells.

MULTIPLE CHOICE QUESTIONS

1. What is the primary purpose of creating transgenic animals?

a. Enhancing natural biodiversity
b. Improving livestock for agricultural purposes
c. Studying plant genetics
d. Exploring marine ecosystems

2. In the context of animal biotechnology, what does the term "knockout" refer to?

a. Enhancing gene expression
b. Inactivating a specific gene
c. Creating hybrid organisms
d. Accelerating growth rates

3. Which biotechnological technique is commonly used for introducing foreign genes into animals?

a. Polymerase Chain Reaction (PCR)
b. Gel Electrophoresis
c. Gene Editing
d. DNA Cloning

4. Which of the following is an application of transgenic animals in medicine?

a. Improved crop yield
b. Production of biopharmaceuticals
c. Enhanced environmental monitoring
d. Aquaculture management

5. What role do genetically modified animals play in xenotransplantation?

a. Producing disease-resistant livestock
b. Providing organs for human transplantation

   c. Enhancing fish populations
   d. Developing novel vaccines

6. **In which field of study are transgenic animals commonly used for disease modeling?**
   a. Environmental Science
   b. Physics
   c. Biomedical Research
   d. Astronomy

7. **What is the significance of using transgenic animals in conservation efforts?**
   a. Enhancing natural habitats
   b. Controlling invasive species
   c. Preserving endangered species
   d. Studying marine ecosystems

8. **What is the primary concern when using transgenic animals in research and applications?**
   a. Environmental impact
   b. Ethical considerations
   c. Economic feasibility
   d. Technological complexity

9. **How do transgenic animals contribute to improving agricultural practices?**
   a. Controlling pest populations
   b. Enhancing soil fertility
   c. Improving crop resistance to diseases
   d. Reducing water consumption in farming

10. **Which of the following is an example of an application of animal biotechnology in aquaculture?**
   a. Development of disease-resistant crops
   b. Enhanced oil recovery
   c. Improving fish growth rates
   d. Restoration of coral reefs

Answer to Multiple Choice Questions

1. (b); 2. (b); 3. (c); 4. (b); 5. (b); 6. (c); 7. (c); 8. (b); 9. (c); 10. (c)

REFERENCES

1. Abrahams, M. V., and A. Sutterlin. 1999. The foraging and antipredator behavior of growth-enhanced transgenic Atlantic salmon. Animal Behavior 58:933–942.
2. Bailone, R. L., Luís, K. De. A., Roberto De, O. R., Ricardo, C. B., Tatiana, C., Helena, J., & Hirla, C. F. S., (2019). Zebrafish as an animal model for food safety research: Trends in the animal research. *Food Biotechnology, 33*(4), 283–302.
3. Carter, B. J., (2004). Adeno-associated virus and the development of adeno-associated virus vectors: A historical perspective. *Mol. Ther., 10*, 981–989.
4. Fiester, A., (2008). Justifying a presumption of restraint in animal biotechnology research. *American Journal of Bioethics, 8*(6), 36–44.
5. Giassetti, M. I., Maria, F. S., Assumpção, M. E., & Visintin, J. A., (2013). Chapter 4. *Genetic Engineering and Cloning: 567 Focus on Animal Biotechnology* (pp. 1–95). 10.5772/56071.
6. Hastie, E., & Samulski, R. J., (2015). Adeno-associated virus at 50: A golden anniversary of discovery, research, and gene therapy success a personal perspective. *Hum. Gene. Ther., 26*, 257–265. doi: 10.1089/hum.2015.025.
7. Nascimento, I. P., & Leite, L. C., (2012). Recombinant vaccines and the development of new vaccine strategies. *Revista Brasileira de Pesquisas Medicas e Biologicas., 45*(12), 1102–1111.
8. Nguyen, B. D., & Bich, T. T. L., (2019). Current research, challenges, and perspectives of biotechnology: An overview. *Vietnam Journal of Agricultural Sciences, 1*(2), 187–199.
9. Parks, K. R., E. J. Eisen, and J. D. Murray. 2000. a. Correlated responses to selection for large body size in oMt1a-oGH mice: Growth, feed efficiency, and body composition. Journal of Animal Breeding Genetics 117:385–405.
10. Rexroad, C. E., Green, R. D., & Wall, R. J., (2007). Regulation of animal biotechnology: Research needs. *Theriogenology, 68*, S3–S8.
11. Shakweer, W. M. E., Hafez, Y. M., El-Sayed, A. A., Awadalla, I. M., & Mohamed, M. I., (2017). Construction of ovine GH-pmKate2N expression vector and its uptake by ovine spermatozoa using different methods. *J. Genet. Eng. Biotechnol., 15*, 13–21.
12. Thompson, P. B., (2008). Current ethical issues in animal biotechnology. *Reproduction, Fertility and Development, 20*(1), 67.
13. Ung-Lanki, S., (2014). Constructing the biotech nonhuman animal: Instrumentalism and ambivalence. *Society & Animals, 22*(5), 439–458.

CHAPTER

6 Industrial Biotechnology

LEARNING OBJECTIVES

After studying this chapter, you will be able to:

- Understand the basic of industrial biotechnology;
- Explain the bioprocess development;
- Define products of industrial biotechnology.

INTRODUCTORY EXAMPLE

Sustainable Biofuel Production through Industrial Biotechnology:

Industrial biotechnology plays a pivotal role in addressing global challenges related to energy sustainability. One compelling example is the production of biofuels as an alternative to traditional fossil fuels. The integration of biotechnological processes within industrial settings has paved the way for the development of sustainable and eco-friendly solutions. This example focuses on how industrial biotechnology is utilized for the production of advanced biofuels, specifically bioethanol, as a cleaner and renewable energy source.

- **Feedstock Selection:** The first crucial step in industrial biotechnology for biofuel production involves the careful selection of feedstock. Lignocellulosic biomass, such as agricultural residues, forestry waste, and dedicated energy crops, is chosen for its abundance and non-competition with food resources. This choice aligns with the principles of sustainability, mitigating concerns associated with food-based biofuel production.
- **Enzymatic Hydrolysis:** The selected feedstock undergoes enzymatic hydrolysis, a process where enzymes break down complex carbohydrates into fermentable sugars. This step is essential for unlocking the potential energy stored in biomass. Advanced enzymes, engineered through biotechnological methods, enhance the efficiency of hydrolysis and reduce processing costs. Enzymes derived from extremophiles, organisms thriving in extreme conditions, have been particularly instrumental in developing robust biocatalysts.
- **Fermentation:** The hydrolyzed sugars are then subjected to fermentation, a key biotechnological process where microorganisms, typically yeast or bacteria, convert sugars into bioethanol. Genetic engineering and synthetic biology enable the modification of microorganisms to enhance fermentation efficiency, increase ethanol yield, and improve resistance to inhibitors present in the biomass. This iterative optimization process ensures that the fermentation step is both economically viable and environmentally sustainable.
- **Downstream Processing:** The bioethanol produced undergoes downstream processing, where separation and purification techniques are employed to obtain a high-purity end product. Membrane technologies, chromatography, and distillation are integrated into the industrial biotechnological process to achieve the desired level of bioethanol purity. This ensures that the final biofuel meets stringent quality standards for use in transportation and other applications.
- **Sustainabil**ity Impact: The adoption of industrial biotechnology in biofuel production contributes significantly to sustainability goals. By utilizing non-food biomass, reducing greenhouse gas emissions, and optimizing processes through biotechnological innovations, the biofuel industry minimizes its environmental footprint. Additionally, the development of biodegradable and energy-efficient enzymes aligns with the principles of green chemistry, fostering a more sustainable approach to industrial processes.

UNIT INTRODUCTION

Industrial biotechnology is a branch of biotechnology that harnesses the power of living organisms, primarily microorganisms and enzymes, to produce valuable products on an industrial scale. Unlike traditional manufacturing processes, which often rely on chemical synthesis and energy-intensive methods, industrial biotechnology offers a more sustainable and environmentally friendly approach. By leveraging the natural capabilities of biological systems, this field enables the production of a wide range of products, including chemicals, materials, fuels, and pharmaceuticals, with greater efficiency and reduced environmental impact.

At the heart of industrial biotechnology is the use of microorganisms, such as bacteria, yeast, and fungi, as microbial factories. These organisms can be genetically engineered to carry out specific tasks, such as producing enzymes or metabolizing raw materials into desired products. The precision of genetic manipulation allows for the optimization of biological processes, leading to increased yields and improved product quality. Additionally, the use of renewable feedstocks, such as plant-based sugars or agricultural waste, further contributes to the sustainability of industrial biotechnology processes.

One of the key advantages of industrial biotechnology is its versatility. The same principles can be applied across various industries, from biofuels and bio-based chemicals to agriculture and healthcare. For example, bio-based polymers can replace traditional petroleum-based plastics, reducing reliance on fossil fuels and minimizing the environmental impact of plastic waste. Similarly, industrial biotechnology plays a crucial role in the production of biofuels, offering a renewable alternative to conventional fossil fuels and contributing to efforts to mitigate climate change.

6.1. THE BASIC OF INDUSTRIAL BIOTECHNOLOGY

→Learning Objectives

- Trace the history of industrial biotechnology;
- Recognize the benefits of industrial biotechnology;
- Identify the tools of industrial biotechnology;
- Understand the impact of industrial biotechnology.

The use of enzymes and microorganisms to produce chemicals (for a wide range of applications) from renewable sources forms the basis of industrial biotechnology. The use of contemporary biotechnology to the sustainable synthesis of materials, compounds, and fuels from living cells and/or enzymes (renewable sources) is known as industrial biotechnology. After the first wave of medical biotechnology, also referred to as red biotechnology, and the second wave of agricultural biotechnology, also known as green biotechnology, the field is thought to be in its third wave.

Industrial biotechnology has attracted a lot of attention and is being widely used because of its advantages, which include a decreased dependency on energy sources, discounted emissions of greenhouse gases, and a decrease in waste generation. The area is growing quickly. As the industry grows, it might force a paradigm change in value-added chemical production from fossil fuel-based to bio-based. This will make it possible for related industries to combat climate change and lower their carbon footprints.

6.1.1. History of Industrial Biotechnology

Industrial biotechnology has a rich history that spans centuries, with roots in ancient fermentation practices and the more recent advancements in molecular biology and genetic engineering. This narrative captures key milestones and developments that have shaped the evolution of industrial biotechnology into a cornerstone of modern manufacturing and sustainable processes.

1. **Ancient Foundations:** The foundations of industrial biotechnology can be traced back to ancient civilizations that unknowingly harnessed the power of

microorganisms for fermentation. Brewing, bread-making, and the production of fermented foods were among the early applications. These processes relied on the metabolic activities of yeast and bacteria, marking the humble beginnings of biotechnological practices.

2. **20th Century Breakthroughs:** The mid-20th century witnessed transformative breakthroughs that laid the groundwork for industrial biotechnology as we know it today. The discovery of the structure of DNA by James Watson and Francis Crick in the 1950s marked a pivotal moment, paving the way for the manipulation of genetic material. This era also saw the emergence of recombinant DNA technology, allowing scientists to splice and recombine genes.

   In 1978, Genentech achieved a groundbreaking feat by producing human insulin through genetic engineering. This achievement marked the birth of the biotechnology industry, as it showcased the potential to use microorganisms as biofactories for the large-scale production of valuable proteins and pharmaceuticals.

3. **Biotechnological Renaissance:** The 1980s and 1990s witnessed a biotechnological renaissance, characterized by rapid advancements in genetic engineering and the commercialization of biotechnological products. Techniques for gene cloning and expression were refined, enabling the production of a wide range of enzymes, hormones, and therapeutic proteins. This period also saw the development of bioremediation technologies, where microorganisms were employed to clean up environmental pollutants.

4. **Omics Revolution:** As the new millennium dawned, the advent of omics technologies, including genomics and proteomics, provided a deeper understanding of microbial systems. This knowledge facilitated the design and optimization of industrial processes, making biotechnological applications more efficient and precise.

5. **Synthetic Biology and Precision Engineering:** The 21st century brought forth the era of synthetic biology, offering unprecedented control over biological systems. Synthetic biology allowed scientists to design and construct new biological components, leading to the creation of tailor-made microorganisms for specific industrial applications. This precision engineering opened avenues for the development of bio-based chemicals, advanced materials, and innovative pharmaceuticals.

6. **Technological Marvels:** The development of CRISPR-Cas9 technology in the 2010s revolutionized genetic engineering. This powerful tool allowed for precise and targeted gene editing, enhancing the efficiency of creating GM organisms for industrial purposes. The ability to edit genes with unprecedented accuracy accelerated progress in optimizing microorganisms for industrial processes.

7. **Sustainability and Renewable Resources:** In recent years, industrial biotechnology has embraced a sustainability paradigm. Researchers and industries are increasingly focused on utilizing renewable resources, such as biomass and agricultural waste, to produce biofuels, bioplastics, and other bio-based products. This shift towards eco-friendly practices aligns with global efforts to reduce environmental impact and address climate change.

8. **Current Landscape and Future Prospects:** Today, industrial biotechnology is a multifaceted discipline with applications spanning diverse sectors, including energy, healthcare, and environmental management. The field continues to evolve as researchers explore innovative solutions to global challenges. From the development of biofuels as an alternative to fossil fuels to the production of bio-based materials as replacements for traditional plastics, industrial biotechnology is at the forefront of sustainable innovation.

Keyword

Renewable Resource – It is a substance of economic value that is replenished naturally over time thereby supporting sustainability despite consumption.

6.1.2. Benefits of Industrial Biotechnology

Industrial biotechnology offers a multitude of benefits across various sectors, contributing to sustainable development and addressing environmental, economic, and social challenges. Some key advantages include:

1. Environmental Sustainability:
   i. **Reduced Environmental Impact:** Industrial biotechnology often utilizes renewable feedstocks and employs microorganisms that can operate under mild conditions, reducing the need for harsh chemicals and energy-intensive processes. This leads to lower greenhouse gas emissions and a decreased overall environmental footprint.
   ii. **Biodegradability:** Bioproducts generated through industrial biotechnology are often biodegradable, minimizing pollution and waste accumulation. This contrasts with some traditional manufacturing processes that produce persistent pollutants and non-degradable materials.
2. Resource Efficiency:
   i. **Optimized Processes:** The ability to genetically engineer microorganisms enables the design of highly efficient and specific biological processes. This optimization results in higher yields, reduced resource inputs, and increased overall production efficiency.
   ii. **Utilization of Renewable Resources:** Industrial biotechnology can utilize renewable resources such as plant-based sugars, agricultural residues, and other biomass, reducing dependence on finite fossil fuels and contributing to a more sustainable resource base.
3. Economic Benefits:
   i. **Innovation and Job Creation:** The adoption of industrial biotechnology fosters innovation and the development

of new products and processes, leading to economic growth. Moreover, the expanding biotechnology sector creates job opportunities in research, development, manufacturing, and related fields.

ii. **Cost-Effective Production:** Bioprocesses can often be more cost-effective than traditional chemical methods, particularly when using inexpensive raw materials. This can lead to more competitive pricing for bio-based products.

4. Diversity of Applications:

i. **Versatility:** Industrial biotechnology finds applications in diverse industries, including pharmaceuticals, agriculture, energy, and chemicals. This versatility allows for the production of a wide range of products, contributing to the resilience and adaptability of the biotechnology sector.

ii. **Bio-Based Alternatives:** The development of bio-based alternatives to traditional products, such as biofuels, bioplastics, and bio-based chemicals, offers sustainable choices that reduce reliance on fossil fuels and mitigate environmental impact.

5. Health and Medicine:

i. **Biopharmaceuticals:** Industrial biotechnology plays a crucial role in the production of biopharmaceuticals, including vaccines, insulin, and monoclonal antibodies. These advanced medical products often have improved efficacy and reduced side effects compared to traditional pharmaceuticals.

ii. **Personalized Medicine:** Advances in biotechnology contribute to the development of personalized medicine, tailoring treatments based on an individual's genetic makeup for more effective and targeted healthcare.

6.1.3. Tools of Industrial Biotechnology

The advances of a wide variety of biological toolsets, including genetic engineering, protein engineering, metabolic engineering, synthetic biology and systems biology ("omics" and in silico approaches) for industrial biotechnology are discussed in the following are:

6.1.3.1. Genetic Engineering

Recombinant DNA (rDNA) technology, another name for genetic engineering, is the process of transferring a desired gene from one organism to another so that it can be expressed there. The rDNA technology uses both in vitro and in vivo methods. Phage particles, which transfer genetic material from one organism's chromosomes to another, are employed in in vivo procedures. The second organism becomes infected with the genetic material from the first host. On the other hand, in vitro methods rely on vectors that gather data from one cell and transfer it to another. It begins with the in vitro manipulation of a gene into the vector, which is followed by the insertion of the carrier and any additional DNA it has into the recipient cell. The issue of gene expression emerges after the target gene has been integrated into the recipient cell. When a bacterium is given a eukaryotic gene, this becomes especially challenging. Because eukaryotic genes differ from prokaryotic genes in that they contain non-coding DNA segments, the expression of eukaryotic genes in prokaryotic cells presents a challenge. Thus, the production of a eukaryotic product by a prokaryotic cell necessitates the incorporation of the genes coding for the product in a form that may be translated by the recipient cell. The most publicized application of recombinant DNA technology in the context of fermentation is the construction of strains capable of synthesizing foreign proteins.

Human growth hormone (hGH), which treats hypopituitary dwarfism, was the first commercially produced heterologous protein. Before it was made by fermentation, it was taken out of human cadavers' brains. The synthesis of recombinant human growth hormone from *E. coli*, which met the demand of the compound and removed the potential hazards of the source. When recombinant human insulin was first made available for purchase in the UK in 1982, it surpassed all other forms of the drug by 1989. Recombinant proteins are the only workable method of producing many other human proteins in large enough quantities for use as therapeutic agents because their synthesis occurs at very low levels. Products like erythropoietin and interferons are examples of this type. Commercial recombinant proteins such as insulin, growth hormone, erythropoietin, G-CSF, GM-CSF, factor VIII, α and β interferon, hepatitis B surface antigen, etc., are produced for therapeutic purposes. A lot of work has gone into improving and increasing the yield of native microbial products using recombinant DNA technology. At an industrial scale, a wide range of recombinant microbial products are produced, including polysaccharides, enzymes, amino acids, vaccines, and antibiotics. The main challenge in using recombinant microbes is determining how stable they can be in large-scale culture, particularly in continuous processes.

6.1.3.2. Protein Engineering

Protein engineering is a key component of industrial biotechnology. A wild-type enzyme found in nature is almost never appropriate for use in an industrial process. For the enzymatic process to be commercially viable, it is necessary to engineer and optimize the performances of the enzymes in terms of activity, selectivity, enantioselectivity, and substrate/product inhibition.

In general, there are two methods for protein engineering: directed evolution and rational design. In rational design, site-directed mutagenesis is a useful tool for achieving desired changes in a protein only if its structure, function, and catalytic mechanism are well understood. Nevertheless, for the majority of interesting proteins, this knowledge is lacking. Furthermore, only patchy results have been shown despite the development of computational protein design algorithms to forecast ideal mutations at particular residue positions in the protein. On the other hand, the directed evolution method just needs the protein sequence to work. Using this method, random mutagenesis and/or gene recombination are repeated cycles followed by selection or screening for positive mutants.

It is frequently like searching for a needle in a haystack when trying to locate an enzyme with desired characteristics among a library of mutants produced by directed evolution. To identify variants of interest, numerous screening and/or selection methods have been developed in recent years. Microtitre plate-based screening is a frequently employed technique. In a 96-well microtitre plate format, a standard screening process starts with the creation of a library of mutants selected and cultured in 96-well plates. When the target proteins are expressed, high throughput assays based on UV absorption, fluorescence, or colorimetric techniques are frequently applied. After that, mutants exhibiting the desired traits are confirmed and sequenced. The best mutant is subsequently chosen to serve as the mutagenesis template for the following cycle. Iteratively repeating the process continues until the objective is met or no more advancements are feasible. The agar plate screen, the cell-in-droplet screen, the cell as microreactor cell surface display, and in vitro compartmentalization are additional screening/selection techniques. Even though many different screening and selection tools are available, their applicability is limited.

Keyword

Protein Design – It is the process of creating new proteins with specific properties by manipulating the sequence of amino acids.

6.1.3.3. Metabolic Engineering and Modeling

As our understanding of microbial metabolism improves, there will be more and more opportunities to modify bacteria, yeasts, and fungi to produce new products and increase yields. Metabolic engineering typically uses rDNA technology to develop microorganisms giving improved product yields or even having new pathways. By manipulation of enzymatic, transport, and regulatory functions in the cell, metabolic engineering redirects precursor metabolic fluxes, changes protein cellular levels, fine-tunes gene expression, and controls gene expression regulation in microorganism hosts such as *E. coli*, *S. cerevisiae*, and actinomycetes. Such "design organisms" can be seen as cell factories and form the cornerstone of industrial biotechnology. As in classical strain development, knowledge of microbial physiology remains essential to the practice of metabolic engineering. Understanding the regulation of genes, in particular, the interaction of promoters and regulatory proteins is also essential. The five key strategies for metabolic engineering can be summarized as: (i) enhance committed step from central metabolism or branch point, (ii) eliminate transcriptional and allosteric regulation, (iii) identify and relieve rate-limiting step(s), (iv) prevent carbon and energy loss to competing pathways, and (v) production of foreign enzymes or metabolites.

An alternative is to metabolically engineer microbes to produce large quantities of valuable metabolites that are too complex or costly to produce chemically and difficult to extract from their natural sources. With an annual sales revenue of US $1 billion, taxol (paclitaxel) is an antimitotic agent used in the treatment of metastatic breast cancer and ovarian cancer. In the beginning, only a very small amount of paclitaxel—roughly 9000 kg of yew bark from 3000 trees—could be obtained from the bark of the yew Taxus brevifolia to produce 1 kg of pure paclitaxel. Therefore, producing taxol in a microorganism is a desirable and cost-effective substitute for extracting it from plant biomass. Recently, yeast was used to synthesize taxadiene, an intermediate in the biosynthesis of taxol, efficiently.

The identification of key locations for metabolic engineering is becoming increasingly dependent on the combination of sophisticated metabolic models and experimental data. For instance, DesignPath® is a method that uses mathematical models and specially created algorithms to predict the ideal gene combination that would allow a cell to maximize the synthesis of a desired biomolecule, like 1,3-propanediol, and minimize the synthesis of undesirable by-products, like, e.g., acetic acid. Since this field's knowledge and research capacities have advanced significantly, the change should now concentrate on developing predictive metabolic models rather than just using a trial-and-error methodology.

The focus areas of research in metabolic engineering and modeling are listed are:

- Better understanding of molecular aspects and microorganism metabolism under industrial conditions.
- Advanced metabolic engineering research for efficient industrial production of products such as bioethanol, biomaterials, bulk chemicals, and specialties including enantiopure molecules.
- Mathematical modeling of microbial metabolism, directed towards both steady state and dynamic models.
- Design of new pathways and synthetic microorganisms focusing on the synthesis of molecules and products new to nature.
- Novel and efficient production systems focussing on protein synthesis and expression in new hosts to develop new or improved enzymes.

6.1.4. Impact of Industrial Biotechnology

An evaluation of the potential of white biotechnology was carried out by a number of prominent industrial biotechnology companies, including DSM, Cargill Dow, BASF, Novozymes, and Genecor, in collaboration with the reputable and independent German Oko Institute, the European and US biotechnology industry associations (EuropaBio and Bio). In-depth case studies and a McKinsey & Company market analysis were utilized to calculate the effect on the three components of sustainable development: profit, planet, and people (Figure 6.1). The outcomes supported an earlier OECD study that found industrial biotechnology has positive social, environmental, and economic effects. In a self-reinforcing cycle involving all parties involved, industrial biotechnology has the potential to produce new equipment, lessen its environmental impact, and even generate economic value.

A further factor in the appeal of biotechnology is the resurgence of interest in the sustainability of industrial processes. The sustainability of all the main pillars of European society and economy—agriculture, environmental conservation, and industry—is being called into question. Industrial biotechnology has a significant potential impact. It can: (i) increase the competitiveness and sustainability of agriculture (in its broadest sense, including forestry) by opening up new non-food markets, (ii) enhance human well-being while lessening environmental impact by producing innovative goods at reasonable prices, and (iii) assist industry in becoming more sustainable, eco-friendly, and efficient in both the economy and the environment while preserving or enhancing its competitive edge and capacity for growth. Moreover, white biotechnology has the potential to improve amiability in all three domains, including the environment, the economy, and society. To put it succinctly, the knowledge-based bioeconomy is built around industrial biotechnology. It increases the value of agricultural products and creates new industrial production plans that aim higher overall.

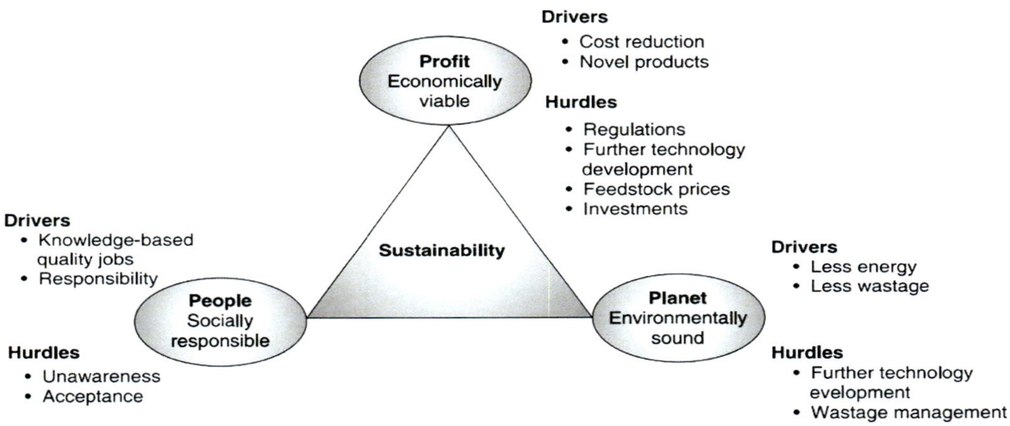

Figure 6.1. *The triple-P bottom line.*

Source: researchgate.net/publication/311576484/figure/fig3/AS:614001801191431@152340084557 0/The-impact-of-industrial-biotechnology-on-the-economy.png.

The environment benefits because biotechnological processes are efficient users of (often renewable) raw materials, creating little end-of-the-pipe waste, which itself can be often used as input into a further biological process.

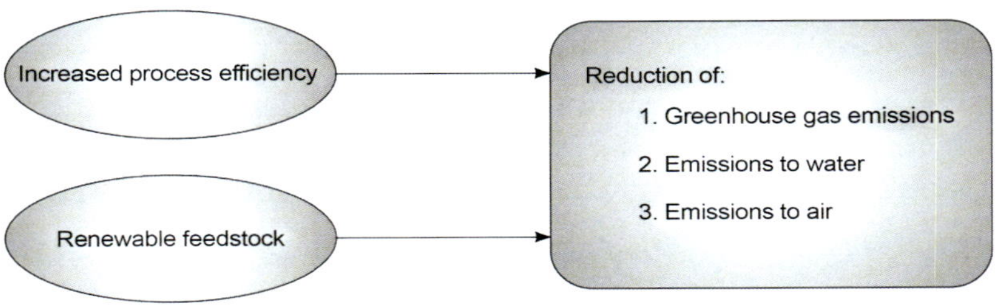

Figure 6.2. *The impact of industrial biotechnology on the environment.*

Source: researchgate.net/publication/311576484/figure/fig3/AS:614001801191431@152340084557 0/The-impact-of-industrial-biotechnology-on-the-economy.png.

At the same time, moving from chemicals to biological processes can lead to a significant reduction in carbon dioxide emissions, energy consumption, and water use. Industrial biotechnology has a dual impact: it increases process efficiency and enables the use of renewable feedstock. Therefore, it is a key tool in the development of sustainable production processes. The impact of industrial biotechnology on the environment is illustrated in Figure 6.2.

Biotechnology facilitates the introduction of less energy-intensive, more efficient processes, which benefits the economy. In the fine chemicals industry, fermentation and enzymatic methods are already widely employed to create flavors, vitamins, and pharmaceutical intermediates. Additionally, they are beginning to penetrate a number of other industrial sectors as well as high-volume markets like biofuels, bulk chemicals, and polymers. The consumer and automotive industries will use new polymers in a variety of ways. Feedstock costs are a growing concern as industrial biotechnology progresses from fine chemicals into the commodities market and ultimately into bulk goods. Figure 6.3 illustrates how industrial biotechnology affects the economy.

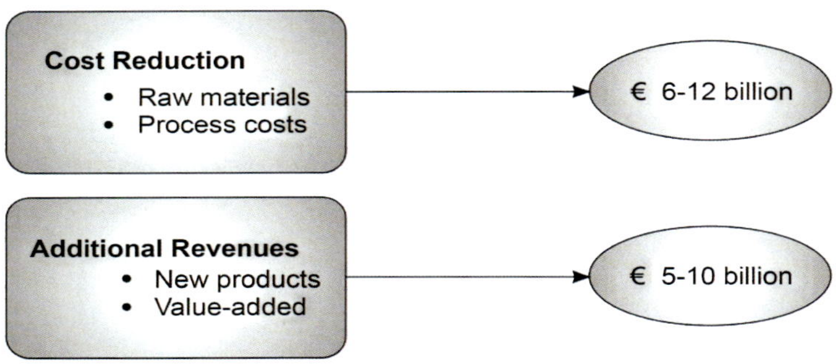

Figure 6.3. *The impact of industrial biotechnology on the economy.*

Source: researchgate.net/publication/311576484/figure/fig3/AS:614001801191431@1523400845570/The-impact-of-industrial-biotechnology-on-the-economy.png.

It is anticipated that the benefits of industrial biotechnology will be felt in a number of crucial areas related to society: the creation and retention of jobs, the creation of new technological platforms, and the lessening of societal reliance on finite fossil fuels, which will preserve them for future generations. Industrial biotechnology can support advanced research and education by creating R&D projects and jobs for highly qualified individuals. Figure 6.4 illustrates how industrial biotechnology has affected society.

The full potential of industrial biotechnology cannot be realized unless all stakeholders work together in concert. Stakeholders should first engage in a dialog to exchange information and facts, as well as to talk about the opportunities and potential drawbacks of this technology. Relevant parties include government agencies, academic institutions, and businesses. Non-government organizations, the financial sector, suppliers, industry users, and observers from organizations like the OECD are additional examples of stakeholders.

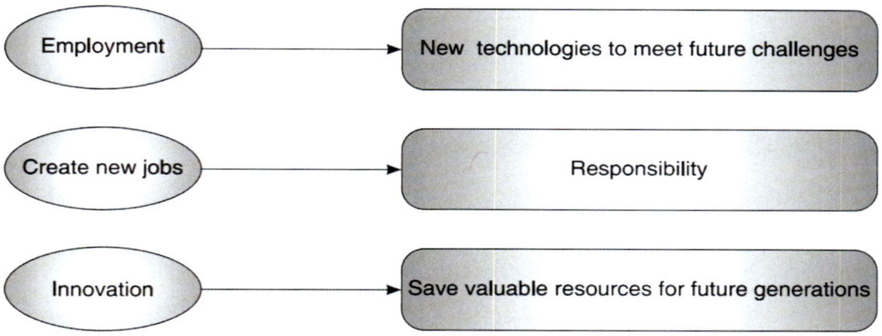

Figure 6.4. *The impact of industrial biotechnology on society.*

Source: Ram Sarup Singh. Industrial Biotechnology: An Overview. https://www.researchgate.net/publication/311576484.

Did you get it?

1. Summarize the key points in the history of industrial biotechnology.
2. Identify and discuss the benefits associated with industrial biotechnology.

6.2. BIOPROCESS DEVELOPMENT

→Learning Objectives

- Define upstream processing (USP);
- Explain the downstream processing (DSP).

Biotechnology has gained global attention due to advancements in the fields of genetics and molecular biology in the 21st century. Our understanding of agriculture, medicine, and environmental management have all undergone significant shifts since DNA manipulation became possible. Innovative new products and services are being offered by the biotechnology industry thanks to advances in gene expression research, protein engineering, and cell fusion. Molecular biology and cell manipulation expertise will be put to use in biotechnology to process biological materials on a large scale. Biotechnology and bioprocess development are combined to achieve this aim. In the food, chemical, and pharmaceutical industries, bioprocess development is crucial. Enzymes, as well as the cells of microorganisms, plants, and animals, are used in bioprocess operations to produce new goods and manage waste.

The utilization of microorganisms for the purpose of transferring biological material to produce fermented foods has its roots in ancient history. Since then, bioprocesses have been created for a vast array of commercial goods, ranging in price from costly specialty chemicals like antibiotics, therapeutic proteins, and vaccines to comparatively inexpensive materials like industrial alcohol and organic solvents. Bioprocesses may find use in additional fields where chemical processes are currently employed as methods and equipment are improved. Because bioprocesses use living cells or enzymes, they offer several advantages over conventional chemical methods of production: (i) they usually require lower temperature, pressure, and pH (the measure of acidity), (ii) they can use renewable resources as raw material, and (iii) greater quantities can be produced with less energy consumption.

Although enzymes can produce certain products quickly and in large quantities, improvements are frequently required before they can be used in industrial processes. Methods like directed evolution, protein engineering, and gene splicing, enzymes are better suited for usage in industrial processes. These instruments facilitate the creation of fresh biocatalysts for entirely new uses. Production processes will operate more efficiently when biocatalytic processes are designed optimally. In industrial practice, promising process interactions are rarely found and utilized because case-based reasoning is the norm. As a result, systematic design technology that can swiftly and accurately develop new high-performance processes is desperately needed. Table 6.1 lists the research focus areas for biocatalytic process design.

Table 6.1. *Focus Areas of Research in Biocatalytic Process Design*

• Expansion of the range of enzymes for industrial applications. • Use of newly generated information on the biochemical pathways and enzymes to search for catalysts capable of novel chemical conversions. • Expanded applicability of high-throughput screening methods for activity and (stereo) selectivity of enzyme and mutant libraries. • Optimization of biocatalysts by directed evolution and/or rational design. • In-depth understanding of catalytic enzyme-substrate interactions to enable new enzyme applications or synthesis of new bioproducts. • Integration of bio- (and chemo-) catalysts into industrial processes, as well as development of robust biocatalysts for industrial processes to ensure successful scale-up. • Direct integration of enzyme production and transformation, including downstream processing of target compounds.
Development of modular and multiphase bioreactors. • Reduction of the number of unit operations in biocatalytic processes by using multi-step bioreactions without isolation of intermediates via the design of both enzymes and bioreactors. • Use of cascades of enzymes in a single unit process and whole cell bioconversion system. • Development of micro- and nano-devices for chemical and biochemical analysis, including biochips for molecular and cellular detection and (automated online) monitoring.

Source: Ram Sarup Singh. Industrial Biotechnology: An Overview: https://www.researchgate.net/publication/311576484.

Progress in the fields of microbiology, biochemistry, and cell physiology has been intimately linked to our capacity to utilize the potential of cells and enzymes. Modern biotechnology tools like recombinant DNA, gene probes, cell fusion, and tissue culture offer ways to create new products or enhance bioprocessing techniques as our understanding in these areas is rapidly growing. Biology is a major factor in the Industrial Revolution because modern biotechnology has enabled us to envision advanced medications,

Keyword

Industrial Alcohol – It is distilled ethyl alcohol (C_2H_5OH), normally of high proof, produced, and sold for purposes other than beverages. It is usually distributed in the form of pure ethyl alcohol, completely denatured alcohol, specially denatured alcohol, and proprietary solvent blends.

the culture of human tissues and organs, biochips for cutting-edge computers, environmentally friendly pesticides, and potent microbes that degrade pollution. The primary challenge, however, lies in implementing these newly developed products—which were conceived and partially developed in laboratories—at an industrial scale. As a result, developing bioprocesses has grown to be crucial to industrial biotechnology. Upstream (USP) and downstream (DSP) processing are components of all bioprocesses. Both of them are very important for the development of an efficient bioprocess.

Remember

Biocatalytic process design must be especially considered when using biological processes in industrial settings, potentially in conjunction with conventional chemistry.

6.2.1. Upstream Processing (USP)

The factors and processes that lead to and include fermentation are all included in upstream processing (Figure 6.5). There are three primary sections. The producer microorganism's aspects are the subject of the first. It comprises the plan, the initial acquisition of a suitable microorganism, the enhancement of productivity and yield through strain improvement, the preservation of strain purity, the creation of an appropriate inoculum, and the ongoing development of specific strains to raise the economic efficiency of the process. The first step in the development of a strain is to locate or isolate an organism that can produce the desired molecule. To satisfy the demands of the production process and the bottom line, the isolated strain is inevitably enhanced through mutagenesis, screening, selection, and genetic engineering. The strain handling, medium formulation, fermentation conditions, product recovery, and application are among the process requirements. It should come as no surprise that process development and strain occur simultaneously. For instance, processes that are cost-sensitive when it comes to the fermentation of raw materials search for strains that are low in carbon. Similar to this, procedures that seek to produce products with high levels of purity must employ fermentations that need the use of little medium and strains that yield few byproducts. It is now possible to rationally engineer a production strain thanks to new genetic techniques. A rational approach entails the targeted engineering of genes through mutagenesis, deletion, or over-expression that determine the production of a specific metabolite. The rate (productivity), yield (carbon conversion efficiency), and titer (concentration) of the product are all determined by the expression of these genes. The result is a production organism that is perfectly suited for both the process and the product. A desired product strain can be kept alive as a stock culture once it has been developed. Genetic alterations must be kept to a minimum in-stock culture maintenance. The best practice is to keep cells or spores (if any are generated) in liquid nitrogen-filled sealed ampoules at extremely low temperatures.

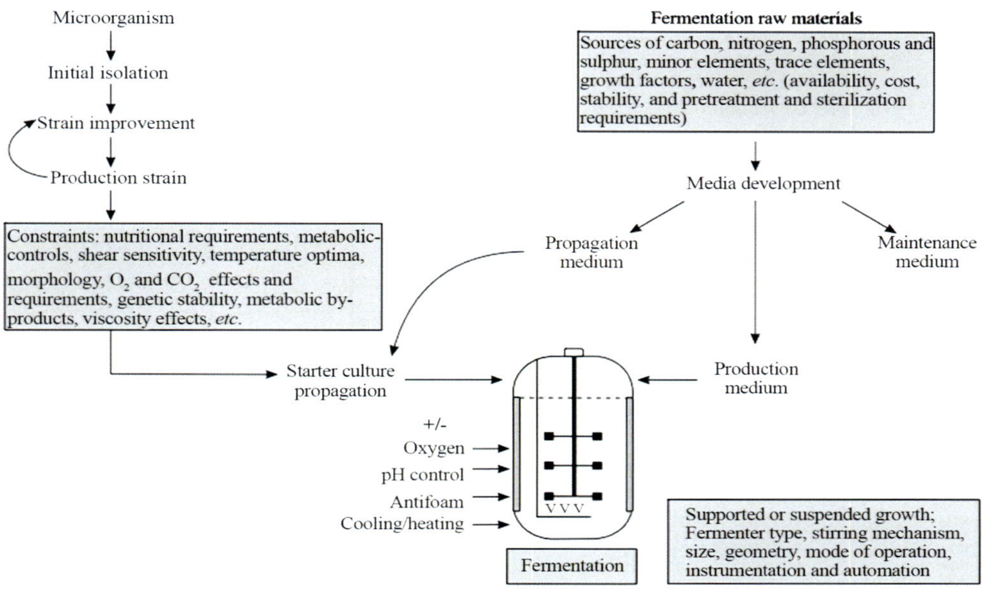

Figure 6.5. *Outline of upstream processing.*

Source: Ram Sarup Singh. Industrial Biotechnology: An Overview. https://www.researchgate.net/publication/311576484.

The second USP component is the fermentation medium, specifically the choice of an appropriate, economical carbon and energy source in addition to other necessary nutrients. Media optimization is, therefore, a crucial component of process development to guarantee yield and profit maximization. In-depth analyzes are required to determine which medium is best for a given bioprocess, but any such medium needs to satisfy a few fundamental requirements. Water, energy sources, carbon, nitrogen, mineral elements, etc., are necessities for all microbes and if the process is aerobic, oxygen is also needed. Making a pure compound medium on a small scale is not too difficult, but the resulting medium might not be appropriate for use in a large-scale process even if it supports satisfactory growth. Large-scale nutrient sources are typically needed to produce a medium that satisfies as many of the requirements listed in Table 6.2. Because they are inexpensive substrates, cane, and beet molasses, cereal grains, starch, glucose, sucrose, and lactose are used as carbon sources; as nitrogen sources, ammonium salts, urea, nitrates, corn steep liquor, soybean meal, slaughterhouse waste, and fermentation residues are used. These materials have a tendency to meet most of the stated criteria for production media. However, it is possible to use simpler procedures; hence, other more costly pure substrates might be selected if the total cost of the entire process can be decreased. Certain elements of a production medium may occasionally aid in controlling product production as opposed to promoting the microorganism's growth. Precursors, inducers, and inhibitors are a few examples of these additives that can be used to control how the process proceeds.

Table 6.2. *Criteria for the Selection of a Medium at an Industrial Scale*

•	It will produce the maximum yield of product or biomass per gram of substrate used.
•	It will produce the maximum concentration of product or biomass.
•	It will permit the maximum rate of product formation.
•	There will be a minimum yield of undesired products.
•	It will be of consistent quality and be readily available throughout year.
•	It will cause minimal problems during media making and sterilization.
•	It will cause minimal problems in other aspects of the production process, particularly aeration and agitation, extraction, purification, and waste treatment.

Source: Ram Sarup Singh. Industrial Biotechnology: An Overview. https://www.researchgate.net/publication/311576484.

It is also necessary to take into account the challenge of developing the process from the laboratory to the pilot scale and then to the industrial scale. In a large fermenter with a low gas transfer pattern, a laboratory medium might not be the best choice. For efficient stirring, a medium with a high viscosity will also require higher power inputs. The medium may affect pH fluctuation, foam formation, the oxidation-reduction potential, and the morphological form of the organism, in addition to fulfilling requirements for growth and product formation. Moreover, precursors or metabolic inhibitors might be required. The effluent treatment and product recovery is also impacted by the medium. Various workers have also reviewed different facets of microbial media.

Foaming is an issue in certain microbiological bioprocesses. It might be caused by a microorganism-produced factor or something in the medium. Proteins in the medium, such as peanut meal, yeast extract, meat extract, corn-steep liquor, etc., are the most frequent cause of foaming. At the air-broth interface, these proteins may denature and form a skin that is difficult to rupture. When foaming happens during the fermentation process, antifoam agents are added. Due to their low solubility, many antifoams require a carrier, such as liquid paraffin, lard oil, or castor oil, which can be metabolized and impact the fermentation process. Antifoam additions should be minimized because, regrettably, the concentrations of many antifoams required to regulate fermentation will lower the oxygen transfer rate by as much as 50%. Other antifoams that speed up the transfer of oxygen are also available. If the antifoam addition reduces the oxygen transfer rate, mechanical foam breakers might need to be taken into consideration as a potential substitute.

Historically, undefined complex natural materials have been used in fermentation processes because they are much cheaper than pure substrates. However, there is often considerable batch variation because of variable concentrations of the component parts and impurities in natural materials, which cause unpredictable biomass and/or product yields. As a consequence of these variations in composition small yield improvements are difficult to detect. Undefined media often make product recovery and effluent treatment more problematical, because not all the components of a complex nutrient source will be consumed by the organism. The residual components may interfere with recovery and contribute to the biological oxygen demand of the effluent. Thus, although manufacturers have been reluctant to use defined media components because they are more expensive, but pure substrates give more predictable yields from batch to batch and recovery. Purification and effluent treatment are much simpler and, therefore, cheaper. Process improvements are also easier to detect when pure substrates are used.

Table 6.3. *Focus Areas of Research of Upstream Processing*

• Studying the physiology of microorganisms under extreme conditions: pH and temperature, slow growth, high concentration of substrates and products to maximize product yield.
• Development of microbioreactors as a screening tool to shorten the process development time, based on realistic large scale production conditions.
• Development of engineering tools to design strategies for process intensification, such as low-cost fermenters, alternative novel reactor concepts, advanced control strategies and simulation tools for modeling fermentation processes on different scales.
• Combination of energy production and bioprocesses for chemical(s) production.

Source: Ram Sarup Singh. Industrial Biotechnology: An Overview. https://www.researchgate.net/publication/311576484.

The third element of USP is fermentation, which is typically carried out under strictly regulated circumstances and designed to maximize the organism's growth or the production of a desired microbial product. From a chemical and physical point of view, analysis of the various industrial fermentation processes reveals that they are common reactions. The reaction mechanisms that turn raw materials into products can be used to categorize fermentation processes. These comprise hydrolyzes, complicated biosyntheses, reductions, simple and complex oxidations, transformations, polymerizations, substrate conversions, and cell formation. With the exception of sterilization, many common operations of chemical engineering units are shared by fermentation processes. Microorganisms, medium, and air are three heterogeneous phases that are "mixed" in aerobic fermentations, for instance. Additional unit operations include "heat transfer" from the fermentation medium and "mass transfer" of oxygen from the air to the organisms. Understanding the behavior of fermentations has been greatly enhanced by analysis using the unit operation technique. Table 6.3 lists the main upstream processing research areas.

6.2.2. Downstream Processing (DSP)

Downstream processing is the separation and refinement of a biotechnological product to a state appropriate for its intended application. DSP includes all post-fermentation procedures. Its main goals are to recover the targeted product safely, efficiently, and to the necessary specifications (biological activity, purity, etc.), even as recovery yield is maximized and expenses are reduced. Depending on whether the target product is extracellular or intracellular, it may be recovered by processing the spent medium or the cells. Most of the time, the particular use of the product dictates the required level of purity.

Every step of the overall recovery process is heavily reliant on the previous fermentation's protocol. The characteristics of microorganisms, specifically their morphology, flocculation characteristics, cell size, and cell wall rigidity, are among the fermentation factors that impact DSP.

These elements significantly affect the efficiency of sedimentation, homogenization, and filterability. DSP procedures and the related product analysis may be hampered by the presence of fermentation additives like antifoams, media impurities, and fermentation byproducts. Consequently, creating a new industrial purification strategy calls for an all-encompassing approach. All aspects of the process, including those upstream and downstream, must be taken into account. For instance, the fermentation substrate selected affects the DSP that comes after. A cheap carbon and energy source with lots of impurities could save money at first, but it might also mean paying more for DSP. Hence, using a more expensive but pure substrate may result in overall cost savings. Additionally, implementing techniques that use already-existing equipment might be less expensive than introducing more efficient techniques that call for the construction of new facilities.

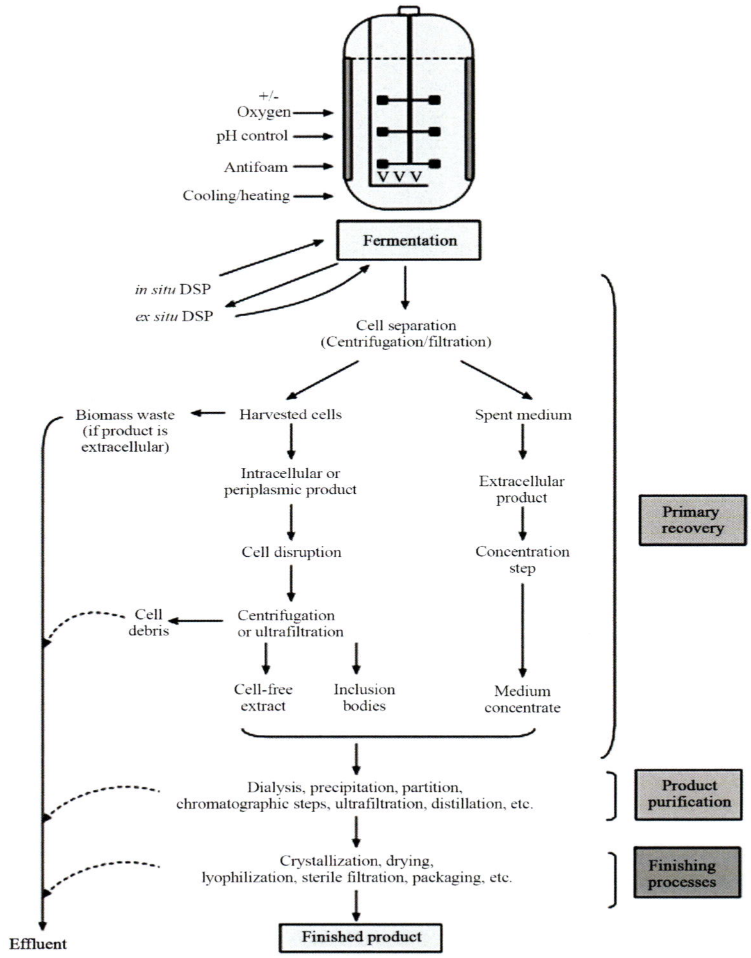

Figure 6.6. *Outline of downstream processing.*

Source: https://www.researchgate.net/figure/Flow-diagram-for-downstream-processing-of-pullulan_fig14_278722442.

The product's concentration, location, and physical and chemical characteristics all play a crucial role in determining the first steps in the separation process as well as the overall purification plan. The target product could be the entire cell or an intracellular product that could be found inside an organelle or as inclusion bodies. As an alternative, the target product might have been secreted into the fermentation medium or the producer cell's periplasm.

The product's stability also affects whether any pretreatment is required to stop the product from degrading or becoming inactive. DSP is a multi-step procedure that can be broken down into several discrete unit operations that are connected to produce a purified product (Figure 6.6). There should be as few stages as possible. This will avoid overall losses during the multistage purification process in addition to being cost-effective. The economics of the process, the necessary product purity, the yield attained at each stage, and safety considerations will all impact the particular unit steps selected. Integration of DSP and fermentation is now preferred in many cases. Integration frequently results in higher output, fewer unit operations, shorter processing times, and lower costs.

Physical process integration can be accomplished by connecting the two systems directly or by putting separation units inside the fermenter. When growth and product formation go hand in hand, i.e., primary metabolites, employing integrated systems that sustain high cell densities by cell retention or recycling may lead to increased productivity.

When products are inhibitory, bioreactors can be partitioned using a variety of techniques to enable the products to be removed quickly in situ by extraction, adsorption, or stripping, thus frequently increasing productivity and yield.

An alternative method of processing is called ex situ, in which the product is taken out of the fermenter, and the medium is processed before being used again in the fermentation. Although extracting certain proteins has become feasible, the removal of alcohols and solvents has proven to be a particularly effective use for these procedures. When there are unfavorable thermodynamic equilibria, i.e., when there is a mode of product removal, it can frequently improve raw material conversion. The elimination of product "pulls" the reaction in the direction of additional product formation. For products susceptible to extended exposure to the fermentation environment—where shear, proteolytic enzymes, oxidation, etc.—early product extraction can also increase yield and might be harmful. Table 6.4 lists the main research focus areas for creating an inventive downstream processing.

Keyword

Bioreactor – It is a large vessel where different cells, such as human or plant, or animal cells can be cultured to obtain new biological products.

Table 6.4. *Focus Areas of Research of Downstream Processing*

•	Development of innovative downstream processing techniques such as computer-aided design systems and continuous fermentation processes with new in situ product removal techniques.
•	Development of a toolbox of generic technique (i.e., different techniques for different groups of compounds) to cover the majority of product recovery questions in bioconversions.
•	Integration of downstream recovery and chemistry to facilitate the further conversion of fermentation products, minimizing energy and water input as well as waste streams.

Source: researchgate.net/publication/311576484.

Did you get it?

1. Define upstream processing (USP).
2. Define downstream processing (DSP).

6.3. PRODUCTS OF INDUSTRIAL BIOTECHNOLOGY

→Learning Objectives

- Define organic acids and polymers;
- Explain the vitamins and neutraceuticals;
- Understand the biopharmaceuticals;
- Discuss biofuels and bioenergy;
- Explain the fine and specialty chemicals.

Employing biotechnology, new products as well as replace current chemical processes can be created. Biotechnology has already produced a wide range of products in industries such as food chemicals, medicines, detergents, textiles, energy, materials, and polymers, among others.

Table 6.5. *Milestones of Bioprocess Industry Development*

Chronolgy	*Milestones*
6000 BC	Brewing (Sumeria, Babylonia)
2400 BC	The first bioprocess complete description (ancient Egyptian)
1680	Yeast under the microscope (van Leeuwenhoek)
1835	Alcoholic fermentation associated with yeasts
1857	Fermentation correlated with metabolism (Pasteur)
1877	Term "enzyme" (in yeast) introduced (Kuhne)
1923	Industrial production of citric acid
1930s	Industrial productin of amino acids
1940s	Industrial production of antibiotics
1979	Monoclonal antibody production by hybridoma cell
1982	Industrial production of rHuman insulin in *E.coli* (Eli Lily)
1984	First commercial production of therapeutic MAb (Anti CD3)
1994	First commercial vaccine from recombinant yeast (hepatitis B vaccine)
1996	Completion of the yeast genome project for *S. cerevisiae*
2000	10 m^3 STR for mammalian cell culture
2002	Disposable bioreactors were used at industrial scale
2007	Bioprocess products market exceeded 700 billion US $, Biopharmaceuticals market reached 70 billion US $ & mammalian cell culture products reached 25 billion US $
2013	Biopharmaceutical products market reached 109 billion US $ and by 2017, it is expected to exceed 166 billion US $

Source: researchgate.net/publication/311576484.

Table 6.6. *Some of the Major Industrial Fermentation Products*

Product	*Annual production (metric tons)*	*Main application*	*Microorganism*
Citric acid	1,200,000	Food	*Aspergillus niger*
Ethanol	31,600,000	Fuel	*Saccharomyces cerevisiae*
Glutamate	1,000,000	Flavouring	*Corynebacterium glutamicum*
Lactic acid	259,000	Food, plastics	*Lactobacillus sp.*
Lysine	1,200,000	Feed	*Corynebacterium glutamicum*
Penicillin	215,000	Pharmaceutical	*Penicillium chrysogenum*
Xanthan gum	110,000	Food, oil drilling	*Xanthomonas campestris*

Source: researchgate.net/publication/311576484.

Table 6.5 lists numerous industry development milestones for the bioprocess sector. Enzymes and cellular processes have advantages in process efficiency due to their high specificity and mild reaction conditions, which yield high-quality products. To increase the use of biological processes in many other fields, however, there are still obstacles to overcome. Table 6.6 shows a selection of the main industrial fermentation products.

6.3.1. Organic Acids and Polymers

Numerous organic acids can be generated through microbial or biochemical processes. Currently, industrial fermentation only produces acetic acid (as vinegar), citric acid, itaconic acid, gluconic acid, 2-keto-gluconic acid, and lactic acid. Previously produced through fermentation or enzymatic processes, other organic acids like fumaric, gallic, malic, and tartaric acids are now produced commercially, mostly through the more cost-effective method of chemical synthesis. Xanthomonas campestris uses aerobic submerged fermentation to produce xanthan gum for commercial use. Novel large-scale polymers like biodegradable polymers and monomers (e.g., 1,3-propane-diol for the innovative manufacture of polyester) have also entered the market. In the future, biotechnology will make it feasible to produce new polymers with enhanced functionality that are not achievable through traditional methods.

6.3.2. Alcohols

This group of organic compounds includes ethanol; 1,3-propanediol; butanol; acetone; 2,3-butanediol; and glycerol. These can all be produced through fermentation. Although fermentation has been used to produce butanol and acetone in industry, synthesis is currently the preferred manufacturing process due to cost considerations. However, there is a renewed interest in looking into the fermentation processes as the cost and availability of ethylene and propylene as feedstocks for the synthesis processes become concerns. An increasing number of people are interested in using biological processes to produce chemical feedstocks from renewable sources, such as agricultural crops, as the price of renewable resources, such as crude oil, keeps rising.

6.3.3. Amino Acids

Methionine is the only material amino acid that is produced chemically in the form of a racemic mixture. The L-enantiomer of all other natural types is created through fermentation. The L-forms of glutamic acid, primarily monosodium glutamate (MSG), lysine, phenylalanine, aspartic acid, threonine, tryptophan, arginine, citrulline, glutamine, histidine, isoleucine, leucine, ornithine, proline, tyrosine, and cystine, are the most frequently produced amino acids in the commercial world. The biological techniques include enzymatic synthesis, fermentation, and extraction from natural sources. Purified proteins can be hydrolyzed chemically or enzymatically to produce amino acids. They can also be created synthetically using organic, enzymatic, or microbiological methods, separated from industrial byproducts, or extracted from plant or animal tissues. These techniques all produce the L-isomer. Commercial aspartic acid and alanine production involves the use of enzymes.

6.3.4. Vitamins and Nutraceuticals

In the commercial market for vitamins, the most significant are ascorbic acid (vitamin C), the D-group (D_2, D_3, ergosterol), folic acid and α-tocophenol (vitamin E). The next important group of vitamins is biotin, cyanocobalamin (B_{12}), inositol, pantothenic acid, and riboflavin (B_2). Microorganisms can synthesize many vitamins of medical importance. Vitamin B_2 (riboflavin) and vitamin B_{12} (cyanocobalamin) are products of fermentation. Vitamin C (ascorbic acid) precursor, 2-keto-L-gluconic acid (2-KLG) is produced microbiologically as well as chemically.

6.3.5. Antibiotics

Since the early 1940s, intensive research for new and useful antibiotics has been undergoing throughout the world. More than 10,000 antibiotics have been discovered from microorganisms. A number of natural antibiotics are produced by microorganisms like *Streptomyces nodosus* (Amphotericin B), *Streptomyces verticillus* (Bleomycin), *Cephalosporium* sp. (Cephhalosporin), *Saccharopolyspora erythrae* (Erthromycin A), *Micromonospora purpurea* (Gentamycin), *Aspergillus terreus* (Lovastatin), *Streptomyces fradiae* (Neomycin), *P. chrysogenum* (Penicillin G), *Streptomyces ambofaciens* (Streptomycin), *Streptomyces aureofaciens* (Tetracycline), *Sarda orientalis* (Vancomycin), etc. Most of them are produced by fermentation or by biotransformations.

6.3.6. Biopharmaceuticals

The term "biopharmaceuticals" has been generally accepted for the large molecule pharmaceuticals (mainly proteins) that have emerged as a result of the modern "biotechnology" from the traditional small molecule drugs. Biopharmaceuticals that are industrially produced by microbial (bacteria or yeast) fermentation include insulin, human growth hormone, hepatitis B surface antigen vaccine, α-interferon, β-interferon, γ-interferon, granulocyte colony stimulating factor and interleukin-2. Some of the major biopharma products are listed in Table 6.7. Some of them are produced by mammalian cell culture.

Table 6.7. *Examples of Biopharmaceutical Products, Applications, and Manufacturers*

Biopharma product	*Application*	*Companies*
Insulin	Diabetes	Eli Lily, Novo Nordisk
Erythropoietin	Anemia	Amgen, Johnson & Johnson
Human growth hormone	Growth deficiency	Genetech, Pharmacia
Interferon α	Hepatitis	Schering, Roche
Interferon β	Multiple sclerosis	Chiron, Biogen
Factor VIII	Hemophilia	Bayer
Tissue plasminogen activator	Bloot clot	Genentech
Glucocerebroidase	Gaucher's disease	Genzyme
Therapeutic antibodies	Cancer	Glaxo, Amgen, Genentech
GCSF	White blood cell	Amgen, Sankyo

Source: researchgate.net/publication/311576484.

6.3.7. Enzymes

Enzymes from prokaryotic systems, such as both Gram-positive and Gram-negative bacteria, to eukaryotic systems, such as yeasts and fungi, are produced by microorganisms. Two significant developments that significantly impacted the enzyme industry occurred in the 1960s: (i) the commercialization of glucoamylase, which catalyzes the synthesis of glucose from starch more efficiently than the chemical process of acid hydrolysis; and (ii) the introduction of the first detergents containing enzymes. The 1980s saw the development of genetic engineering, which made it possible to produce and market new enzymes. This led to a second, equally explosive expansion of the enzyme industry, which is currently worth billions of dollars.

Table 6.8. *The Industrial Applications of Some Enzymes*

Industry	*Enzyme*	*Application/s*
Laundry detergents	Proteinase (91%)	Used in pre-soaks to remove protein based stains.
	Lipase (6%)	Now commonly included to digest oils and fats.
	Amylase (2%)	Removes resistant starch residues.
	Cellulase (1%)	Digests the cotton "fuzz" which accumulates with excessive washing.
Starch industry	Amylases, amyloglucosidases and glucoamylases	Converts starch to glucose and other sugar syrups.
	Glucose isomerase	Converts glucose syrups into fructose syrups.
Dairy industry	Rennin from the stomachs of young ruminant animals	Manufacture of cheese.
	Lipases	Enhances ripening of blue-mold cheeses.
	Lactases	Break down lactose to glucose and galactose.
Textile industry	Amylase	Now widely used to remove starch from woven fabrics. Starch is used as an adhesive (or 'size') on the threads of many fabrics to prevent damage during weaving. Traditionally, chemicals were favoured but now bacterial amylases are commonly used.

Brewing industry	Amylases, glucanases, proteinases	Splits polysaccharides and proteins in the malt.
	Proteinases	Reduces clouding of the beer.
	Amyloglucosidase	Low calorie beer production.
	β-glucanase	Improves filtration characteristics.
Baking industry	α-amylase	Catalyses the breakdown of starch in flours. Used in the manufacture of bread.
	β-xylanase	Improves the characteristics and rising of bread.
	Proteinases	Reduces the protein in flour. Used in biscuit manufacture.
Leather industry	Proteinase (trypsin)	The process known as bating treats the leather with proteinases to make it more pliable. Trypsin isolated from both slaughterhouses and microorganisms replaces the old method of using dog and pigeon faeces.
Pulp and paper industry	β-xylanases	Emerging technology for enhancing pulp-bleaching.
	Lipases	Reduces 'pitch' which causes paper to stick to rollers and tear.

Source: researchgate.net/publication/311576484.

Table 6.9. *Some of the Enzyme Manufacturing Companies*

Company	*Country*
Amano Pharmaceutical Co.	Nagoya, Japan
BASF	Ludwigshafen, Germany
Biocon India	Banglore, India
Biozyme Laboratories	South Wales, UK
Danisco	Copenhagen, Denmark
DSM	Delft, The Netherlands
Finnzymes	Espoo, Finland
Novozymes	Bagsvaerd, Denmark
Genzyme	Cambridge, MA, USA
Gist Brocades	Delft, The Netherlands
New England Biolabs	Beverly, MA, USA
Prodigene	College Station, TX, USA
Rhone-Poulene	Cedex, France
Roche Molecular Biochemicals	Indianapolis, IN, US
Worthington Biochemical Corporation	Lakewood, NJ, USA

Source: researchgate.net/publication/311576484.

Table 6.8 enumerates the numerous industrial uses for enzymes. The study and manipulation of enzyme structure and function are being made possible by recent developments in X-ray crystallography and other analytical techniques in the field of protein chemistry, as well as the ever-increasing amounts of biological information available from genomic programs and molecular techniques like directed evolution and gene and genome shuffling. It is reasonable to assume that the quest for bettering currently used enzyme-catalyzed processes, the requirement to create new technologies, and the growing awareness of the responsible use and reuse of raw materials will encourage both the development of new enzymes with completely unique properties as well as the logical modification of existing enzymes to meet specific requirements. Many different enzymes, including oxidoreductases, nucleases, glucose isomerase, pullulanases, xylanases, cellulases, lipases, pectinases, inulinases, amylases, glucoamylase, and β-galactosidase are made commercially employing various microorganisms. Table 6.9 lists a few of the businesses that manufacture enzymes for the commercial market.

6.3.8. Biofuels and Bioenergy

Reducing reliance on energy imports will be facilitated by the development of indigenous energy sources. Although the European Union uses less biofuels overall, it has set lofty goals to replace 20% of its gasoline and diesel with biofuel by 2020. Both a biological method for producing biodiesel and a biotechnological method for using cellulosic waste materials like straw and corn cobs as a fermentation source for bioethanol would be huge advancements. Methane can already be produced from biomass, and in the future, it may also be used as a source of hydrogen.

6.3.9. Fine and Specialty Chemicals

Specialized high-value chemicals often require complex and inefficient chemical processes for their production. Therefore, the simpler and more efficient biological processes are most important in this sector.

6.3.10. New Materials

In the long term, completely new materials and manufacturing techniques will be inspired by nature, e.g., controlled transport phenomena to create a solar cell that is more efficient. The characteristics of bio-based performance and nano-composite materials will come from their unique micro- or nano-scale structure, or they will be created utilizing the ideas of natural self-organization.

Did you get it?

1. What are examples of organic acids and polymers?
2. Provide examples of vitamins and nutraceuticals.

FOCUS ON CAREERS

BIOMEDICAL ENGINEERING

Biomedical engineering is a multidisciplinary field that applies principles and problem-solving techniques from engineering to biology and medicine. It involves the development and application of technologies to improve healthcare, diagnosis, and treatment. Here's an overview of biomedical engineering:

1. Interdisciplinary Nature:
   - **Integration of Engineering and Biology:** Biomedical engineering combines principles from engineering disciplines, such as electrical, mechanical, chemical, and materials engineering, with biological and medical sciences.
   - **Collaboration with Healthcare Professionals:** Biomedical engineers often collaborate with healthcare professionals, including physicians, surgeons, and therapists, to develop solutions to medical challenges.
2. Key Focus Areas:
   - **Medical Imaging:** Development of advanced imaging technologies, such as MRI, CT, and ultrasound, for improved diagnostics and visualization.
   - **Biomechanics:** Study of the mechanical aspects of living organisms, including movement, materials properties, and structural analysis, often applied in prosthetics and orthopedics.
   - **Biomedical Instrumentation:** Design and development of medical devices and instruments for monitoring, diagnosis, and treatment.
   - **Biological Signal Processing:** Analysis of physiological signals, such as EEG and ECG, to extract meaningful information for medical diagnosis and monitoring.
   - **Tissue Engineering:** Construction of artificial tissues and organs to replace or repair damaged biological structures.
   - **Biomedical Materials:** Development of materials compatible with the human body for use in implants, prosthetics, and medical devices.
   - **Biopharmaceuticals:** Application of engineering principles to the design and production of pharmaceuticals and therapeutic agents.

Biomedical Engineer Job Responsibilities:

- Supports patient diagnosis and treatment by installing, testing, calibrating, and repairing biomedical equipment; training users; maintaining safe operations.
- Approves new equipment by conducting tests, ensuring adherence to codes, and making modifications.

- Installs new equipment by establishing, adjusting, calibrating, and testing performance.
- Maintains equipment by completing preventive maintenance schedules, conducting tests, and troubleshooting and repairing malfunctions.
- Calls for special equipment service, evaluates service contracts and maintains equipment inventories.
- Directs biomedical equipment operation by providing equipment operation demonstrations, instructing other healthcare staff in the use and care of equipment, and answering questions.
- Evaluates proposed equipment by conducting tests and evaluations.
- Improves equipment performance by studying machine-patient interaction, conferring with equipment users, developing modifications, and collaborating with manufacturers.
- Maintains supplies inventory by checking stock, anticipating needs, placing, and expediting orders, and verifying receipt.
- Prepares biomedical reports by collecting, analyzing, and summarizing information and trends.
- Maintains patient confidence by keeping information confidential.
- Maintains a safe and healthy working environment by conducting safety tests, recommending, and complying with procedures, training, and guiding medical and healthcare personnel, and complying with codes.

Biomedical Engineer Skills and Qualifications:

- Documentation skills
- Analyzing information
- Creating a safe, effective environment
- Infection control
- Training management
- Lab environment
- Clinical lab testing
- Confidentiality
- Equipment calibration
- Sterilization
- Conceptual skills

SUMMARY

- The basis of industrial biotechnology is the use of enzymes and microorganisms to generate chemicals (for numerous applications) from renewable sources. Industrial biotechnology is the application of modern biotechnology to the sustainable production of materials, chemicals, and fuels from living cells and/or enzymes (renewable sources).
- The foundations of industrial biotechnology can be traced back to ancient civilizations that unknowingly harnessed the power of microorganisms for fermentation. Brewing, bread-making, and the production of fermented foods were among the early applications. These processes relied on the metabolic activities of yeast and bacteria, marking the humble beginnings of biotechnological practices.
- Genetic engineering, also known as recombinant DNA (rDNA) technology, involves the incorporation of a desired gene from one organism to another, where it may be expressed. Both in vivo and in vitro techniques are used in rDNA technology.
- Metabolic engineering of microbes to produce large amounts of valuable metabolites that are difficult to extract from their natural sources and too complex or expensive to produce via chemical synthesis, is an alternate option. Taxol (paclitaxel) is an antimitotic agent used in the treatment of ovarian cancer and metastatic breast cancer, with annual sales revenue of US $ 1 billion.
- The developments in the field of genetics and molecular biology in the 21st century have brought world-wide interest in biotechnology. The ability to manipulate DNA has radically changed our perceptions of medicine, agriculture, and environmental management.
- Enzymes can produce specific products rapidly and with high yield, but developments are often needed to make them suitable for use in industrial processes. Techniques such as protein engineering, gene shuffling and directed evolution make enzymes better suitable for use in industrial processes.
- Isolation and purification of a biotechnological product to a form suitable for its intended use is called downstream processing. DSP encompasses all processes following the fermentation. Its primary aim includes efficiency, reproducibility, and safe recovery of the targeted product to the required specification (biological activity, purity, etc.), while maximizing recovery yield and minimizing costs.
- Biotechnology can both replace existing chemical processes and allow the production of new products. Already, there are many products of biotechnology in various areas like food chemicals, pharmaceuticals, detergents, pulp & paper, textiles, energy, materials, and polymers, etc.
- The organic chemicals that fall into this category and can be produced by fermentation include ethanol, 1,3-propanediol, butanol, acetone, 2,3-butanediol, and glycerol. Butanol and acetone have been produced industrially by fermentation, but currently for economic reasons, synthesis is the manufacturing method of choice.

REVIEW QUESTIONS

1. Explain the key principles and components encompassing industrial biotechnology.
2. Discuss the stages and considerations in the development of bioprocesses.
3. Provide an in-depth overview of various products produced through industrial biotechnology.
4. How does industrial biotechnology contribute to sustainable practices?
5. Compare and contrast traditional manufacturing with industrial biotechnology processes.

MULTIPLE CHOICE QUESTIONS

1. What is the primary purpose of using genetically modified microorganisms in industrial biotechnology?

a. To increase biodiversity
b. To produce pharmaceuticals and biofuels
c. To reduce carbon emissions
d. To control pest populations

2. When did the field of Industrial Biotechnology emerge?

a. 19th century
b. 20th century
c. 21st century
d. 18th century

3. In the mid-20th century, the development of recombinant DNA technology marked a significant advancement in Industrial Biotechnology. Who were the scientists credited with this breakthrough?

a. Watson and Crick
b. Mendel and Darwin
c. Pauling and Franklin
d. Cohen and Boyer

4. Which of the following is a common application of industrial biotechnology in the food and beverage industry?

a. Textile production
b. Vaccine development
c. Cheese and yogurt production
d. Plastic manufacturing

5. What is the role of enzymes in industrial biotechnology processes?

a. To provide energy for microorganisms

b. To speed up chemical reactions

c. To store genetic information

d. To regulate temperature in bioreactors

6. What is a common product of Industrial Biotechnology in the pharmaceutical industry?

a. Synthetic polymers

b. Vaccines

c. Fertilizers

d. Textiles

7. Which biofuel is often produced through Industrial Biotechnology processes?

a. Natural gas

b. Diesel

c. Ethanol

d. Coal

8. In the production of enzymes for industrial use, what is a typical application?

a. Flavor enhancers

b. Detergents and cleaning agents

c. Paints and coatings

d. Plastics

9. Which of the following is a common product obtained through fermentation processes in Industrial Biotechnology?

a. Computer software

b. Air purifiers

c. Bioplastics

d. Solar panels

10. What is the significance of metabolic engineering in industrial biotechnology?

a. It involves the study of ecosystems

b. It focuses on the design and improvement of metabolic pathways in microorganisms

c. It deals with the extraction of minerals from ores

d. It is related to the synthesis of artificial polymers

Answer to Multiple Choice Questions

1. (b); 2. (a); 3. (d); 4. (c); 5. (b); 6. (b); 7. (c); 8. (b); 9. (c); 10. (b)

REFERENCES

1. Bailey, J. E., (2001). Reflections on the scope and the future of metabolic engineering and its connections to functional genomics and drug discovery. *Metabolic Engineering, 3*, 111–114.
2. Chen, G. Q., & Patel, M., (2012). Plastics derived from biological sources: Present and future – a technical and an environmental review. *Chemical Reviews, 112*, 2082–2099.
3. Chotani, G., Dodge, T., Hsu, A., Kumar, M., LaDuca, R., Trimbur, D., Weyler, W., & Sanford, K., (2000). Commercial production of chemicals using pathway engineering. *Biochimica et Biophysica Acta, 1543*, 434–455.
4. Duarte, N. C., Herrgård, M. J., & Palsson, B. O., (2004). Reconstruction and validation of *Saccharomyces cerevisiae* IND750, a fully compartmentalized genome-scale metabolic model. *Genome Research, 14*, 1298–1309.
5. Engels, B., Dahm, P., & Jennewein, S., (2008). Metabolic engineering of taxadiene biosynthesis in yeast as a first step towards Taxol (paclitaxel) production. *Metabolic Engineering, 10*, 201–206.
6. Feist, A. M., Henry, C. S., Reed, J. L., & Krummenacker, M., (2007). A genome-scale metabolic reconstruction for *Escherichia coli* K-12 MG1655 that accounts for 1260 ORFs and thermodynamic information. *Molecular Systems Biology, 3*, 121.
7. Gibson, D. G., Benders, G. A., Andrews-Pfannkoch, C., Denisova, E. A., Baden-Tillson, H., Zaveri, J., Stockwell, T. B., et al., (2008a). Complete chemical synthesis, assembly, and cloning of a *Mycoplasma genitalium* genome. *Science, 319*, 1215–1220.
8. Gibson, D. G., Benders, G. A., Axelrod, K. C., Zaveri, J., Algire, M. A., Moodie, M., Montague, M. G., et al., (2008b). One-step assembly in yeast of 25 overlapping DNA fragments to form a complete synthetic *Mycoplasma genitalium* genome. *Proceedings of the National Academy of Sciences USA, 105*, 20404–20409.
9. Johnson, K., Jiang, Y., Kleerebezem, R., Muyzer, G., & Van, L. M. C. M., (2009). Enrichment of a mixed bacterial culture with a high polyhydroxyalkanoate storage capacity. *Biomacromolecules, 10*, 670–676.
10. Lee, J. W., Kim, H. U., Choi, S., Yi, J., & Lee, S. Y., (2011). Microbial production of building block chemicals and polymers. *Current Opinions in Biotechnology, 22*, 758–767.
11. Luetz, S., Giver, L., & Lalonde, J., (2008). Engineered enzymes for chemical production. *Biotechnology and Bioengineering, 101*(4), 647–653. https://pubmed.ncbi.nlm.nih.gov/18814289/ (accessed on 04 April 2024).
12. Martí, S., Andrés, J., Moliner, V., & Silla, E., (2008). Computational design of biological catalysts. *Chemical Society Reviews, 37*, 2634–2643.
13. McLachlan, M., Sullivan, R. P., & Zhao, H., (2009). Directed enzyme evolution and high throughput screening. In: Tao, J., Lin, G., & Liese, A., (eds.), *Biocatalysis for the Pharmaceutical Industry – Discovery, Development, and Manufacturing* (pp. 45–64). John Wiley and Sons, Singapore.

14. Oh, Y. K., Palsson, B. O., Park, S. M., Schilling, C. H., & Mahadevan, R., (2007). Genome-scale reconstruction of metabolic network in *Bacillus subtilis* based on high-throughput phenotyping and gene essentiality data. *Journal of Biological Chemistry, 282*, 28791–28799.
15. Park, J. H., Lee, K. H., Kim, T. Y., & Lee, S. Y., (2007). Metabolic engineering of *Escherichia coli* for the production of L-valine based on transcriptome analysis and in silico gene knockout simulation. *Proceedings of National Academy of Sciences, USA, 104*, 7797–7802.
16. Picataggio, S., (2009). Potential impact of synthetic biology on the development of microbial systems for the production of renewable fuels and chemicals. *Current Opinions in Biotechnology, 20*, 325–329.
17. Sindelar, G., & Wendisch, V. F., (2007). Improving lysine production by *Corynebacterium glutamicum* through DNA microarray based identification of novel target genes. *Applied Microbiology and Biotechnology, 76*, 677–689.
18. Tang, W., & Zhao, H., (2009). Industrial biotechnology: Tools and applications. *Biotechnology Journal, 4*(12), 1725–1739. https://onlinelibrary.wiley.com/doi/pdfdirect/10.1002/biot.200900127 (accessed on 04 April 2024).
19. Ward, T. R., (2008). Artificial enzymes made to order: Combination of computational design and directed evolution. *Angewandte Chemie International Edition, 47*, 7802–7803.
20. Wendisch, V., Bott, M., & Eikmanns, B., (2006). Metabolic engineering of *Escherichia coli* and *Corynebacterium glutamicum* for biotechnological production of organic acids and amino acids. *Current Opinion in Microbiology, 9*(3), 268–274.

CHAPTER 7

Environmental Biotechnology

LEARNING OBJECTIVES

After studying this chapter, you will be able to:

- Understand overview of environmental biotechnology;
- Define bioremediation;
- Explore specific environmental biotechnology applications in cleaning up pollution;
- Discuss the environmental biotechnology for medical waste management.

INTRODUCTORY EXAMPLE

Bioremediation for Environmental Cleanup: Harnessing Environmental Biotechnology:

Environmental biotechnology plays a pivotal role in addressing contemporary challenges related to pollution and ecosystem degradation. One compelling application is bioremediation, a process that utilizes living organisms to eliminate or neutralize pollutants in the environment. This example focuses on the use of environmental biotechnology in bioremediation, showcasing how microorganisms and their metabolic capabilities can be harnessed for the cleanup of contaminated sites.

- **Contaminated Site Identification:** The first step in employing environmental biotechnology for bioremediation involves identifying contaminated sites. These areas may be polluted with various substances, such as petroleum hydrocarbons, heavy metals, or industrial chemicals. Advanced analytical techniques, including molecular biology tools, aid in precisely characterizing the nature and extent of contamination. Once identified, a tailored bioremediation strategy is designed based on the specific pollutants present.

- **Biostimulation and Bioaugmentation:** Biostimulation involves enhancing the activity of indigenous microorganisms at the contaminated site. This can be achieved by providing essential nutrients, oxygen, or other growth-promoting factors. Environmental biotechnology contributes by formulating custom microbial consortia that thrive in the given conditions, accelerating the natural degradation processes. Bioaugmentation, on the other hand, introduces exogenous microorganisms with specialized metabolic capabilities to augment the existing microbial community and enhance pollutant degradation.
- **Phytoremediation:** Environmental biotechnology extends beyond microorganisms to include plants in bioremediation strategies. Phytoremediation exploits the natural ability of certain plants, known as hyperaccumulators, to absorb and accumulate pollutants in their tissues. This method is particularly effective for removing heavy metals from soil. Biotechnological advancements facilitate the genetic engineering of plants to express enhanced pollutant uptake and tolerance, expanding the range of contaminants that can be remediated through phytoremediation.
- **Microbial Fuel Cells (MFCs):** Innovative technologies within environmental biotechnology, such as Microbial Fuel Cells (MFCs), offer a sustainable approach to remediate contaminated water while simultaneously generating electricity. MFCs utilize the metabolic activity of microorganisms to break down organic pollutants, producing electrons that can be harnessed as a source of renewable energy. This dual-purpose application demonstrates the potential of environmental biotechnology not only in pollutant removal but also in resource recovery.
- **Monitoring and Ad**aptive Management: Environmental biotechnology relies on continuous monitoring to assess the effectiveness of bioremediation strategies. Molecular tools, sensors, and remote sensing technologies enable real-time tracking of microbial populations and pollutant concentrations. This information facilitates adaptive management, allowing for the modification of bioremediation approaches based on the dynamic conditions of the contaminated site.

UNIT INTRODUCTION

Environmental biotechnology is an interdisciplinary field that leverages the power of biological processes to address pressing environmental challenges and promote sustainable solutions. As the global population continues to grow, industrialization expands, and the demands on natural resources increase, the impact on the environment becomes more pronounced. Environmental biotechnology emerges as a crucial tool in mitigating the adverse effects of human activities on ecosystems, air, water, and soil.

Environmental biotechnology draws upon the principles of biology, microbiology, chemistry, and engineering to develop innovative and eco-friendly approaches to various environmental issues. The field encompasses a diverse range of applications, from waste management and pollution control to renewable energy production and ecosystem restoration. By harnessing the capabilities of microorganisms, plants, and genetic engineering techniques, environmental biotechnology seeks to find sustainable and efficient solutions to the complex challenges posed by environmental degradation.

Waste management is a fundamental aspect of environmental biotechnology, offering alternatives to traditional disposal methods that often result in pollution and resource depletion. Through processes like bioremediation and composting, microorganisms are employed to break down pollutants and organic waste, turning them into less harmful substances or valuable compost. This not only reduces the environmental impact of waste but also contributes to the conservation of natural resources.

In the realm of pollution control, environmental biotechnology plays a vital role in developing strategies to combat air, water, and soil pollution. Biofilters utilize microorganisms to remove airborne pollutants, while microbial-based processes in water treatment contribute to the purification of water resources. These approaches present sustainable alternatives to conventional methods, aligning with the growing global emphasis on reducing pollution and safeguarding environmental quality.

Environmental biotechnology, with its diverse applications, holds the promise of fostering a more sustainable coexistence between human activities and the natural environment. As we confront the challenges of climate change, resource depletion, and environmental pollution, the role of environmental biotechnology becomes increasingly significant in charting a path towards a resilient and harmonious future. The field's ability to offer practical, science-based solutions positions it as a key player in the broader efforts to achieve environmental sustainability on a global scale.

7.1. OVERVIEW OF ENVIRONMENTAL BIOTECHNOLOGY

→Learning Objectives

- Define the scope of environmental biotechnology;
- Assess the impact of environmental biotechnology.

Environmental biotechnology refers to biotechnology that is applied to and used to understand the environment. The term "environmental biotechnology" may also imply attempts to manipulate biological processes for financial gain. The International Society for Environmental Biotechnology defines environmental biotechnology as "the development, use, and regulation of biological systems for remediation of contaminated environments (land, air, and water), and for environment-friendly processes (green manufacturing technologies and sustainable development)."

- **Definition:** "The best use of nature, in the form of plants, animals, bacteria, fungi, and algae, to produce renewable energy, food, and nutrients in a synergistic integrated cycle of profitable processes where the waste of each process becomes the feedstock for another process," is the most basic definition of environmental biotechnology.."

7.1.1. Scope of Environmental Biotechnology

The pressure of a growing population has also increased the need to meet each individual's needs for clothing, food, and shelter. To meet these demands, industry sector development is steadily rising. This heavily altered the ecosystem while also contaminating every area, i.e., earth, water, and air. An environmentally friendly method that could shield the environment from the pollutants released from various sources is urgently needed at this time. Under such circumstances, environmental biotechnology must be a key component of the effective solution to this issue and for future safety. Molecular biology, ecology, environmental engineering, microbiology, biochemistry, and other fields are all integrated into the field of environmental biotechnology. The application of environmental biotechnology in environmental safety is quite broad. Figure 7.1 presents some of the main applications of environmental biotechnologies.

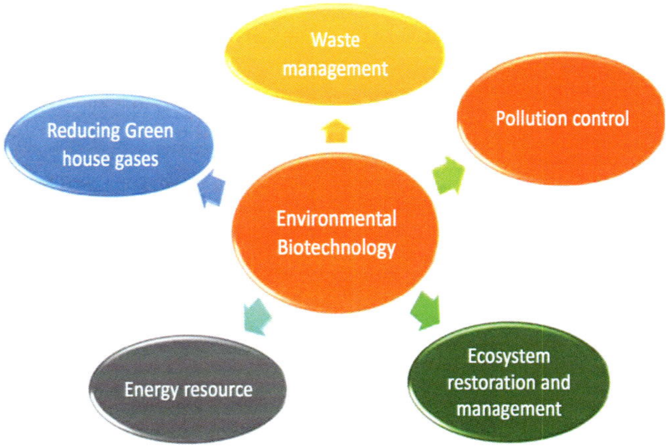

Figure 7.1. *Major scopes of environment biotechnology.*

Source: https://egyankosh.ac.in/bitstream/123456789/95583/1/Unit-1.pdf.

Keyword

Sustainable Development – It is an organizing principle that aims to meet human development goals while also enabling natural systems to provide necessary natural resources and ecosystem services to humans.

7.1.2. Impact of Environmental Biotechnology

The impact of environmental biotechnology on society, ecosystems, and the planet as a whole is profound, contributing to sustainable development, improved resource management, and the mitigation of environmental challenges. Here are key aspects of the impact of environmental biotechnology:

1. Waste Reduction and Remediation:
   - **Land and Water Cleanup:** Environmental biotechnology, particularly bioremediation, has significantly reduced the environmental impact of contaminants in soil and water. This has implications for human health and the restoration of ecosystems affected by industrial pollution, oil spills, and hazardous waste.
   - **Organic Waste Management:** The use of biotechnology in composting and anaerobic digestion has led to more efficient organic waste management, minimizing the volume of waste sent to landfills and reducing greenhouse gas emissions.
2. Pollution Control and Air Quality:
   - **Biofilters for Air Purification:** The implementation of biofilters in industries has led to improved air quality by utilizing microorganisms to break down pollutants. This has contributed to reduced emissions of volatile organic compounds (VOCs) and other harmful substances.

- **Water Treatment:** Microbial-based water treatment processes have played a crucial role in purifying water sources, ensuring access to clean water for communities, and protecting aquatic ecosystems from contamination.

3. Renewable Energy Production:
   - **Biogas and Biofuels:** The adoption of anaerobic digestion and microbial fermentation for biogas and biofuel production has diversified the energy portfolio. This not only contributes to reducing dependence on fossil fuels but also helps in mitigating climate change by lowering greenhouse gas emissions.
4. Ecosystem Restoration and Biodiversity:
   - **Microbial Inoculants:** The use of microbial inoculants and mycorrhizal fungi has enhanced ecosystem restoration efforts by promoting plant growth, improving soil health, and fostering biodiversity in degraded areas.
   - **Phytoremediation:** Genetic engineering and phytoremediation techniques have facilitated the restoration of contaminated sites by using plants to absorb and mitigate pollutants, aiding in the recovery of ecosystems.
5. Sustainable Agriculture:
   - **Biofertilizers and Biological Pest Control:** Environmental biotechnology has contributed to sustainable agriculture through the development and use of biofertilizers and microbial pesticides. These alternatives reduce the reliance on chemical inputs, promoting environmentally friendly farming practices.
6. Human Health and Biomedical Applications:
   - **Bioremediation for Health Protection:** The cleanup of contaminated sites through bioremediation has direct implications for human health, preventing exposure to harmful substances in the environment.
   - **Biotechnological Medicines:** Techniques developed in environmental biotechnology have paved the way for advancements in biopharmaceuticals, including the production of therapeutic proteins using genetically engineered microorganisms.
7. Innovations in Monitoring and Assessment:
   - **Biosensors:** The development of biosensors for environmental monitoring enables real-time detection of pollutants, allowing for proactive responses to environmental changes and pollution incidents.
8. Job Creation and Economic Growth:
   - **Emergence of New Industries:** The growth of environmental biotechnology has led to the establishment of new industries, creating job opportunities and contributing to economic development.

- **Technology Transfer:** The transfer of environmental biotechnology knowledge and technologies between countries has facilitated international collaborations, supporting global efforts to address shared environmental challenges.

Did you get it?

1. Explain the scope of environmental biotechnology and its applications.
2. How would you assess the impact of environmental biotechnology on modern society and ecosystems?

7.2. BIOREMEDIATION

→Learning Objectives

- Explain the microorganisms used in bioremediation;
- Classify types of bioremediations;
- Understand the bioremediation of various pollutants;
- Discuss the advantages and disadvantages of bioremediation;
- Identify limitations of bioremediation.

Emerging contaminant issues are also thought to have solutions in bioremediation and natural reduction; microbes play a major role in cleaning up contaminated environments. Numerous microorganisms, such as fungi, aerobic, and anaerobic bacteria, are involved in the bioremediation process. Through the universal action of microorganisms, bioremediation is highly involved in the degradation, eradication, immobilization, or detoxification of various chemical wastes and physically hazardous materials from the surrounding environment. The fundamental idea is to break down and transform pollutants into less harmful forms. Biotic and abiotic conditions are the two categories of factors that impact degradation rate. Various techniques and approaches are currently used in the bioremediation process.

- **Definition of Bioremediation:** Biological organisms are employed in the bioremediation process to eliminate or neutralize environmental pollutants through metabolic processes. The term "biological" organisms refers to both the "remediation" process, which involves treating the situation, and microscopic organisms like fungi, algae, and bacteria.

Microorganisms grow in the most diverse range of habitats found in the Earth's biosphere. They can grow in the deep sea, on freezing ice, in plants, animals, soil, and water. Microorganisms are the ideal choice to take care of our environment because of their sheer quantity and their hunger for a diverse array of substances.

> *"Bioremediation is a waste management technique that includes the use of living organisms to eradicate or neutralize pollutants from a contaminated site."*
>
> *"Bioremediation is a 'treatment techniques' that uses naturally occurring organisms to break down harmful materials into less toxic or non-toxic materials."*

Technologies for bioremediation were widely used and are still expanding exponentially today. Because bioremediation uses a microbial process, it is an eco-friendly method of cleaning up polluted sites that has been found to be reliable and effective. Recent advances in bioremediation techniques over the last 20 years have made it possible to successfully restore contaminated environments cost-effectively and in a environmentally responsible manner. Different bioremediation techniques that restore contaminated environments have been developed by researchers. Both native and non-native microorganisms can be introduced to the contaminated site during the bioremediation process. Native bacteria found in contaminated areas are the key to resolving the majority of problems pertaining to pollutant biodegradation and bioremediation. Among the main benefits of bioremediation over chemical and physical remediation techniques are its cost-effectiveness and environmental friendliness.

One method of bioremediation involves reducing, detoxifying, degrading, mineralizing, or changing more hazardous pollutants into less hazardous ones. Pesticides, agrochemicals, chlorinated compounds, heavy metals, xenobiotic compounds, organic halogens, greenhouse gases, hydrocarbons, nuclear waste, dyes, plastics, and sludge are among the substances whose nature determines the process of pollutant removal. Cleaning methods are used to get rid of toxic waste from contaminated areas. Through the all-encompassing action of microorganisms, bioremediation is heavily involved in the degradation, eradication, immobilization, or detoxification of various chemical wastes and physical hazardous materials from the surrounding environment (Figure 7.2).

Keyword

Nuclear Waste – It refers to the byproducts generated from nuclear reactions, primarily in nuclear power plants, medical facilities, and research institutions. This waste can remain hazardous for thousands of years due to the long half-lives of some radioactive isotopes.

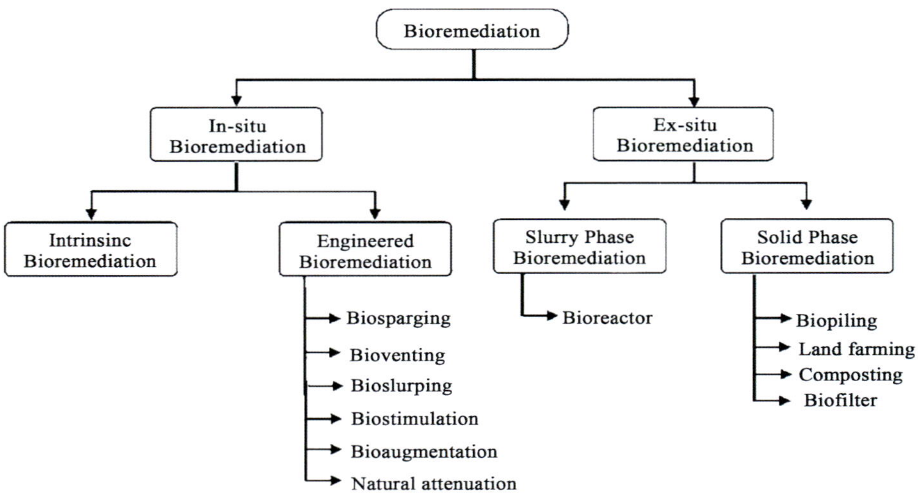

Figure 7.2. *Bioremediation approaches for environmental clean-up.*

Source: Indu, S., (2020). Bioremediation Techniques for Polluted Environment. doi: 10.5772/intechopen.90453.

7.2.1. Microorganisms Used in Bioremediation

Microorganisms are essential to nutritional chains, which are crucial to maintaining the biological balance of life. Through the use of bacteria, fungi, algae, and yeast, contaminated materials are removed through the process of bioremediation. In the presence of hazardous compounds or any kind of waste stream, microbes can grow in temperatures below zero as well as extremely high temperatures. The biological system and adaptability of microbes make them appropriate for the remediation process. The primary element needed for microbial activity is carbon. The bioremediation process was carried out by microbial consortiums in different environments. These microorganisms comprise *Achromobacter*, *Arthrobacter*, *Alcaligenes*, *Bacillus*, *Corynebacterium*, *Pseudomonas*, *Flavobacterium*, *Mycobacterium*, *Nitrosomonas*, *Xanthobacter*, etc.

There are groups of microbes that are used in bioremediation, such as:

1. **Aerobic:** These bacteria have degradative capacities to degrade the complex compounds such as *Pseudomonas*, *Acinetobacter*, *Sphingomonas*, *Nocardia*, *Flavobacterium*, *Rhodococcus*, and *Mycobacterium*. These microbes have been reported to degrade pesticides, hydrocarbons, alkanes, and polyaromatic compounds. Many of these bacteria use the contaminants as carbon and energy source.
2. **Anaerobic:** These bacteria are not as regularly used as aerobic bacteria. There is an increasing interest in aerobic bacteria used for bioremediation of chlorinated aromatic compounds, polychlorinated biphenyls, and dechlorination of the solvent trichloroethylene and chloroform, degrading, and converting pollutants to less toxic forms.

7.2.1.1. Factors Affecting Microbial Bioremediation

Through the action of bacteria, fungi, algae, and plants, the bioremediation process breaks down, removes, alters, immobilizes, or detoxifies a variety of chemicals and physical pollutants from the environment. Microorganisms' enzymatic metabolic pathways speed up the biochemical reactions that aid in the pollutant's breakdown. Microorganisms only react to pollutants when they come into contact with substances that provide them with nutrients and energy to divide into new cells. Numerous variables affect the efficacy of bioremediation, such as the pollutants' chemical makeup and concentration, the environment's physicochemical properties, and their accessibility to already-existing microorganisms.

The primary factors contributing to the degradation of pollutants are the microbial population, the accessibility of contaminants to the microbial population, and environmental factors such as soil type, pH, temperature, oxygen content, and nutrient availability.

7.2.1.2. Biotic or Biological Factors

Biotic factors facilitate antagonistic interactions between microorganisms, such as protozoa and bacteriophages, and the degradation of organic compounds by microorganisms with insufficient carbon sources. The concentration of the contaminant and the amount of catalyst present in the biochemical reaction are often related to the rate of contaminant degradation. Enzyme activity, interactions (predation, succession, and competition), mutation, horizontal gene transfer, population size and composition, growth for biomass production, and mutation are among the main biological factors.

7.2.1.3. Abiotic or Environmental Factors

The way environmental pollutants interact with the metabolic processes and physicochemical characteristics of the microorganisms being targeted in the process. The conditions of the surroundings affect how well the bacteria and pollutants interact. Temperature, pH, moisture, soil structure, water solubility, nutrients, site conditions, oxygen content, redox potential, resource scarcity, physico-chemical bioavailability of pollutants, concentration, chemical structure, type, solubility, and toxicity all affect microbial growth and activity. The aforementioned elements govern the kinetics of degradation.

The pH range of 6.5 to 8.5 is generally ideal for pollutant biodegradation in most aquatic and terrestrial environments. Because it depends on the type and quantity of soluble constituents that are accessible as well as the pH and osmotic pressure of terrestrial and aquatic systems, moisture has an impact on the metabolism of contaminants (Table 7.1).

Table 7.1. *Critical Factors for Microbial Bioremediation*

Factors	Remarks
Biological factors	Soil microorganisms compete for carbon sources, or bacteriophages and protozoa prey on each other, all of which can affect organic compound degradation. Derivatization rates are influenced by contaminants and catalyst levels. Expressed enzymes can speed up or slow contaminant degradation.
Oxygen	Biodegradation rates can be improved by using organisms that don't require oxygen. Anaerobic decomposition occurs as most living organisms need oxygen to survive. In most cases, hydrocarbon metabolism can be boosted by the addition of oxygen.
Moisture content	Microorganisms require a sufficient amount of water to achieve their growth goals. When the soil is too wet, the biodegradation agents also don work.
Nutrients	Nutrients can influence microbial growth and reproduction, as well as biodegradation rate and effectiveness. Optimizing the bacterial C:N:P ratio can improve biodegradation efficiency, especially when essential nutrients like N and P are supplied. Carbon, phosphorous, and nitrogen are just a few of the nutrients microorganisms need to survive. In low concentrations, hydrocarbon degradation is also limited. Adding nutrients to cold environments can increase microorganisms' metabolic activity and, thus, the biodegradation rate. Aquatic biodegradation is limited by nutrient availability. Oil-eating microbes require nutrients to thrive. These essential nutrients are found in small amounts in nature.

Temperature	The most important physical factor influencing microorganism survival and hydrocarbon composition is temperature. In cold climates like the Arctic, natural oil degradation is slow, putting more pressure on microbes to clean up spilled oil. Here, the sub-zero water freezes the microbial transport channels, rendering them unable to perform their metabolic functions. Temperature affects the metabolic turnover of enzymes involved in degradation. Also, each compound's degradation requires a specific temperature. Temperature affects microbial physiological properties and thus speeds up or slows down bioremediation. Increased microbial activity occurs at higher temperatures. It started to drop suddenly as the temperature increased or decreased, and then eventually stopped.
pH	A compound's acidity, alkalinity, and basicity affect microbial metabolism and the removal process. Microbial growth can be predicted by the pH of the soil. Even minor pH shifts have a significant impact on metabolic processes.
Site characterization and selection	Before proposing a bioremediation remedy, it is necessary to conduct adequate remedial investigation work to characterize the extent of the contamination. Site selection procedures include determining the horizontal and vertical extent of contamination, defining parameters, and sampling locations, and describing sampling and analysis methods.
Metal ions	Metals are essential for bacteria and fungi, but excessive amounts inhibit cell metabolism. Degradation rates are influenced by metal compounds on both a direct and indirect basis.
Microorganisms	High concentrations of some toxic compounds can harm microorganisms and slow the decontamination process. Toxicity varies with the toxicant, concentration, and microorganisms exposed.

Source: Saroj, B., Diksha, G., Banjagere, V. T., (2022). Recent Strategies for Bioremediation of Emerging Pollutants. doi: 10.3390/toxics10080484.

7.2.2. Types of Bioremediations

There are far more than nine types of bioremediation, but the following are the most common ways in which it is used.

1. **Biopile:** Excavated polluted soil is piled above ground for bioremediation; aeration and nutrient addition are then added to enhance bioremediation through microbial metabolism. This method includes treatment bed systems, leachate collection, irrigation, nutrients, and aeration. A growing number of studies are being conducted on this particular ex-situ technique because of its cost-effectiveness and practical features, which enable efficient control of temperature, aeration, pH, and nutrients during biodegradation. In addition to being used to treat volatile low molecular weight pollutants, biopiles are also an effective way to clean up polluted extreme cold environments. Heating systems can be incorporated into biopile designs to enhance microbial activity and contaminant availability, thereby accelerating the rate of biodegradation, and remediation duration can be reduced due to the biopile's flexibility. To promote improved bioremediation, heated air can also be injected into the design of the biopile to supply heat and air simultaneously. In a biopile structure, bulking agents like bark, wood chips,

sawdust from straw, and other organic materials have been added to improve the remediation process. The absence of power at remote sites prevents biopile systems from being connected to additional field ex-situ bioremediation techniques like land farming, bioventing, biosparging, robust engineering, maintenance, and operation costs, and constant air circulation in contaminated piled soil through an air pump. Furthermore, excessive air heating can cause soil to dry out during bioremediation, suppressing microbial activity and promoting volatilization rather than biodegradation.

2. **Windrows:** To improve bioremediation by increasing the microbial degradation activities of native and transient hydrocarbonoclastic present in polluted soil, windrow bioremediation techniques rely on periodically rotating the piled polluted soil. The rate of bioremediation can be effectively accelerated through acclimatization, biotransformation, and mineralization. The regular turning of polluted soil increases aeration with the addition of water, uniform distribution of nutrients, pollutants, and microbial degradation activities. Although windrow treatment was more effective than biopile treatment at removing hydrocarbons from the soil, the rate of hydrocarbon removal from the soil was still higher with windrow treatment. Periodic turning in conjunction with windrow treatment is not, however, the ideal selection technique for bioremediating soil contaminated with volatile toxic compounds. Because windrow treatment often results in reduced aeration due to the formation of an anaerobic zone inside piled polluted soil, it has been linked to the release of greenhouse gases (CH_4).

3. **Land Farming:** It is the simplest, outstanding bioremediation techniques due to its low cost and less equipment requirement for operation. It is mostly observed in ex-situ bioremediation, while in some cases of in-situ bioremediation techniques. This consideration is due to the site of treatment. Pollutant depth is important in land farming, which can be carried out ex-situ or in-situ. Inland farming, polluted soils are regularly excavated and tilled and site of treatment speciously regulates the type of bioremediation. When excavated polluted soil is treated on-site, it is ex-situ as it has more in common than other ex-situ bioremediation techniques. Generally, excavated polluted soils are carefully applied on a fixed layer of support above the ground surface to allow aerobic biodegradation of pollutants by autochthonous microorganisms. Overall, the land farming bioremediation technique is very simple to design and

Remember

Enzymes must also be involved in contaminant metabolism to have an affinity for the contaminant and availability. The major biological factors: interaction (competition, predation, and succession), population size, and composition.

implement, requires low capital input, and can be used to treat large volume of polluted soil with minimal environmental impact and energy requirement.

4. **Bioreactor:** It is a container where a sequence of biological reactions transforms raw materials into one or more targeted products. Bioreactors can operate in a variety of ways: batch, fed-batch, sequencing batch, continuous, and multistage. The ideal growing environment for bioremediation is provided by a bioreactor. Polluted samples are placed in a bioreactor for the remediation process. There are a number of benefitsof treating contaminated soil using a bioreactor instead of an ex-situ bioremediation process. The bioremediation process using bioreactors effectively shortens the bioremediation time due to their excellent control over pH, temperature, agitation, and aeration, substrate, and inoculum concentrations. The ability to regulate and alter process parameters in a bioreactor suggests that biological reactions are occurring. The flexible nature of bioreactor designs allows maximum biological degradation while minimizing abiotic losses.

- **Advantages of *Ex-Situ* Bioremediation:**
  - Suitable for a wide range of contaminants.
  - Suitabile relatively simple to assess from site investigation data.
  - Biodegradation is greater in a bioreactor system than or in solid-phase systems because the contaminated environment is more manageable, controllable, and predictable.

Disadvantages:

- Not applicable to heavy metal contamination or chlorinated hydrocarbons such as trichloroethylene.
- Non-permeable soil requires additional processing.
- The contaminant can be stripped from the soil via soil washing or physical extraction before being placed in a bioreactor.

5. **In-Situ Bioremediation Techniques:** These methods include handling contaminated materials on the pollution scene. It requires little to no disturbance of the soil's construction and no excavation. Perfectly speaking, these methods ought to be less expensive than ex-situ bioremediation methods. Certain in-situ bioremediation methods, such as intrinsic bioremediation or natural attenuation, may advance without any kind of improvement, while others, such as bioventing, biosparging, and phytoremediation, may be improved. Polluted sites containing hydrocarbons, heavy metals, dyes, and chlorinated solvents have been successfully treated using in-situ bioremediation techniques.

- **Types of In-Situ Bioremediation:** In-situ bioremediation is two types; these are intrinsic and engineered bioremediation.

  i. **Intrinsic Bioremediation:** Natural reduction, another name for intrinsic bioremediation, is an in-situ bioremediation technique that involves passively cleaning up contaminated areas without the use of outside force (human intervention). The stimulation of native or naturally occurring microbial populations is the focus of this process. The procedure for biodegrading pollutants, including those that are resistant, using both anaerobic and aerobic microbial processes. In comparison to other in-situ techniques, the technique appears to be less costly due to the lack of an external force.

ii. **Engineered In-situ Bioremediation:** The second strategy entails introducing specific microorganisms to the contaminated area. The in-situ bioremediation process is accelerated by genetically engineered microorganisms because they improve the physicochemical conditions that promote microorganism growth.

Did you know?

The concept of bioremediation gained prominence in the late 20th century as a strategy to address environmental pollution. Researchers began exploring the use of microorganisms to degrade or detoxify pollutants in soil and water. The Exxon Valdez oil spill in 1989 highlighted the potential of bioremediation for cleaning up oil-contaminated environments.

6. **Bioventing:** These techniques to boost the activity of native microbes for bioremediation, include carefully regulating airflow by supplying oxygen to the unsaturated (vadose) zone. To boost bioremediation, amendments are made in bioventing by adding moisture and nutrients. This will cause pollutants to undergo microbial transformation and become safe. Among the various in-situ bioremediation methods, this one has grown in favor.

7. **Bioslurping:** This method achieves soil and groundwater remediation by indirectly supplying oxygen and promoting the biodegradation of contaminants. It combines vacuum-enhanced pumping, soil vapor extraction, and bioventing. This method is intended for product recovery from unsaturated and saturated zones, light non-aqueous phase liquids (LNAPLs), and remediating capillaries. This method is used to clean up soils contaminated with organic compounds that are volatile or semi-volatile. By spreading into the free product layer, the technique employs a "slurp" to draw liquids up from this layer. By moving upward, the pumping machine carries LNAPLs to the surface, where they separate from water and air. By limiting air permeability and decreasing the rate of oxygen transfer, soil moisture in this method lowers microbial activity. This method is not appropriate for remediating low permeable soil, but because it uses less groundwater and requires less storage, treatment, and disposal, it is an economical process to use.

8. **Biosparging:** This method is comparable to bioventing in that it involves injecting air beneath the soil to enhance microbial activity, which in turn promotes the removal of pollutants from contaminated

areas. However, when air is injected into the saturated zone during bioventing, it can aid in the upward movement of volatile organic compounds to the unsaturated zone, which can accelerate the biodegradation process. The two main variables that determine the effectiveness of biosparging are the biodegradability of pollutants and the permeability of the soil. While in-situ air sparging (IAS), which depends on high air-flow rates for pollutant volatilization, and biosparging, which promotes biodegradation, are closely related techniques in bioventing and soil vapor extraction (SVE). Aquifers tainted with kerosene and diesel have typically been treated via biosparging.

9. **Phytoremediation:** The contaminated soils are being cleaned up via phytoremediation. This method reduces the toxic qualities of pollutants in contaminated sites by utilizing plant interactions with physical, chemical, biological, microbiological, and biochemical elements. Phytoremediation involves a number of processes, including extraction, degradation, filtration, accumulation, stabilization, and volatilization, and is dependent on the type and quantity of pollutants present. Extraction, transformation, and sequestration are frequently used techniques to eliminate pollutants like radionuclides and heavy metals. Degradation, rhizoremediation, stabilization, and volatilization are the main methods used to eliminate organic pollutants like hydrocarbons and chlorinated compounds. Mineralization is a possibility when certain plants, like alfalfa and willow, are utilized.

   A ability of a plant to function as a phytoremediator depends on a number of critical elements, including its root system, which can be tap or fibrous depending on the pollutant's depth; above-ground biomass; the pollutant's toxicity to plants; the plant's existence and ability to adapt to its surroundings; plant growth rate; site monitoring; and, most importantly, the amount of time required to reach the desired level of cleanliness. Furthermore, the plant needs to be pest- and disease-resistant. Pollutant removal in phytoremediation involves uptake and translocation from roots to shoots. Furthermore, transpiration, and partitioning are necessary for translocation and accumulation. However, other variables, like the type of contamination and the plant could alter the process. Most plants that are found growing in contaminated areas are effective phytoremediators.

   Therefore, the success of any phytoremediation method mainly depends on improving the remediation potentials

Keyword

Soil Moisture – It is the water content of the soil. It can be expressed in terms of volume or weight. Soil moisture measurement can be based on in situ probes (e.g., capacitance probes, neutron probes) or remote sensing methods.

of native plants growing in polluted sites either by bioaugmentation with endogenous or exogenous plant. One of the major advantages of using plants to remediate polluted site is that some precious metals can bioaccumulate in some plants and are recovered after remediation, a process known as phytomining.

10. **Permeable Reactive Barrier (PRB):** This approach is frequently used as a physical remediation strategy for contaminated groundwater. On the other hand, the PRB method uses biological mechanisms for pollutant removal through sorption and precipitation degradation. The biotechnology and bioremediation aspects of the technique have been accommodated by the suggestion of alternative terms such as biological PRB, bio-enhanced PRB, and passive bioreactive barrier. PRB is a general in-situ method for cleaning up groundwater pollution that contains heavy metals and chlorinated compounds.

- Advantages of In-Situ Bioremediation:
  - In-situ bioremediation methods do not required excavation of the contaminated soil.
  - This method provides volumetric treatment, treating both dissolved and solid contaminants.
  - The time required to treat sub-surface pollution using accelerated in-situ bioremediation can often be faster than pump and treat processes.
  - It may be possible to completely transform organic contaminants to innocuous substances like carbon dioxide, water, and ethane.
  - It is a cost effective method because there is minimal site disruption.
- Limitation of In-Situ Bioremediation:
  - Depending on the specific site, some contaminants may not be absolutely transformed to harmless products.
  - If transformation stops at an intermediate compound, the intermediate may be more toxic and/or mobile than the parent compound; some are recalcitrant contaminants are not biodegradable.
  - When incorrectly applied, injection wells may become blocked by profuse microbial growth due to the addition of nutrients, electron donor, and electron acceptors.
  - Heavy metals and organic compounds concentration inhibits the activity of indigenous microorganisms.
  - In-situ bioremediation usually requires acclimatization of the mocrobe, which may not develop for spills and recalcitrant compounds.

7.2.3. Bioremediation of Various Pollutants

7.2.3.1. Bioremediation for Organic Pollutants

Due to their extensive and pervasive use in the environment, organic compounds (OCs), such as flame retardants and biocides, are now regarded as a threat to almost all forms of life on Earth. Microbes have the ability to break down the majority of organic compounds (OCs), including polychlorinated biphenyls (PCBs), polybrominated biphenyl ethers (PBEs), and polycyclic aromatic hydrocarbons (PAHs). The process by which microbes convert organic compounds into residues that are either completely non-toxic or less toxic is known as biodegradation. The microbes eat the organic substrate to obtain energy and organic carbons. An individual microbial species typically thrives in a community and does not break down any organic substrate when isolated from other microbes.

The exchange of genetic information among microbial species results in the conferment of resistance, chemical-degrading ability, and tolerance as a result of community microbe interactions. Due to a lack of globally accepted microbial identification protocols and methods, many OC-degrading microorganisms are misidentified. This emphasizes the importance of researching microbial consortiums with traditional genetic engineering techniques and metagenomics tools. Depending on their chromosomal genes and extracellular enzymatic activity (in the case of bacteria), bacteria and other microorganisms can break down a broad variety of organic compounds (fungal degradation process). These processes are made more complex by the different environmental factors that influence the pattern of microbe growth.

The right species and strains must be identified for every substrate in order for a bioengineered microbe to succeed. Biodegradation by microbes utilizing easily accessible organic carbon and energy sources in the surrounding environment is a feasible substitute for the recombinant degradation of resistant organic compounds. Microbes identify the best growing environments by observing changes in the chemical gradients in their surroundings.

As a result, they can flourish in the ideal setting. Two recent innovations in microbiological bioremediation are being utilized to break down resistant organic compounds: microbial consortia and microbial fuel cells (MFCs and bioreactors). Fungi are a better option for remediating toxic organics than bacteria because the former cannot grow in high concentrations of toxic organics. For instance, the lignocellulosic compound metabolism of the white rot fungus Phanerochaete chrysosporium is facilitated by the activity of the enzymes laccase (LAC), lignin peroxidase (Lip), and manganese decarboxylase (MDA).

7.2.3.2. Bioremediation for Inorganic Pollutants

Among the most prevalent inorganic contaminants are toxic heavy metals and their compounds that come from mining, metallurgy, power plants, and chemical manufacturing processes. Toxic elemental pollution is a major worry for environmentalists because it poses intolerable health risks when toxic metals are disposed of in soils, waters, or subsurface formations. It's important to understand that microbes can only stabilize metals by altering their oxidation states; they cannot break down metal ions. Like any other nutrient in the cells, they are able to metabolize and detoxify metals. It has been reported that a number of microorganisms can bioremediate both organic and inorganic pollutants (Table 7.2). The bioavailability of metal ions is increased by microbes that release chelating agents and acids, and change physicochemical properties like redox potential in their surroundings.

These microbes have the potential to significantly alter the environment. The initial stages of the interaction between metals and microbial cells are physical adsorption, biosorption, and ion complexation. The biochemical transformation of metal ions by microorganisms involves enzymes for oxidation, methylation, reduction, precipitation, and dealkylation. A multidrug-resistant isolate of pseudomonas aeruginosa T-3 from tannery effluent showed how bacteria can adapt to heavy metals in textile waste, including iron, zinc, chrome, magnesium, mercury, and barium. This demonstrates how bacteria are able to adjust to their surroundings. Additionally, it has been found that the bacteria *Pseudomonas putida* PhCN has a copper and cadmium metal resistance gene encoded in a plasmid.

The metallothionein gene (Neurospora crasa) for Cd uptake was expressed in recombinant *Escherichia coli* through the use of plasmid-encoded biochemical data and genetic engineering techniques, leading to a noticeably faster uptake of Cd than the donor microbe. *Staphylococcus xylosus* and *Staphylococcus carnosus* were modified to express a poly-histidyl peptide, which encoded genes that enabled the microbes to bind nickel.

Table 7.2. *Potentially Hazardous Organic and Inorganic Pollutants and Their Degrading Microbes (Bacteria, Fungi, and Algae)*

Substrate	Compound	Microorganisms
Organic substrate	Chlorobenzenes	P. putida (GJ31)
	N, N-dimethyl-pphenylenediamine	Klebsiella pneumonia (RS-13)
	Polycyclic aromatic hydrocarbons	Burkholderia sp., Myceliophthorathermophila
	Remazol Black B	Kluyveromyces marxianus (IMB3)
	Sulfonate benzene	A. radiobacter (S2)
	4,4-dibromodiphenyl ether	Phanerochaete chrysosporium
	Aromatic hydrocarbons	Acinetobacter sp., Microbacterium sp., Pseudomonas sp. and Ralstonia sp.
	Phenol	Alcaligenes odorans, Corynebacterium propinquum, B. subtilis, and P. aeruginosa
	Toluene and its derivatives	P. putida (F1), Penicillium chrysogenum
	Methyl parathion and chlorpyrifos	Acinetobactor sp., Pseudomonas sp., Enterobacter sp., and Photobacterium sp.
	Endosulfan	Bacillus, Staphylococcus
	Azo dyes effluents	Exiguobacterium indicum, B. cereus, E. aurantiacums and A. baumanii

	Vat dyes	Bacillus firmus, Staphylococcus aureus, B. macerans, and K. oxytoca
	Oil-based based paints	B. subtilis strain NAP1, NAP2, NAP4
	Crude oil	Aspergillus niger, Candida krusei, C. glabrata, and Saccharomyces cerevisiae
	Diesel oil	P. cepacia, B. coagulans, B. cereus, B. cereus A and Serratia ficaria
	Oils	Alcaligenes odorans, Corynebacterium propinquum, P. aeruginosa and Fusarium sp.
Inorganic substrate	Heavy metals, mercury nickel and lead	Saccharomyces cerevisiae and, Cunninghamella elegans
	Cr^{6+}	Pseudomonas putida
	Cobalt, chromium, copper, and lead	Lysinibacillus sphaericus CBAM5
	Cadmium	A. versicolor, Paecilomyces sp., A. fumigatus, Paecilomyces sp., Terichoderma sp. and Cladosporium sp.
	Uranium, copper, nickel, chromium	P. aeruginosa, Aeromonas sp.
	Lead, chromium, and cadmium	Aerococcus sp., Rhodopseudomonas palustris
	Hg^{2+}	Cyclotella cryptica, Chlamydomonas reinhardtii, Pseudochlorococcum typicum, Spirogyra hyaline
	$Cr_2O_7^{22}$	Chlorella spp. Spirulina sp. (HD-104)
	Cr^{51}	Spirulina sp., Ulothrix tenuissima and C. reinhardtii
	Pb^{21}	Oscillatoria laete-virens, Arthrospira platensis, Pseudochlorococcum typicum and Spirogyra insignis

Source: Saroj, B., Diksha, G., Banjagere, V. T., (2022). Recent Strategies for Bioremediation of Emerging Pollutants. doi: 10.3390/toxics10080484.

7.2.4. Advantages and Disadvantages

All bioremediation techniques have their own specific applications attributed to their advantages and disadvantages because.

- **The Advantage of Bioremediation:**
  - It is a natural process; it takes a little time, as an adequate waste treatment process for contaminated material such as soil. Microbes can degrade the contaminant, and the biodegradative populations become reduced. The treatment products are commonly harmless, including cell biomass, water, and carbon dioxide.
  - It needs only little effort and can commonly carry out on-site, regularly without disturbing normal microbial activities. This also eradicates the transport amount of waste off-site and the possible threats to human health and the environment.
  - It is functional in a cost-effective process in comparison to other conventional methods that are used for the clean-up of toxic hazardous waste regularly for the treatment of oil-contaminated sites. It also supports the complete degradation of the pollutants; many of the toxic hazardous compounds can be transformed to less harmful products and the disposal of contaminated material.
  - It does not use any dangerous chemicals. Nutrients, especially fertilizers added to make active and fast microbial growth. Bioremediation changes harmful chemicals into water and harmless gases, and the harmful chemicals are completely destroyed.
  - Simple, less labor intensive and cheap due to their natural role in the environment.
  - Contaminants are destroyed, not simply transferred to different environments.
  - Nonintrusive, possibly allowing for continued site use.
  - The current approach of remediating the environment of large contaminates and acts as an ecofriendly sustainable opportunity.
- **The Disadvantage of Bioremediation:**
  - It is restricted to biodegradable compounds. Not all compounds are disposed to quick and complete degradation process.
  - There are particularly new products of biodegradation that may be more toxic than the initial compounds and persist in the environment.
  - Biological processes are highly specific and ecofriendly, and include the presence of metabolically active microbial populations, suitable environmental growth conditions, and availability of nutrients and contaminants.
  - It is demanding to encourage the process from bench and pilot-scale to large-scale field operations. Contaminants may

be present as solids, liquids, and gases. It often takes longer than other treatment preferences, such as excavation and removal of soil or incineration.

– Research is needed to develop and engineer bioremediation technologies that are appropriate for sites with complex mixtures of contaminants that are not evenly dispersed in the environment.

7.2.5. Limitations of Bioremediation

Biodegradable compounds are the only ones that can be treated with bioremediation. This technique is prone to total and quick breakdown. The environmental byproducts of biodegradation could be more hazardous or persistent than the original compound.

1. **Specificity:** The biological processes are very particular. The presence of metabolically capable microbial populations, ideal environmental growth conditions, and suitable concentrations of contaminants and nutrients are important site factors essential for success.
2. **Scale Up Limitation:** It is difficult to scale up the bioremediation process from batch and pilot scale studies applicable to large-scale field operations.
3. **Technological Advancement:** Modern engineer bioremediation technologies that are appropriate for sites with composite combinations of contaminants that are not equally distributed in the environment need more research. It could exist in the form of liquids, solids, or gases.
4. **Time Taking Process:** Bioremediation takes longer time compare to other treatment options, such as excavation and removal of soil from contaminated sites.
5. **Regulatory Uncertainty:** Since there is no agreed-upon definition of what constitutes clean, we cannot be certain that remediation is 100% finished. Because of this, evaluating the effectiveness of bioremediation is challenging, and there is no recognized endpoint for bioremediation procedures.

Did you get it?

1. What are the different types of bioremediation, and how do they differ in their approaches?
2. What are the primary advantages and disadvantages associated with the use of bioremediation in environmental cleanup?

7.3. ENVIRONMENTAL BIOTECHNOLOGY APPLICATIONS IN CLEANING UP POLLUTION

→Learning Objectives

- Define environmental biotechnology and alternative solutions;
- Explain the waste water treatment;
- Discuss biomonitoring.

The ability of environmental biotechnology to reduce pollution is one of its main uses. This field uses bio-augmentation and bio-stimulation to accelerate certain natural processes in order to combat pollution. While bio-stimulation increases the number of already-existing microbial populations by providing nutrients, bio-augmentation entails the introduction of particular strains of bacteria that are capable of degrading pollutants.

7.3.1. Environmental Biotechnology and Alternative Solutions

7.3.1.1. Biomass Energy Production

One of the main sources of pollution in the environment is fossil fuels. Various air pollutants, such as soots, CO, HC, and PAHs are discharged as a result of using fossil fuels. The environmental biotechnology approach, applying several biomass sources, has aided in the production of alternative fuels. Any type of living thing or its leftovers with the ability to produce energy can be considered biomass. Ethanol, when combined with other fuels, is one of the major alternatives used to combat fossil fuels. An excellent source of environmentally friendly biogas is the biomass used in biotechnology. Biogas is produced from organic matter, kitchen waste, courtyard waste, sugarcane waste, and many other carbon sources. Methane and CO_2 make up the majority of the gas. Again, this biogas is a valuable energy source that can be applied to a variety of industries, which lessens the demand for fossil fuels.

Keyword

Fossil Fuel – It is a generic term for non-renewable energy sources such as coal, coal products, natural gas, derived gas, crude oil, petroleum products and non-renewable wastes.

7.3.1.2. Biopesticides

The conventional agricultural system uses a vast amount of chemicals, including fertilizers and insecticides, to produce consistently high yields. However, due to concerns about environmental preservation and issues pertaining to human health, alternatives to this system are required. Biopesticides are an excellent alternative to traditional chemical pesticides. Biopesticides are groups of substances that can be used to suppress, inhibit, damage, or kill pests directly or indirectly. They are derived from natural sources such as plants, animals, microbes, or particular minerals. Thus, biotechnology is a rapidly developing field in biology with many applications in sustainable agriculture, beginning with the development of biopesticides that don't require chemicals.

Pesticides that are highly effective, species-specific, and eco-friendly are made using microorganisms. These microbes play a crucial role in integrated pest management strategies. Several microorganisms, such as *Bacillus thuringiensis*, *Bacillus phaericus*, *Pseudomonas aeruginosa*, *Nostoc piscinale*, *Chlamydopodium fusiforme*, and *Chlorella vulgaris*, are utilized as biopesticides and are deemed safe.

7.3.1.3. Biofertilizers

Biofertilizers are biological products that use microorganisms to naturally increase plant growth and improve soil fertility such as the production of plant growth hormones, the solubilization of phosphorus, and the fixation of nitrogen is set off by applying these kinds of biofertilizers. Both free-living and symbiotic nitrogen-fixing bacteria and cyanobacteria are the microorganisms used to make biofertilizers. The biofertilizers *Rhizobium*, *Azolla pinnata*, *Clostridium*, *Azotobacter*, and *Azospirillum* are frequently used. Additional biofertilizers include phosphate-soluble microorganisms like *Erwinia*, *Serratia*, *Alcaligenes*, and *Rhizobacteria*, which stimulate plant growth. Biotechnology facilitates the large-scale culture of these organisms for use as biopesticides, and genetic engineering further increases their efficiency.

7.3.1.4. Bio-Enzymes

An efficient biological tool that promotes environmental cleanliness in a variety of ways is the bioenzyme. Bioenzymes are employed in agriculture, soil stabilization, water body treatment, air purification, pollutant monitoring, and waste management for various kinds of waste. Bioenzymes have the ability to totally change the toxicological structure of a contaminant or convert a dangerous chemical into an innocuous inorganic end product. The bio-enzymes have the ability to effectively transform and detoxify certain types of contaminants because they are specific to them. A few enzymes effective against xenobiotics, such as herbicides, pesticides, organophosphorus compounds, nitrile compounds, and other polymers, are hydrolases, esterases, nitrilases, lipases, and cutinases, among others. It has been reported that a wide variety of enzymes derived from different microorganisms are crucial in wastewater treatment applications. Enzymes have the ability to selectively target and eliminate resistant pollutants through precipitation and conversion into other products. They can also change a waste's characteristics to make it easier to treat or assist in converting garbage into products with higher value. Through immobilization, the possibility of enzymes leaking into the solution is decreased while their mechanical and thermal stability is boosted.

7.3.2. Waste Water Treatment

Numerous industries and the residential sector continuously produce a significant amount of wastewater. To prevent contamination of the receiving body, this wastewater must be treated before being released into the environment. A large number of treatment procedures in industries rely on biotechnological methods. Membrane bioreactors, trickling filters, activated sludge, oxygenation ponds, anaerobic treatment, and membrane bioreactors are some of the tools used in this process. Every technique is used in conjunction with biological treatment to convert wastewater's organic waste into carbon dioxide, water, and other byproducts. It shields the public and the environment from the negative effects of wastewater discharge. Furthermore, biotechnology is a useful tool for turning sludge into compost and biogas. Biotechnology is a key component of several processes used in the secondary wastewater treatment process. Major of these are discussed below.

Did you know?

The use of microorganisms for wastewater treatment began in the early 20th century. The activated sludge process, developed in the United Kingdom in 1913, marked a crucial advancement in using microbes to break down organic matter in sewage.

7.3.2.1. Activated Sludge Process

Wastewater is biologically broken down by bacteria, algae, fungi, and protozoa in the activated sludge process. Microorganisms are seeded in the aeration chamber through the passage of wastewater. The flow of liquid waste, known as activated sludge, is one aspect of this process; it settles in a clarifier tank bottom alongside the solid wastes. The activated sludge is then reinserted into the aeration chamber, and the process is repeated with the next batch of sewage.

7.3.2.2. Tricking Filters

A fixed-bed bioreactor is used in the aerobic wastewater treatment process known as trickling filtering. Stones, bricks, or any other rough material are used to create a biological filter, onto which a slimy layer of microorganisms is seeded. Because of its porosity, air can easily flow through this filter. Wastewater flows through a rotary distributor that distributes it evenly over the filter bed, where organic waste is broken down by microorganisms. This method reduces the biological oxygen requirement as well as suspended solids.

7.3.2.3. Membrane Bioreactors

It is an aerobic suspended growth process where microorganisms remain suspended in the reactor with the wastewater. This mixture is continuously stirred, and high compressed air is sparged under pressure from below to keep the aggregates in suspension.

7.3.2.4. Anaerobic Treatment

The degradation of organic waste with the help of microorganisms without using oxygen is known as anaerobic treatment. Mostly, anaerobic degradation of wastes releases carbon dioxide and methane, which are used as biogas.

7.3.2.5. Oxidation Pond

Oxidation ponds are large, shallow ponds where organic matter from wastewater is stabilized or degraded by the influence of sunlight, bacteria, and algae. Lagoons and stabilization ponds are other names for these. Algae in the pond produce oxygen through photosynthesis, which uses sunlight. The organic waste from the wastewater is broken down by aerobic bacteria in the oxidation pond with the help of this oxygen. In the ponds, the majority of the waste is broken down and settled, resulting in well-treated effluent.

7.3.3. Biomonitoring

It is critical to identify the pollutants compromising the integrity of the environment to maintain safety and sustainability. In addition to identification, it is critical to regularly monitor them in order to limit their value to an acceptable level. Numerous biological techniques are available thanks to environmental biotechnology, which can identify and track the concentration of pollutants in the environment. These devices are known as biological sensors or biosensors. Biosensors are analytical tools with the ability to identify alterations in biological processes and translate them into an electrical signal that is then measured. Materials such as enzymes, microorganisms, cells, tissues, etc., can be a part of biological processes. Biomarkers and bioindicators are primarily used in biomonitoring.

7.3.3.1. Biomarker

Biochemical, cellular, physiological, or behavioral alterations in tissue, bodily fluids, or an organism's entire body that act as markers of exposure to chemical pollutants and may or may not also indicate the existence of a detrimental outcome are referred to as "biomarkers." A few of the frequently used biomarkers are metallothiones, DNA integrity 16, acetylcholinesterase inhibition, pigments, and cytochrome P4501A enzyme induction.

7.3.3.2. Bioindicator

A bioindicator is an organism, species, or community whose status, function, or population can reveal information about the condition of the surrounding environment. There will be a correlation between the indicator species and a small variation in the ecosystem's variable. Bioindicators, which are analogous to instrument measurements, use living organisms to measure pollutants. Examples of these bioindicators include lichens, mosses, plants, plankton, animals, and microbes. Bioindicators provide information on the quality of the environment as well as the health of an ecosystem or organism. Bioindicators include, but are not limited to, earthworms, cyclops, Vogesella Indigofera, and Hylocomium splendens.

Did you get it?

1. Briefly explain how environmental biotechnology provides alternative solutions to environmental challenges.
2. What is biomonitoring, and how does environmental biotechnology contribute to this practice? Provide a concise example of biomonitoring in action.

7.4. ENVIRONMENTAL BIOTECHNOLOGY FOR MEDICAL WASTE MANAGEMENT

→Learning Objectives

- Identify key impact of medical waste on the environment;
- Define biotechnology in managing medical waste.

Since improper medical waste management can pose a serious risk to the environment and the public, it is a major concern in the healthcare sector. Different approaches to treatment and disposal are required depending on whether medical waste is classified as infectious, dangerous, or non-hazardous. The World Health Organization (WHO) estimates that the healthcare industry produced 87,000 tons of PPE (personal protective equipment), 0.73 million liters of chemical waste, and 0.15 million tons of additional medical waste totaling over 8 billion vaccine doses in 2020–2021. According to WHO reports, of waste, about 85% is non-hazardous and general, similar to household waste, and the remaining 15% is classified as hazardous material, which could include chemicals, radioactive materials, or infectious materials. If not properly handled and disposed of, these wastes may have negative effects on both the environment and public health. Dioxins, furans, and particulate matter, for instance, are carcinogenic substances that can cause respiratory ailments when they are released into the atmosphere during the open burning or incineration of medical waste.

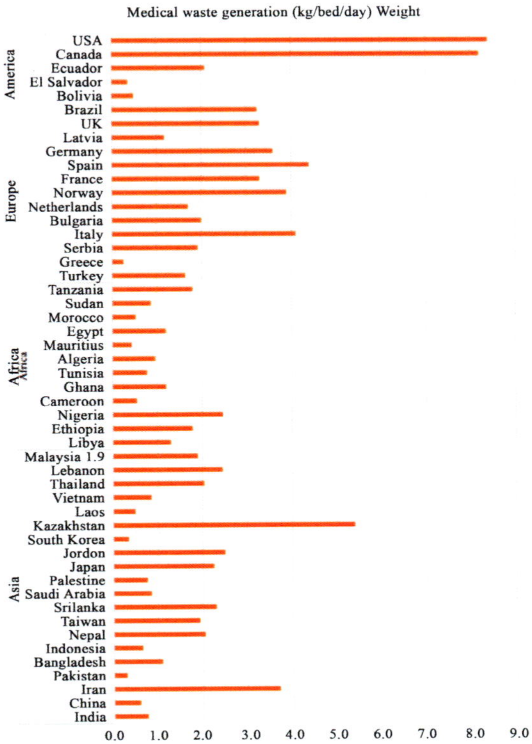

Figure 7.3. *An average medical waste generation in kg per bed in one day of the top 50 countries.*

Source: https://www.researchgate.net/figure/An-average-medical-waste-generation-in-kg-per-bed-in-one-day-of-the-top-50-countries_fig1_372518718.

Furthermore, health professionals and waste handlers may sustain injuries or infections as a result of inappropriate disposal of sharps waste, which includes syringes, needles, disposable scalpels, and blades. Moreover, pharma waste (i.e., tainted, unused, and expired medications and vaccines) can contaminate soil and water sources, harming crops and aquatic life. Moreover, there is a long-term risk of radiation exposure for humans and animals due to radioactive waste (i.e., products contaminated by radionuclides, such as radioactive diagnostic material or radiotherapeutic materials). Furthermore, the type of treatment (conventional or alternative medicine), the level of development (high-income or low-income countries), and the source (hospitals, laboratories, and blood banks) all affect the quantity and makeup of healthcare wastes. Therefore, there isn't a single waste management solution that works for everyone (Figure 7.3).

7.4.1. Impact of Medical Waste on the Environment

The management of medical waste is a crucial issue that calls for ongoing investigation and improvement. The development of medical technology and the rise in pandemic and epidemic diseases like SARS-CoV-2 in 2019, SARS-CoV in 2003, and influenza A (H1N1) in 2009–10 have led to an increase in the type and quantity of medical waste. As a result, it's imperative that medical waste removal procedures and guidelines be updated on a regular basis. Otherwise, the pollution of the air, water, and land would cause severe harm to the environment and pose health risks to all living things, including humans and animals. Biological processes are employed in the field of environmental biotechnology to lessen pollution, repair the environment, and encourage sustainable development. It has demonstrated potential in reducing the harm that medical waste causes to the environment. On the basis of this, the review paper provides a current, thorough overview of medical waste management as well as the part environmental biotechnology plays in lessening the harm that medical waste causes to the environment. It discusses generation, segregation, collection, transportation, treatment, and disposal, among other aspects of waste management, and emphasizes the significance of environmental biotechnology at every turn.

Additionally, the review looks at various environmental biotechnologies, including microbial fuel cells, composting, anaerobic digestion, and phytoremediation, that can be used to manage medical waste. It explores each technology's benefits, limitations, and possible applications. It also examines the financial and social effects of managing medical waste with environmental biotechnology. Overall, this review explored the use of environmental biotechnology to lessen the impact of medical waste on the environment and examines at both the role that biotechnology can play in resolving this issue and current management practices (Figure 7.4).

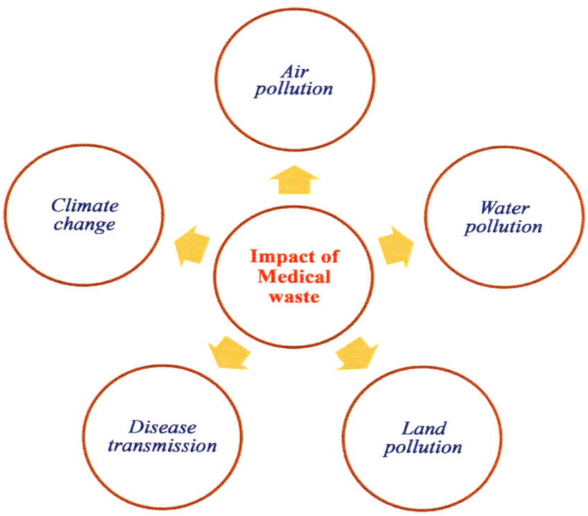

Figure 7.4. *Impact of medical waste on the environment.*

Source: https://www.researchgate.net/publication/372518718/figure/fig2/AS:11431281176100154@1690026186554/Impact-of-medical-waste-on-the-environment.png.

If handled improperly, medical waste, which includes any kind of waste produced by healthcare facilities like hospitals, clinics, and laboratories, can have a substantial negative influence on the environment. The following are some ways that medical waste may harm the environment:

- **Air Pollution:** When medical waste is burned, hazardous chemicals like dioxins, furans, and mercury can be released into the atmosphere, causing air pollution and perhaps endangering the health of adjacent communities.
- **Water Pollution:** If medical waste is not properly disposed of, it may contaminate water sources, posing major health dangers to humans as well as wildlife.
- **Land Pollution:** Medical waste can contaminate the soil and landfills, endangering the environment and endangering human health.
- **Disease Transmission:** Infectious agents, including viruses, germs, and parasites can be carried by medical waste and spread illness to both people and animals.
- **Climate Change:** By releasing greenhouse gases that cause global warming, improper medical waste disposal can exacerbate climate change

7.4.2. Biotechnology in Managing Medical Waste

Utilizing living things, cells, or molecular analogs to create goods and services is known as biotechnology. Among its many potential applications is waste management, where biological techniques can be used to treat and get rid of various types of waste, including chemical, pharmaceutical, infectious, and radioactive waste. By stopping dangerous substances like dioxins, furans, and heavy metals from being released into the air, water, and soil, these techniques can also lessen the negative effects of waste on the environment and human health. However, there are certain difficulties and restrictions in the use of biotechnology for waste management, including moral, societal, legal, and financial concerns. The study found that medical waste management (26.6%), operational management concerns (21.7%), training for medical waste management procedures (17.8%), raising awareness (17.5%), and environmental assessment (16.4%) were the most crucial factors for reducing medical waste. One of the key factors of medical waste management, is explained in detail below.

Keyword

Polycyclic Aromatic Hydrocarbons (PAHs) – These are a class of chemicals that occur naturally in coal, crude oil, and gasoline.

7.4.2.1. Bioremediation

Bioremediation is the process of using living organisms or their byproducts to break down or remove toxins from the environment. Recombinant enzymes can be used in bioremediation to treat medical waste because they have undergone genetic alteration to increase their stability or activity. One example of a recombinant enzyme that can be used for the bioremediation of medical waste is horseradish peroxidase (HRP). Medical waste frequently contains organic contaminants like phenols and polycyclic aromatic hydrocarbons (PAHs), which can be catalyzed to oxidize by HRP because it contains heme. Genetic engineering techniques can be used to create recombinant HRP, which can then be immobilized on a solid support to enable its use in bioremediation applications. Another example of a recombinant enzyme that can be used for medical waste bioremediation is laccase. Laccase is an enzyme that contains copper that can aid in the oxidation of many organic molecules, including lignin, a complex polymer found in plant material and a large amount of some medical waste streams. The environmental impact of medical waste can be reduced by breaking down the lignin through the use of genetic engineering techniques to produce recombinant laccase. Other recombinant enzymes that can be used for the bioremediation of medical waste are lipases, proteases, and cellulases; these enzymes can break down lipids, proteins, and cellulose, respectively. Mixing these enzymes allows it to bioremediate a range of medical waste streams with success.

7.4.2.2. Composting

Composting is a natural and sustainable way to manage medical waste. It entails the microbial decomposition of organic waste materials to produce a nutrient-rich soil supplement that can be used in gardening and farming. When it comes to medical waste, composting can be used to manage a variety of items, such as food waste from clinics, hospitals, and labs, as well as obsolete pharmaceuticals and other organic waste products. Composting functions by creating an atmosphere that is ideal for the growth of microorganisms. Decomposition is carried out by microorganisms like actinomycetes, fungi, and bacteria. These bacteria convert the organic material into simpler compounds like carbon dioxide, water, and organic acids. Any bacteria that might be in the waste are also killed by the heat of the procedure. One type of bacteria used in composting is *Bacillus subtilis*. This bacterium is commonly found in soil and is capable of breaking down a wide variety of organic materials. In composting, it's widely used to speed up decomposition and produce high-quality end products.

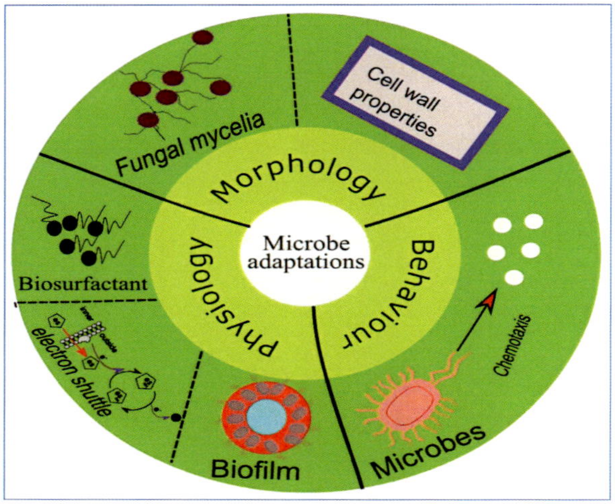

Figure 7.5. *Microbial adoptations in medical waste composting.*

Source: Vishal, C., & Kumari, A., (2023). Environmental Biotechnology for Medical Waste Management. doi: 10.20944/preprints202306.2232.v1.

Another example is the fungus *A. niger*, which is widely used in the food and pharmaceutical industries because of its ability to produce enzymes. A. By breaking down complex organic molecules like lignin and cellulose into simpler forms that are easier for other bacteria in the compost pile to use, *A. niger* can help with the composting process. Medical waste can be composted to produce an essential soil amendment that can be used to improve the soil's fertility and health while also reducing the amount of waste disposed of in landfills. When carried out correctly, it is a safe, natural way to treat medical waste that can help prevent the release of harmful toxins into the surrounding environment (Figure 7.5).

Medical waste can be composted with the help of recombinant enzymes, such as cellulase. The enzyme cellulase breaks down cellulose, which is one of the primary components of medical waste, which includes bandages, gauze, and cotton swabs. Utilizing cellulase can speed up the breakdown process, lowering the volume of waste produced and creating compost that is high in nutrients. This enzyme degrades the fats and lipids in medical waste, including surgical drapes and soiled bandages. These materials can break down more quickly when lipase is used, which lowers their volume and the possibility of environmental contamination. Protease is a third enzyme that can be used in addition to lipase and cellulase to break down proteins in medical waste, including biological fluids like blood. The use of proteases can facilitate the breakdown of these compounds and reduce the potential for infection.

7.4.2.3. Phytoremediation

Phytoremediation, a type of bioremediation, uses plants and the microorganisms that live in and on them to remove or detoxify contaminants from the environment. Environmental biotechnology has shown tremendous promise in the cleanup of medical waste through the use of phytoremediation. Medical waste is one type of hazardous waste, along with other pollutants, germs, and various chemicals. These pollutants can endanger both human health and the environment if they are not properly controlled. Phytoremediation offers a safe, cost-effective, and long-lasting substitute for cleaning up medical waste. Vetiver grass is phytoremediation agent that has been used to clean up medical waste. Vetiver grass has been shown to be effective in eliminating pathogens, organic pollutants, and heavy metals from medical waste. It can absorb and retain pollutants in its tissues, preventing them from evaporating into the environment, thanks to its extensive root system. Furthermore, vetiver grass may promote the growth of beneficial soil microbes, hastening the cleanup process even more. Willow is another phytoremediation agent that has shown promise in the treatment of medical waste. Willow has been utilized to purify contaminated soil, water, and air. It has a strong tolerance to toxins and is able to absorb and store them in its tissues. Willow can also promote the growth of beneficial soil microbes, which can expedite the detoxification process.

7.4.2.4. Genetic Engineering

Genetic engineering, in particular, plays a significant role in managing medical waste in the field of bioremediation, which uses microorganisms to convert pollutants, particularly medical waste, into non-toxic chemicals. This approach is affordable, environmentally friendly, and applicable to specific pollutants. Thanks to genetic engineering, we may now be able to increase the effectiveness and efficiency of microorganisms' ability to break down toxins. One example of how genetic engineering is used in the bioremediation of medical waste is the biodegradation of polyethylene terephthalate (PET), a common plastic used in medical equipment. PET has an extended lifespan in the environment due to its resistance to biodegradation. However, a bacterium *Ideonella sakaiensis* has been found by scientists to be capable of degrading PET. By modifying this bacterium's DNA, scientists have increased its ability to degrade PET, making it a more effective bioremediation tool (Figure 7.6).

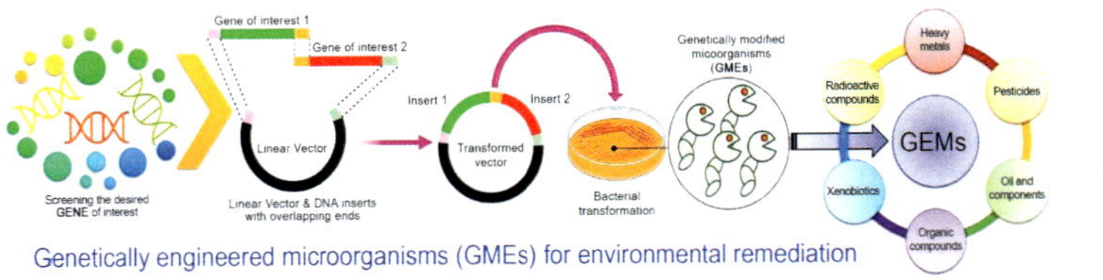

Figure 7.6. *Typical process of medical waste bioremediations by genetically engineered microorganism.*

Source: Vishal, C., & Kumari, A., (2023). Environmental Biotechnology for Medical Waste Management. doi: 10.20944/preprints202306.2232.v1.

An additional instance is the utilization of genetically modified (GM) fungi to degrade PAHs, a class of contaminants found in medical waste. PAHs are dangerous chemicals that can cause cancer in humans. It has been discovered that a fungus called *Phanerochaete chrysosporium* is capable of degrading PAHs. Through genetic engineering, they have increased this fungus' ability to degrade PAHs, making bioremediation a more successful and efficient method.

7.4.2.5. Waste to Energy (WTE) Conversion of Medical Waste

Medical waste is converted into a renewable energy source through the WTE process. Medical waste is generated, including used needles, expired medications, and biological materials. Improper disposal of this waste poses a serious risk to public health and the environment. One effective way to get rid of medical waste is to use WTE conversion, which uses a variety of technologies to turn waste into energy. A common method is incineration, which entails burning waste at high temperatures to produce steam and heat that can be used to generate electricity or heat. Gasification is the process of heating trash in the absence of oxygen to create a gas that can be utilized to generate electricity (Figure 7.7).

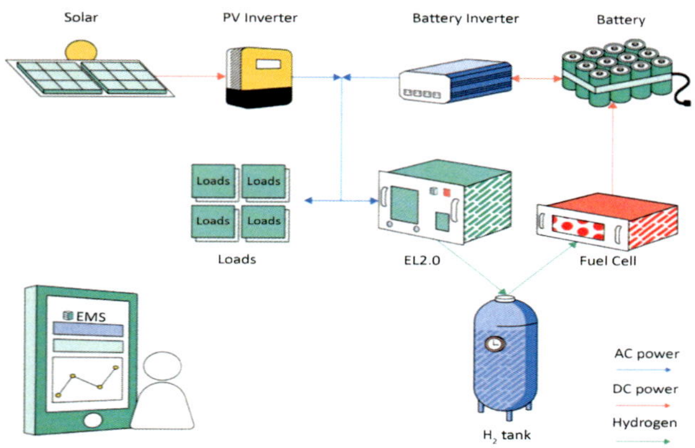

Figure 7.7. *Component of waste-to-hydrogen micrograms. It is one of the most popular methods employed for the conversion of waste material into hydrogen gas which is further used in various applications in day-to-day life.*

Source: Vishal, C., & Kumari, A., (2023). Environmental Biotechnology for Medical Waste Management. doi: 10.20944/preprints202306.2232.v1.

There are many benefits to converting medical waste with WTE. It reduces the quantity of waste disposed of in landfills, reducing the possibility of pollution and environmental damage. Additionally, it might help meet the growing energy demands by generating power from a renewable source. WTE conversion also reduces greenhouse gas emissions by stopping the release of methane and other hazardous gases produced when waste breaks down in landfills. Medical waste conversion to WTE is not without its challenges, though. The possible release of dangerous chemicals during the combustion or gasification process is one thing to be concerned about as it could seriously affect the health of the local populace. Another challenge is the cost of setting up and maintaining the infrastructure and technologies needed for WTE conversion.

Did you get it?

1. Outline the impact of medical waste on the environment in a few sentences.
2. Summarize the role of biotechnology in managing medical waste and its potential benefits in a brief statement.

FOCUS ON CAREERS

- **Environmental Biotechnologist:**

  Environmental biotechnologists combine biology and engineering to develop and use processes that remediate contaminated sites. For example, there are a variety of microbes, fungi, and bacteria capable of consuming pollutants and breaking them down into harmless components over time. Environmental biotechnologists identify, use, and develop appropriate microbes for remediating a particular area and its unique pollutants. Contaminated soil may be remediated on-site or placed in containers and hauled away for treatment.

- What Does an Environmental Biotechnologist Do?

  Environmental biotechnologists apply biological processes to technology to create more sustainable societies. For instance, they may:

  - Convert plants into biofuels
  - Create plant-based bioplastics
  - Discover or engineer plants or microbes optimized for processing environmental contaminants, or for treating toxic waste or sewage
  - Use geographic information systems (GIS) to map contaminated sites and the distribution of pollutants
  - Develop remediation plans for particular sites that adhere to environmental regulations
  - Create processes to turn waste to biogas or other cleaner energy sources
  - Create cleaner industrial activities by replacing chemicals with biological processes, such as replacing chemical bleaching agents with biological enzymes

- **Where Does an Environmental Biotechnologist Work?**

  Most environmental biotechnologists work in private industry for companies providing bioremediation services, or for environmental consulting companies with clients in the manufacturing sector. They may also be involved in research and development for biofuel firms.

  These professionals may also work in government for federal and state agencies involved in land management and restoration, such as the United States Environmental Protection Agency. Those employed by colleges and universities teach and conduct research.

  Environmental biotechnologists often conduct both field and lab work. They may spend time outdoors testing and assessing a site or overseeing the remediation process. They may work long or irregular hours when doing fieldwork.

  Environmental scientists and specialists, including biotechnologists, are employed nationwide. Higher numbers of jobs are available on the East Coast, and in California and Texas.

- **Environmental Biotechnology Jobs & Job Description:**

  Environmental biotechnologists are cross-disciplinary experts, who create new products at the molecular level and then analyze the relationship they develop with our environment. Candidates seeking an environmental biotechnology job may pursue teaching, research, or quite often, work in business applications. While duties may vary from job to job, much of what an environmental biotechnologist does is listed below:

  - Carry out complex analytical, biological testing over time following field and laboratory best practices
  - Manage biological and project timelines
  - Be prepared to present results in an understandable format to various stakeholders, both within the discipline and those with non-technical backgrounds
  - Identify challenges and assess risk data to elements of the project
  - Build professional relationships to support the advancement of knowledge and best practice
  - Ensure that equipment is well-maintained for safety and redundancy purposes
  - Document and analyze all lab and field results
  - Act as a peer mentor to other team members and their reports in the spirit of best scientific process
  - Review peers' data analyzes for accuracy and scientific integrity
  - Alert the project leader of unusual results or potential mishaps

At the management level, the job of a senior environmental biotechnologist involves more budgetary, managerial, and administrative tasks. The job may have these or similar additional responsibilities:

- Apply and communicate understanding of environmental biotechnology to other businesses, stakeholders, and the public
- Improve technical leadership by training colleagues and administrative support staff, as well as communicating effectively with business partners and external stakeholders
- Seek global recognition as a technologist by maintaining a deep understanding of advances in your field
- Communicate effectively with global and cross-disciplinary partners
- Oversee process improvements and data quality to help workgroup meet business and scientific benchmarks
- Create and deliver presentations to internal and external stakeholders
- Establish a leadership role in related professional activities

- **What Are Environmental Biotechnologist Education Requirements?**

  Education and training requirements vary greatly depending on the type of position. While careers in biotechnology have typically required Master's and doctoral degrees, the rapid growth of the field has opened up opportunities for those with bachelor's and associate's degrees. These workers are generally employed as technicians.

  Students following this career path typically pursue degrees in ecology or environmental engineering. Coursework may include classes in microbial ecology, ecosystems biology, soil microbiology, and biogeochemistry. Knowledge of mathematics and computers, as well as communication skills, are also important to careers in this industry.

SUMMARY

- Environmental biotechnology is an interdisciplinary field that leverages the power of biological processes to address pressing environmental challenges and promote sustainable solutions. As the global population continues to grow, industrialization expands, and the demands on natural resources increase, the impact on the environment becomes more pronounced.
- Environmental biotechnology, particularly bioremediation, significantly reduces the impact of contaminants in soil and water. This has implications for human health and the restoration of ecosystems affected by industrial pollution, oil spills, and hazardous waste.
- The development of biosensors for environmental monitoring enables real-time detection of pollutants, allowing for proactive responses to environmental changes and pollution incidents.
- Bioremediation and natural reduction are also seen as a solution for emerging contaminant problems; microbes are very helpful to remediate the contaminated environment. A number of microbes, including aerobic and anaerobic bacteria, and fungi, are involved in bioremediation process.
- Bioremediation is a process where biological organisms are used to remove or neutralize an environmental pollutant through metabolic processes. The "biological" organisms include microscopic organisms, such as fungi, algae, and bacteria, and the "remediation"—treating the situation.
- Microorganisms act on the pollutants only when they are in contact with the compounds, which help them to generate energy and nutrients to multiply cells. The effectiveness of bioremediation depends on many factors, including the chemical nature and concentration of pollutants, the physicochemical characteristics of the environment, and their accessibility to existing microorganisms.
- Bioremediation includes above-ground piling of dug polluted soil, followed by aeration and nutrient amendment to improve bioremediation by microbial metabolic activities. This technique comprises aeration, irrigation, nutrients, leachate collection, and treatment bed systems.
- A bioreactor is a vessel in which raw materials are converted to specific product(s) following a series of biological reactions. There are different operational modes of bioreactors, which include: batch, fed-batch, sequencing batch, continuous, and multistage. A bioreactor provides optimal growth conditions for bioremediation.
- Intrinsic bioremediation, also known as natural reduction is an in-situ bioremediation technique, which involves passive remediation of polluted sites, without any external force (human intervention). This process stimulates indigenous or naturally occurring microbial population.
- Bioventing techniques involve controlled stimulation of airflow by delivering oxygen to unsaturated (vadose) zone to increase activities of indigenous microbes for bioremediation. In bioventing, amendments are made by adding nutrients and moisture to increase bioremediation.

- Phytoremediation is depolluting the contaminated soils. This technique is based on plant interactions like physical, chemical, biological, microbiological, and biochemical, in contaminated sites to diminish the toxic properties of pollutants.
- Fossil fuels are one of the major contributors of pollutants in the environment. Different air pollutants like CO, HC, PAH, and Soots, etc., are released due to use of fossil fuels. The environmental biotechnology approach has helped in the production of alternative fuels through different biomass.
- Bio-enzymes are an effective biological instrument that support environmental cleanliness in a number of ways. The bio-enzymes are used in agriculture and soil stabilization, for treating the water bodies, to purify the air, monitoring the pollutants, and for managing different types of waste.
- The trickling filter process is an aerobic wastewater treatment with a fixed-bed bioreactor. A biological filter is prepared with stones, bricks, or any rough media on which a slimy layer of microorganisms is seeded. This filter is porous so that air can easily pass through it.
- A bioindicator is an individual, species, or community whose function, population, or status can provide information about the state of the environment's quality. A slight change in the ecosystem's variable will show a correlation with the indicator species.

REVIEW QUESTIONS

1. What is the role of environmental biotechnology in environmental clean up?
2. Explain the importance of biofertilizer in the environmental safety.
3. How biotechnology is playing role in waste water treatment?
4. Discuss the limitations of bioremediation and potential challenges that may arise in its application.
5. How does biotechnology contribute to the efficient management of medical waste?

MULTIPLE CHOICE QUESTIONS

1. What is the primary goal of Environmental Biotechnology?

a. Industrial production

b. Environmental conservation and remediation

c. Medical research

d. Agricultural improvement

2. Which of the following is a common application of environmental biotechnology?

a. Mobile app development

b. Water treatment

c. Space exploration
d. Fashion design

3. **What is the role of microorganisms in environmental biotechnology?**
a. Increase pollution
b. Enhance biodiversity
c. Accelerate environmental degradation
d. Aid in pollution remediation and waste treatment

4. **Which process involves the use of living organisms to break down pollutants into less harmful substances?**
a. Photosynthesis
b. Bioremediation
c. Combustion
d. Desalination

5. **What is phytoremediation?**
a. Using bacteria for pollution control
b. Using plants to remove pollutants from the environment
c. Recycling plastic waste
d. Generating renewable energy from wind

6. **Which enzyme is commonly used in biodegradation processes?**
a. Lipase
b. Amylase
c. Catalase
d. DNA polymerase

7. **What is the purpose of anaerobic digestion in environmental biotechnology?**
a. Oxygen production
b. Energy generation
c. Carbon sequestration
d. Soil aeration

8. **Which of the following is an example of a biopesticide used in environmental biotechnology?**
a. Synthetic chemical pesticides
b. Antibiotics
c. Genetically modified organisms (GMOs)
d. Bacillus thuringiensis (Bt)

9. **What is the main advantage of using microbial fuel cells in environmental applications?**
a. Efficient solar energy conversion

b. Carbon sequestration
c. Electricity generation from organic matter
d. Water desalination

10. What is the significance of bioaugmentation in bioremediation processes?

a. Enhancing the genetic diversity of a population
b. Introducing genetically modified organisms
c. Adding specific microorganisms to improve pollutant degradation
d. Increasing the temperature of the environment

Answer to Multiple Choice Questions

1. (b); 2. (b); 3. (d); 4. (b); 5. (b); 6. (a); 7. (b); 8. (d); 9. (c); 10. (c)

REFERENCES

1. Adams, G. O., Fufeyin, P. T., Okoro, S. E., & Ehinomen, I., (2015). Bioremediation, biostimulation and bioaugmention: A review. *International Journal of Environmental Bioremediation & Biodegradation, 3*, 28–39.
2. Aislabie, J., Saul, D. J., & Foght, J. M., (2006). Bioremediation of hydrocarbon contaminated polar soils. *Extremophiles, 10,* 171–179.
3. Angelim, A. L., Costa, S. P., Farias, B. C. S., Aquino, L. F., & Melo, V. M. M., (2013). An innovative bioremediation strategy using a bacterial consortium entrapped in chitosan beads. *J. Environ. Manage., 127*, 10–17.
4. Cases, I., & De Lorenzo, V., (2005). Genetically modified organisms for the environment: Stories of success and failure and what we have learned from them. *International Microbiology, 8*, 213–222.
5. Chauhan, A., Fazlurrahman, Oakeshott, J. G., & Jain, R. K., (2008). Bacterial metabolism of polycyclic aromatic hydrocarbons: Strategies for bioremediation. *Ind. J. Microbiol., 48*, 95–113.
6. Corretto, E., Antonielli, L., Sessitsch, A., Kidd, P., Weyens, N., & Brader, G., (2015). Draft genome sequences of 10 *Microbacterium* spp., with emphasis on heavy metal-contaminated environments. *Genome Announc., 3*, e00432–15.
7. Coulon, F., Al Awadi, M., Cowie, W., Mardlin, D., Pollard, S., Cunningham, C., et al., (2009). When is a soil remediated? Comparison of biopiled and windrowed soils contaminated with bunker-fuel in a full-scale trial. *Environmental Pollution, 158*, 3032–3040.
8. Dang, T. C. H., Nguyen, D. T., Thai, H., Nguyen, T. C., Tran, T. T. H., Le, V. H., et al., (2018). Plastic degradation by thermophilic *Bacillus* sp. BCBT21 isolated from composting agricultural residual in Vietnam. *Adv. Nat. Sci.: Nanosci. Nanotech., 9*, 015014.
9. Deng, X., & Jia, P., (2011). Construction and characterization of a photosynthetic bacterium genetically engineered for Hg^{2+} uptake. *Bioresour. Technol., 102*, 3083–3088.

10. Dias, R. L., Ruberto, L., Calabró, A., Balbo, A. L., Del Panno, M. T., & Mac Cormack, W. P., (2015). Hydrocarbon removal and bacterial community structure in on-site biostimulated biopile systems designed for bioremediation of diesel-contaminated Antarctic soil. *Polar Biology, 38*, 677–687.

11. Folch, A., Vilaplana, M., Amado, L., Vicent, R., & Caminal, G., (2013). Fungal permeable reactive barrier to remediate groundwater in an artificial aquifer. *Journal of Hazardous Materials, 262*, 554–560.

12. Frascari, D., Zanaroli, G., & Danko, A. S., (2015). In situ aerobic cometabolism of chlorinated solvents: A review. *Journal of Hazardous Materials, 283*, 382–399.

13. Frutos, F. J. G., Pe'rez, R., Escolano, O., Rubio, A., Gimeno, A., Fernandez, M. D., et al., (2012). Remediation trials for hydrocarbon-contaminated sludge from a soil washing process: Evaluation of bioremediation technologies. *Journal of Hazardous Materials, 199*, 262–271.

14. Gidarakos, E., & Aivalioti, M., (2007). Large scale and long term application of bioslurping: The case of a Greek petroleum refinery site. *Journal of Hazardous Materials, 149*, 574–581.

15. Gomez, F., & Sartaj, M., (2014). Optimization of field scale biopiles for bioremediation of petroleum hydrocarbon contaminated soil at low temperature conditions by response surface methodology (RSM). *Integrated Journal of Biodeterioration Biodegradation, 89*, 103–109.

16. Gong, T., Xu, X., Dang, Y., Kong, A., Wu, Y., Liang, P., et al., (2018). An engineered *Pseudomonas putida* can simultaneously degrade organophosphates, pyrethroids and carbamates. *Sci. Total. Environ., 628*, 1258–1265.

17. Hobson, A. M., Frederickson, J., & Dise, N. B., (2005). CH_4 and N_2O from mechanically turned windrow and vermincomposting systems following in-vessel pre-treatment. *Waste Management, 25*, 345–352.

18. Huang, K., Chen, C., Shen, Q., Rosen, B. P., & Zhao, F. J., (2015). Genetically engineering *Bacillus subtilis* with a heat-resistant arsenite methyltransferase for bioremediation of arsenic-contaminated organic waste. *Appl. Environ. Microbiol., 81*, 6718–6724.

19. Kuiper, I., Lagendijk, E. L., Bloemberg, G. V., & Lugtenberg, B. J. J., (2004). Rhizoremediation: A beneficial plant-microbe interaction. *Molecular Plant Microbe Interactions, 7*, 6–15.

20. Kulshreshtha, A., Agrawal, R., Barar, M., & Saxena, S., (2014). A review on bioremediation of heavy metals in contaminated water. *IOSR Journal of Environmental Science Toxicology and Food Technology (IOSR-JESTFT), 8*(7), 44–50.

21. Lee, J. H., (2013). An overview of phytoremediation as a potentially promising technology for environmental pollution control. *Biotechnology and Bioprocess Engineering, 18*, 431–439.

22. Maila, M. P., & Colete, T. E., (2004). Bioremediation of petroleum hydrocarbons through land farming: Are simplicity and cost-effectiveness the only advantages? *Review Environmental Science Biotechnology, 3*, 349–360.

23. Meagher, R. B., (2000). Phytoremediation of toxic elemental organic pollutants. *Currant Opinion Plant Biology, 3*, 153–162.

24. Miguel, A. S., Ravanel, P., & Raveton, M., (2013). A comparative study on the uptake and translocation of organochlorines by *Phragmites australis*. *Journal of Hazardous Materials, 244*, 60–69.

25. Mohan, S. V., Sirisha, K., Rao, N. C., Sarma, P. N., & Reddy, S. J., (2004). Degradation of chlorpyrifos contaminated soil by bioslurry reactor operated in sequencing batch mode: Bioprocess monitoring. *Journal of Hazardous Materials, 116*, 39–48.

26. Obiri-Nyarko, F., Grajales-Mesa, S. J., & Malina, G., (2014). An overview of permeable reactive barriers for in situ sustainable groundwater remediation. *Chemosphere, 111*, 243–259.

27. Ortet, P., Gallois, N., Piette, L., Long, J., Berthomieu, C., Armengaud, J., et al., (2017). Crossm draft genome sequence of bacterium isolated from chernobyl radionuclide-contaminated soil. *Genome Announc., 5*, 18, 19.

28. Roy, M., Giri, A. K., Dutta, S., & Mukherjee, P., (2005). Integrated phytobial remediation for sustainable management of arsenic in soil and water. *Environment International, 75*, 180–198. doi: 10.1016/j.envint.2014.11.010.

29. Sanscartier, D., Zeeb, B., Koch, I., & Reimer, K., (2009). Bioremediation of diesel-contaminated soil by heated and humidified biopile system in cold climates. *Cold Regions Science and Technology, 55*, 167–173.

30. Silva-Castro, G. A., Uad, I., Gonzalez-Lopez, J., Fandino, C. G., Toledo, F. L., & Calvo, C., (2012). Application of selected microbial consortia combined with inorganic and oleophilic fertilizers to recuperate oil-polluted soil using land farming technology. *Clean Technology of Environmental Policy, 14*, 719–726.

31. Singh, B., Kaur, J., & Singh, K., (2011). 2, 4, 6-trinitrophenol degradation by *Bacillus cereus* isolated from a firing range. *Biotechnol. Lett., 33*, 2411–2415.

32. Smith, E., Thavamani, P., Ramadass, K., Naidu, R., Srivastava, P., & Megharaj, M., (2015). Remediation trials for hydrocarbon-contaminated soils in arid environments: Evaluation of bioslurry and biopiling techniques. *Integrated Journal of Biodeterioration Biodegradation, 101*, 56–65.

33. Thiruvenkatachari, R., Vigneswaran, S., & Naidu, R., (2007). Permeable reactive barrier for groundwater remediation. *Journal of Industrial and Engineering Chemistry, 14*, 145–156.

34. Whelan, M. J., Coulon, F., Hince, G., Rayner, J., McWatters, R., Spedding, T., et al., (2015). Fate and transport of petroleum hydrocarbons in engineered biopiles in polar regions. *Chemosphere, 131*, 232–240.

Index

N

O

P

R

FSC
www.fsc.org
MIX
Papier | Fördert gute Waldnutzung
FSC® C083411

Kolibri 360 GmbH
Ferdinand-Jühlke-Straße 7
99095 Erfurt, Deutschland
produktsicherheit@kolibri360.de